Imagining Contagion in Early Modern Europe

Also by Claire L. Carlin
PIERRE CORNEILLE REVISITED
THEATRUM MUNDI
WOMEN READING CORNEILLE

Imagining Contagion in Early Modern Europe

Edited by

Claire L. Carlin
University of Victoria
Canada

First published 2005 by
PALGRAVE MACMILLAN
Houndmills, Basingstoke, Hampshire RG21 6XS and
175 Fifth Avenue, New York, N. Y. 10010
Companies and representatives throughout the world

PALGRAVE MACMILLAN is the global academic imprint of the Palgrave
Macmillan division of St. Martin's Press, LLC and of Palgrave Macmillan Ltd.
Macmillan® is a registered trademark in the United States, United Kingdom
and other countries. Palgrave is a registered trademark in the European
Union and other countries.

ISBN-13: 978-1-4039-3926-5 hardback

This book is printed on paper suitable for recycling and made from fully
managed and sustained forest sources.

A catalogue record for this book is available from the British Library.

Library of Congress Cataloging-in-Publication Data
Imagining contagion in early modern Europe / edited by Claire L. Carlin.
     p. cm.
  Includes bibliographical references and index.
  ISBN 978-1-4039-3926-5 (cloth)
     1. Communicable diseases–Europe–History. 2. Epidemics–Europe–History.
I. Carlin, Claire L.

RA643.7.E85I53 2005
614.4'94'09031–dc22   2005045403

10   9   8   7   6   5   4   3   2   1
14 13 12 11 10 09 08 07 06 05

Transferred to Digital Printing 2011

Contents

Acknowledgements

This book would not have been possible without early, enthusiastic support from the Centre for Studies in Religion and Society at the University of Victoria, its founding Director, Harold Coward, the current Director, Conrad Brunk, and staff members Moira Hill, Leslie Kenny and Susan Karim. Major funding was provided by the Social Sciences and Humanities Research Council of Canada and a Hannah Development Grant from Associated Medical Services.[1] Several other offices and departments at the University of Victoria contributed the additional funding that was essential to our success: the Office of the President, David Turpin, the Office of the Vice-President Academic, Jamie Cassels, the Office of the Vice-President for Research, Martin Taylor, the Office of the Dean of Humanities, Andrew Rippin, the Office of the Dean of Human and Social Development, Anita Molzahn, along with the Departments of French and English, the Medieval Art Research Group and the European Studies Program. I am also very grateful to Bruce D. Wonder, who made a substantial contribution to the September 2003 meeting of the project team in Victoria.

Among the contributors to the volume, Hélène Cazes deserves special thanks for her help with the bibliography, as well as the initial conception of the project. Donald Beecher also did exceptional service, having written the Afterword as well as one of the chapters. Indeed, Don's work led to my own interest in the topic of contagion.

Present at the team meeting but not able to contribute to the volume were Philippe Desan, Guy Spielmann and Colette Winn. Their participation was invaluable in the evolution of the project.

The enormous task of doing a first English version of the six chapters submitted in French fell to graduate research assistant Rachel Warrington. Her efforts are greatly appreciated, as is the editorial assistance provided by Daniel Bunyard at Palgrave Macmillan.

Note

1 Associated Medical Services Inc. (AMS) was established in 1936 by Dr Jason Hannah as a pioneer, prepaid, not-for-profit health care organization in Ontario, Canada. With the advent of Medicare, AMS became a charitable organization supporting innovations in academic medicine and health services, specifically the history of medicine and health care, as well as innovations in health professional education and bioethics.

Notes on the Contributors

Donald Beecher is Professor of English at Carleton University in Ottawa. He has published on Renaissance medicine, as well as some twenty critical editions of Renaissance plays, early English prose fiction and early music. Currently, he is preparing a two-volume collection of sixteenth-century Italian comedies in translation for the Da Ponte Library. His most recent articles deal with cognitive approaches to Renaissance authors and sex changes in early medical literature.

Dominique Bertrand, Professor of French Literature at the Université Blaise Pascal Clermont-Ferrand, specializes in the representation of laughter and burlesque poetics in the sixteenth and seventeenth centuries. She is the author of *Dire le rire à l'Age classique* (1995) and *Poétiques du burlesque* (1998), and is preparing a critical edition of the *Aventures* of Dassoucy.

Claire Carlin is Associate Professor of French and Associate Dean of Humanities at the University of Victoria. She is the author of *Pierre Corneille Revisited* (1998) and *Women Reading Corneille* (2000), and is the editor, among other volumes, of *Le mariage sous l'Ancien Régime*. Her work on early modern marriage led to her interest in the history of medicine and the idea of contagion.

Hélène Cazes teaches French Renaissance literature at the University of Victoria. A contributor to the volume *Henri Estienne, Imprimeur et Ecrivain* (2003), her current research topic is 'Children and Childhood in Renaissance France'. She came to the history of medicine during her research on the representation of infants and exploration of the textual transmission of Galen's theories by the humanists.

Frédéric Charbonneau, William Dawson Scholar and Professor of Eighteenth-Century French Literature at McGill University, has been working on seventeenth- and eighteenth-century French literary history in three specific domains: memoirs, prose genres and the representation of the body. He is the author of *Les Silences de l'Histoire* (2001; Raymond-Klibansky Prize Finalist); and co-edited, with Real Ouellet, an anthology of *Nouvelles françaises du XVIIe siècle* (2000).

Marianne Closson, Maître de conférences en littérature de la Renaissance at the Université d'Artois, is the author of *L'Imaginaire démoniaque en France (1550–1650). Genèse de la littérature fantastique* (2000). Her research focuses on demonology and demonic possession through the nineteenth century, viewed from both anthropological and literary perspectives. She has also published on baroque theatre.

Michel Fournier is an Assistant Professor in the Department of French of the University of Ottawa. His publications centre on the anthropology of fiction, history of reading and seventeenth-century French literature.

Nancy M. Frelick is Associate Professor of French at the University of British Columbia. She is the author of *Délie as Other: Toward a Poetics of Desire in Scève's 'Délie'* (1994) as well as articles on a number of early modern writers, including Scève, Rabelais, Montaigne, La Boétie, Gournay, Flore and Marguerite de Navarre.

Claude Gagnon publishes his work in *Chrysopoeia*, a periodical dedicated to the history of alchemy, and regularly contributes to *Aries (Journal for the Study of Western Esotericism)*. His principal field of interest is the history of Western theories of the soul from Aristotle to the present day. In 1979, he founded *Horizons philosophiques*, a journal devoted to popular philosophy, published by the Collège Édouard-Montpetit in Montreal.

Nicole Greenspan is a PhD candidate at the University of Toronto. She is completing a dissertation on news culture and the politics of anti-popery in mid-seventeenth-century Britain and has published articles on the Stuart courts-in-exile in the 1650s. She teaches at Sheridan College in Ontario, Canada.

Mitchell Lewis Hammond is an Assistant Professor of History at the University of Victoria. His research interests in early modern Europe include the activities of medical practitioners, poor relief and public health, and changing perceptions of illness and the body. His current project is a comparative study of medicine and society in the imperial cities of the Holy Roman Empire in the sixteenth and early seventeenth centuries.

Daniel Lindmark is Professor of History and History Didactics in the Department of Historical Studies and the Teacher Education Faculty, Umeå University, Sweden. He has published in various fields of educational and religious history, including emigration and Saami studies. His most recent books are *Reading, Writing, and Schooling: Swedish Practices of Education and Literacy, 1650–1880* (2004), and *Ecclesia Plantanda: Swedishness in Colonial America* (2005).

Isabelle Pantin is Professor of Renaissance Literature at the Université of Paris X-Nanterre, and associate researcher at the Observatoire de Paris (CNRS, SYRTE) on the history of astronomy. Her publications include *La* *poésie du ciel en France* (on cosmological poetry in the Renaissance, 1995), *Les Fréart de Chantelou: une famille d'amateurs au XVIIe siècle* (1999), *La Poésie du XVIe siècle, ouvroir et miroir d'une culture* (2003), and critical editions (with French translations) of Galileo's *Sidereus nuncius* and Kepler's *Dissertatio cum nuncio sidereo.* She has also collaborated in the recent edition of Montaigne's *Essais* under the direction of Jean Céard (2001).

Guy Poirier teaches French Renaissance Literature and French Canadian Literature in the Department of French Studies of the University of Waterloo, and is the Associate Editor of the Canadian journal *Renaissance & Reformation/Renaissance et Réforme.* He is the author of *L'Homosexualité dans l'imaginaire de la Renaissance* (1996), and edited in 2002 the collection of articles entitled *La Renaissance, hier et aujourd'hui.*

Rose Marie San Juan teaches in the Department of History of Art at University College, London. She is the author of *Rome: A City out of Print* (2001) and has published widely on European Renaissance art.

David Shuttleton teaches eighteenth-century literature at the University of Wales Aberystwyth and has published widely on literature and medical culture. He recently co-edited *Women and Poetry, 1660–1750* (2003) and is completing a monograph entitled *Smallpox and the Literary Imagination, 1650–1820.* As contributing editor to the *Cambridge Correspondence of Samuel Richardson,* he is editing correspondence between the fashionable physician George Cheyne and the novelist.

Preface

Seldom does a day go by that contagion is not in the news. Avian influenza, Norwalk virus, SARS, West Nile virus, BSE ('mad cow' disease) and a new strain of AIDS are preoccupations of the twenty-first century. We consider ourselves much better informed than our early modern counterparts about the transmission of disease, but that does not mean that our beliefs are always based in scientific accuracy or that the discourse of desire, of fear (even of panic) does not stimulate our imagination when it comes to this topic. Our very survival can be at stake, and the postmodern imaginary responds at times in ways not dissimilar to what we have found in exploring contagion in the early modern era. Spreading like a virus, the discourse of bodily contagion invades religion, politics, literature and the visual arts, generating powerful metaphors. But in a shared position, the authors of this volume insist that from the fifteenth to the eighteenth centuries, the material and the metaphorical are inseparable. As medical science advanced rapidly, coexisting with traditional beliefs, metaphor is more often than not embodied, and it is this emotionally charged intersection of mind and body that is explored in *Imagining Contagion in Early Modern Europe*.

The General Bibliography at the end of this volume highlights the attention that the notion of contagion has received from scholars in recent years. Our goal is to contribute to this discussion with work that is both interdisciplinary and coherent. In September 2003, twenty scholars came together at the University of Victoria, British Columbia (Canada) to shape this book before the chapters were written. Working from 1,000-word abstracts, we spent two days examining how each chapter would contribute a piece of the puzzle of the early modern imaginary of contagion. As you can see in the Notes on the Contributors, the team includes social and cultural historians, literary scholars, a philosopher and an art historian. Most of us do not claim to be historians of medicine, but all team members have worked in the area of medical history from the perspective of their own discipline.

With Fracastoro, Paracelsus and Ficino as patron saints, Part I sets out the debates that were taking place in the context of new scientific paradigms, including theories of demonic possession. Part II considers the impact of these theories on the daily lives of the citizens of early modern Europe. Part III suggests how the discourse of contagion will continue to project itself into the modern era. Issues of gender and race come in to play throughout the volume as mind and body remain joined in these projections more intensely than will be the case in the nineteenth and twentieth centuries.

In his 'Afterword on Contagion', Donald Beecher weaves together these strands of argument in a synthesis that reaffirms the unity of the volume, and shows how early modern anxieties are often reflected in our own epistemological questioning.

A note on translation

Six of the chapters presented here were translated from the French, first by a graduate research assistant, Rachel Warrington, with revisions by the volume editor. Unless a reference to an existing English translation is indicated in the notes, all translations were done by us.

Part I
Theory

1
Fracastoro's *De Contagione* and Medieval Reflection on 'Action at a Distance': Old and New Trends in Renaissance Discourse on Contagion

Isabelle Pantin

In the Renaissance, the horrors and mysteries of contagion[1] were manifested in their most terrible form by two mysterious diseases: syphilis, recently introduced in Europe,[2] and the plague, which had reappeared in the West at the end of the fourteenth century.[3] The works dedicated to syphilis were still relatively few and did not constitute a popular genre, but the latter had already inspired a considerable literature which expressed both a desperate feeling of impotence in the face of a curse which defied all human resistance, and a renewed effort to understand it. Thus, depending on what we are looking for, we can find in these works either remnants of archaic notions, such as explanations in terms of divine wrath or the use of analogical and magical thinking, or the first manifestations of a medical revolution in progress.

Murderous plagues and evasive causes

The plague provides the most striking and the best documented example of the complexity of the response to contagion. A discouraging disease, the plague was also stimulating in that it challenged the Galenic conception of pathology.

It could still be analysed using Galenic theory, provided that this theory was carried to its limits. In this view, it was a 'putrid fever': its victims inhaled corrupt air that attacked the organism *in tota substantia*, by some

3

occult quality, and putrefied it as would an instantaneous poison.[4] But first of all, the 'corrupt air' explanation, already problematic in itself,[5] did not fit the evidence: medics referred to this thesis, yet, at the same time, they issued sanitary rules (such as burning the patient's clothes and mattress[6]) that reflected a different conception of the problem. Moreover, the Galenic doctrine viewed disease as a totally natural process – giving this expression a meaning that excluded the occult[7] – and studied it only through observable experience. However, the plague defied observation. Medics often noted that it could be present without any external symptoms.[8] And finally, as the result of a kind of 'Copernican revolution', the plague incited medics to centre their reflection on the disease itself and not on the complexion of the patient, as the Galenic approach required. Confronted with epidemics, they perceived the terrible nature of the disease and sensed that it would be more effective to base their therapy directly on knowledge of it.

When considering the propagation of the disease throughout the population, the recurrent – and sometimes suspicious – problem of 'action at a distance' (also involved in magnetism and astrology[9]) was encountered. It is significant that, in the case of the plague, an astrological explanation was traditionally given: a conjunction between Saturn and Mars. According to Avicenna, this contributed to the corruption of the air.[10] The Black Death had been preceded, in March 1345, by a conjunction of the three superior planets (Mars, Jupiter and Saturn) in Aquarius,[11] which had been extensively discussed by the doctors of the University of Paris,[12] followed by many theologians, philosophers and physicians.[13] A treatise entitled *De peste*, composed in Avignon by Raimundus a Vinario, a physician attached to the papal court and published by Jacques Dalechamps in 1552, stresses the relations between celestial *affluxus* and occult properties integrated by the Creator into certain substances (for example, the magnet): these effects are all explained by a kind of sympathy based on similitude, analogy or favourable predisposition (*conveniens aptitudo*) between the agent and the recipient subject.[14] By a comparable process, certain malignant stellar influences can give us the plague when they correspond to a specific disposition of our bodies (*peculiari potestate, seu mavis specifica forma, qua ea valent in corpora nostra*).[15] The spread of syphilis inspired similar speculations.[16]

There was another type of explanation – one that was different, but quite compatible with the first: the idea of invisible particles transmitted by breath and bodily fluids. It had been marginally expressed in Antiquity and the Middle Ages[17] to justify the fact that contagion could take place not only through direct contact, but also at a distance through the air, and through more or less porous materials touched by sick people;[18] these materials were called *fomites*, a somewhat ambiguous expression that referred literally to the kindling feeding a fire: it designated primarily the receptacles of the infectious particles,[19] and sometimes, by extension, the particles themselves. In the Renaissance, this obscure notion encountered that of *pestifera semina* mentioned in the *De natura rerum*, when the poem evokes the pestilence in Athens (first described by Thucydides). According to Lucretius, these invisible *semina* move randomly in space and putrefy the air by amassing together.[20] We can observe that there was nothing in the Lucretian description that could perplex a medic trained in the medieval tradition, except perhaps the use of the very suggestive term *semina*: he had, by other mediations, acquired a similar conception of the causes of the corruption of the air.

Fracastoro adopted this term (and a variant: *seminaria*) and extended its meanings in his *De contagione*, written in 1538,[21] revised around 1542, and published in 1546, in Venice, with the *De sympathia et antipathia*.[22] The work concerned all forms of contagious diseases, but the plague and, to a lesser extent, syphilis offered Fracastoro the most extreme cases, allowing him to progress as far as possible with his own ideas. He had already dealt with these two diseases: his great poem, *Syphilis*, had been published in 1530 (Verona, Da Sabbio),[23] and he had composed a short prose treatise on the same disease, as well as observations on the plague of 1534–35.[24]

De contagione develops the idea that invisible living seeds (*seminaria*), capable of reproducing themselves, are the main agents of the propagation of pestilential epidemics. Thus, in the history of medicine, Fracastoro is sometimes viewed as a forerunner announcing the modern approach to pathology.[25] I do not intend to challenge this interpretation, for the history of science provides many similar examples: the first, seemingly fortuitous, appearance of imprecisely formulated and insufficiently demonstrated ideas, which reappear much later in the form of illuminating theories. Is there a link between the two stages, and what is the nature of this possible link? This is a major question.

I would simply like to explore another pathway – a complementary rather than a contradictory one – by examining in *De contagione* the continuation of medieval reflection on 'action at a distance' and, more precisely, the theory of *species*.[26]

The theory of *species*

The theory of *species* reached its full development in the work of Robert Grosseteste (1168–1253) and Roger Bacon (*ca.*1214–94); it took its inspiration from two main sources, which were closely linked: the Neoplatonic metaphysics of emanation, which was expressed most particularly in the conception of light, and the geometry and physics of optics which had been elaborated first by the Greeks, then the Arabs,[27] before returning to the West in the thirteenth century.[28]

According to this Neoplatonic conception, an infinite number of rectilinear rays emanate spherically from every point of everything in the universe carrying the powers of what they radiate from.[29] Thus, natural forces are propagated until they find an appropriate receiver in which they can express themselves, according to the nature of the receiver.[30] The term *species* is used to designate that which emanates;[31] its meaning is 'aspect', 'image'[32] or 'form', as well as 'likeness'. The sources, or agents, that produce *species* are many: substances, celestial and terrestrial (but not pure matter), and qualities: heat, cold, humidity, dryness, light, odour, taste and sound (in other words, the 'proper sensibles', *sensibilia propria*).[33]

Through the *species*, the agent always seeks to print its likeness on the receiver, but this is possible only if the latter has a certain degree of resemblance with the former,[34] which brings us to the question of sympathy. In short, *species* are what allows natural agents to 'multiply', although through a different sort of reproduction from that of physical generation, as it is carried out without material contact and by the simple activation of a certain potentiality already present in the receiver, and in the medium between agent and receiver.[35] The *species* are not transported but successively generated ('multiplied') in the medium, without discontinuity, but with a progressive attenuation.[36]

For medieval philosophers, the theory of *species* was considerably advantaged by its association with optics: it thus acquired the possibility of complete geometrization, as David C. Lindberg has demonstrated. However, the example of optics also shows that the theory was ambiguous from the very beginning: the light *species*, which transmitted the image of visible objects, conformed to the laws of geometry as long as they travelled through transparent or semi-transparent media; in other words, right to the interior of the eye. But when they were conducted through the dark corridors of the nerves to the various chambers of the brain, they adopted another form of behaviour, similar to that of the *spiritus* which wandered among the humours. The same occurred in all the *species* involved in sensation. The Renaissance philosophers, and Fracastoro in particular, inherited this combination – which appears strange to our eyes – of a geometrical and almost mechanical physics, and a physiological approach, apparently as unpredictable as life itself.

Fracastoro's synthesis

The theory of *species* lasted at least until the beginning of the seventeenth century, thanks primarily to its importance to optics. It is not surprising that Fracastoro was familiar with it. What is more remarkable is that he made an in-depth study of its possible application in solving the problem of contagion, and, with this goal in mind, searched to improve and complete it.

It is significant that he appended his treatise on contagion to a broader study on sympathy and antipathy. Like Fernel and Paracelsus, he thus situated his medical reflection inside a much larger philosophical framework, without limiting himself to the physical world, in the narrow sense of the expression, since he was also interested in sensations, passions, imagination and intellectual knowledge.[37] His purpose was to examine natural changes which could not be fully explained within the framework of strict Aristotelianism.

Enrico Peruzzi[38] has shown, by a thorough examination of some chapters of the *De sympathia* and by a comparison with their sources, that Fracastoro was well acquainted with the philosophical literature dealing with the *species* (especially the work of Bacon), and that he accommodated the standard theory to his own purposes. His chief object was to demonstrate that the 'occult' had no place in philosophy, which had to deal with medium particular causes, excluding what concerned, on the one hand, metaphysics and first causes, and, on the other, pure matter.[39] All natural phenomena had natural agents, which might remain 'latent'. These natural agents were necessarily either substances or qualities.[40] The first finality of all beings was their own conservation[41] which was assured by the first natural cause: sympathy, or *latens consensus rerum*.[42] This sympathy manifested itself by a universal connection between bodies (*nexus universalis*), which excluded the void,[43] and, strictly speaking, action at a distance.[44] Thus, in this new conception of sympathy, a new explanation of phenomena like magnetism had to be elicited. It was developed as follows.

As the iron body and the magnet are not in contact, they must be linked by something that is sent from the one to the other.[45] This 'something' is not material atoms:[46] Fracastoro demonstrates that atomism could explain some limited effects of magnetism, but not all of them; it cannot constitute *principium commune omnibus*.[47] Only *species spirituales* provide a general solution to the problem.[48] In the traditional conception, the *species*, being immaterial, have nothing to do with local movement (according to Bacon, as we have seen, they are not 'transported' but successively and 'spiritually' generated in the medium):[49] they cannot be involved in the process of attraction, except by introducing a new quality in the attracted body, that is, by altering it.[50]

Fracastoro wanted to remain in the general frame of explanations by local movement. Thus, he slightly modified the definition of the *species*: it was necessary that they be not only qualities, but also perfect substances.[51] Thus he assumed that there was a substantial identity between the *species spirituales* and the embodied forms from which they emanated; the only difference was that the former had no spatial limits and were somewhat attenuated or degraded.[52] In other respects, these 'representations' or *simulacra* of the substances[53] had the same capacities and powers as their sources, and by their propagation they were the main agents of the universal sympathy and cohesion in nature.

In the case of magnetism, the *species* of iron and the magnet were mingled, and a new 'whole' (and a new *continuum*) was formed, constituted by the two bodies and by their *species*. Thus, the attraction could be explained by the general theory of *motus partium in toto*, according to the principle that parts always tend to be reunited in the whole.[54] Iron and magnet, which were mixed bodies, were attracted by each other because they were somehow similar, even if the similitude was far from complete and consisted of a latent *quid* that escaped sensory perception.[55] This *quid* was not their exclusive property, but a general principle, common to different substances.

Contagion was another, more complex case of the so-called 'action at a distance'. In this approach, to touch on one of the keys to understanding the universal order, Fracastoro followed a similar path to Bacon's, while at the same time appearing to us an incarnation of the perfect Renaissance philosopher.

As we have seen, he accepted the same premises as his medieval colleagues: the need for sympathy – which involved similitude – between agent and subject in order for one to act on the other, and the possible existence of 'spiritual' intermediaries, which, although invisible, did not escape natural laws, since they too entered the confining logic of sympathy.

As in his explanation of magnetism, his first tool was the theory of *species*. And the main model he used to understand the propagation of disease was the propagation of light and sound, which also had two phases, first in the air or in a fluid, then in the living body. In this way, he was led to combine two models: one mainly geometrical,[56] the other based more on physiological and biological conceptions.

He finds no better comparison to describe the spread of contagion than that of *species*. Thus, it diffuses itself *in orbem*, and 'we see that it imitates the movement of spiritual *"species"*, whereas the bodies that possess known qualities have only one movement upwards or downwards'.[57] Chapter 7 of Book I ('Quomodo seminaria contagionis ad distans ferantur et in orbem') describes the same 'spherical' propagation; when it evokes the circulation of contagion in 'animals', it is the circulation of *spiritus* in bodily humours that becomes the model, but Fracastoro uses the term *soboles* (offspring), which in *De sympathia* express the emanation of *species* from their source (in the discussion on magnetism already mentioned).

> *De sympathia*, cap. 5: 'embodied forms' tenuem et superficialem sui vel partem vel gradum producunt, quem Epipolim vocant qui … momento gignitur, ac propagatur *ceu soboles quædam* …[58]

> *De contag.* I, cap. 7: … unus penetrationis modus est per propagationem et *quasi sobolem*, prima enim seminaria, quæ adhæserunt e vicinis humoribus ad quod habent analogiam, consimilia sibi alia generant et propagant, et hæc alia, donec tota humorum massa et moles efficiatur …[59]

In the first case, the offspring is only an attenuated simulacrum of the substance, in the second, it consists of living beings, the *seminaria*.

Atomism, as we have seen, for Fracastoro was a theory that had to be surpassed, without however being rejected altogether: rather than material atoms, he preferred spiritual *species*, which provided him with a more universal tool of explanation; but the possible existence of atoms is never denied in his book. Moreover, his conception of the generation of *species* from their source certainly has something in common with the Epicurean notion of *simulacra*: he presents the two terms as synonyms, obviously with philosophical reasoning.[60] This *soboles*, we might say, is the offspring of both Bacon and Lucretius.

In fact, his rejection of material atoms did not keep Fracastoro from recuperating the Lucretian *pestifera semina* and developing all their biological potential. Instead of replacing these Epicurean *semina* by something completely different, he improved on them by giving them all the properties of the *species spirituales*, and as well, those of the *spiritus* that lived in animated bodies. His *species spirituales* already possessed some of the powers of living beings; at least, as simulacra of all substances and qualities, they were the agent of everything in nature, in sensation and in cognition:

Harum vero spiritualium quod actio multa sit, et vis in natura : similiter est manifestum (ut iis diximus quæ de rerum sympathia et antipathia diximus) nam et sensus et intellectus movent, et principia sunt motionum in animalibus, deinde et motus locales videntur facere attractionem et fugam: nonnullæ etiam et primas qualitates producere, ut lumen quod calorem gignit ...[61]

The only, but decisive, difference was that the spiritual *species* did not survive when their origin disappeared.[62] Thus, something else had to be involved in contagion. The infectious agents seemed to have a life of their own since, when their first source had vanished, they could subsist in *fomites*, in the air, and be transported from place to place, even across vast seas; thus they were necessarily corporeal, and not only spiritual beings.[63]

The traditional definition of contagion required that the same infectious principle would be transmitted from the first patient, to the second, the third, and so on.[64] Spiritual *species* had the power to transmit an infection, even a lethal one, as in the case of the *catablephas* (the basilisk that kills with its eyes); but they could not 'generate' in another body the power that they possessed, so that a third person – then a fourth by his intermediary, and so on – could be affected.[65] Contagion did not merely consist of the transmission of a putrefaction, it involved the generation of the seeds of the disease in other bodies. The *spiritus* that assured the functions of life in animals could operate in a similar manner (the *spiritus* present in the blood possessed in themselves the power of generating other *spiritus*), but not the spiritual *species*.[66] Since Antiquity, the *spiritus* had allowed medics to resolve satisfactorily the problems posed by the connection between the material and the immaterial, the animate and the inanimate, and contagion was certainly one of these problems. We see that Fracastoro's *seminaria*, which were alive and proliferated through generation,[67] thus providing the only possible explanation of pestilential contagions,[68] were the result of a quite complex synthesis.

In *De sympathia* and *De contagione* Fracastoro had planned to deal with natural phenomena deprived of observable causes, notably different forms of the so-called 'action at a distance'. His intention had been to explain these phenomena in the philosophical framework he had conceived: the occult causes were excluded, but not the 'latent' ones. As he believed in the coherence of the natural world (governed by the laws of sympathy), his method of explanation could not lack in logic and unity. In fact, throughout his treatise he used similar theories and modes of reasoning. But the specificity of each particular phenomenon received due attention. Contagion was more complex a problem than magnetism (which involved only local movement). To resolve it, Fracastoro chose the same point of departure (the theory of *species*) but added other elements, taken from the Epicurean representation of the spreading of diseases and from the medical conception of the *spiritus*. The result was assuredly new, but it would be hazardous to assert that it was decidedly 'modern'.

Notes

1   On contagion, see V. Nutton, 'Seeds of Disease. An Explanation of Contagion and Infection from the Greeks to the Renaissance', *Medical History*, 27 (1983) 1–34.

2   See J. Arrizabalaga, A. Cunningham and R. French, eds, *Medicine from the Black Death to the French Disease* (Aldershot: Ashgate, 1998).

3   See J.-N. Biraben, *Les Hommes et la peste en France et dans les pays européens et méditerranéens*, 2 vols (Paris: Mouton, 1975–76). The plagues described in the Bible and in other ancient sources seem to have been various forms of typhus or pox. The plague appeared in Europe in AD 541 and disappeared at the end of the eighth century. It returned in 1347 (the 'Black Death'). Then, over four centuries, it remained the most lethal disease in Europe.

4   See M. Ficino, *Consilio contra la pestilenza* (Firenze, 1481): 'La pestilentia è uno vapore velenoso concreato nell'aria inimico dello spirito vitale ... quello vapore pestilente non proprio per calidita, frigidita, siccita, humidita è inimico, ma perche la proportione sua è quasi a punto contraria alla proportione, nella quale consiste lo spirito vitale del cuore', ch. 1, 2r°–v°. Ficino here follows the *Consilium contra pestilentiam* composed by Gentile da Foligno during the Black Death (Firenze, bibl. Laurent., Plut. 90 supra Cod. 20, fol. 65r°–v°); trans. in A. M. Campbell, *The Black Death and Men of Learning* (1931; New York: AMS Press,

1966), pp. 37–8; see also R. French, *Canonical Medicine: Gentile da Foligno and Scholasticism* (Leiden: Brill, 2001). Ambroise Paré assumes that the patient must be given an antidote (preferably theriaca) as soon as possible 'pour contrarier et resister au venin, non en tant qu'il soit chaud ou froid, sec ou humide, mais comme ayant une propriété occulte'; *Traité de la peste, verolle et rougeolle, avec une breve description de la lepre* (Paris: Wechel, 1568), ch. 24. See also Lauent Joubert: 'il n'y a propriété aucune de ses qualitez manifestes, qui rende l'air pestilent ... mais ... c'est une propriété que l'on dit specifique et occulte'; *De peste*, trans. G. Des Innocents (Toulouse: Lertout, 1581), p. 9.

5 An element in itself is always pure (see Joubert, *De peste*, ch. 1, pp. 2–3), but the air can contain infectious and poisonous particles. See Jean Jasme (*Traicté de la peste*, 1376; printed in Paris at the end of the fifteenth century): 'ceste pestilencieuse maladie est contagieuse car de corps infectz yssent humeurs et desfluent fumees venimeuses corrumpans et causans infections de lair'; cited in E. Droz and A. C. Klebs, *Remèdes contre la peste* (Paris: Droz, 1925), p. 34.

6 See A. Carmichael, 'Contagion Theory and Practice in XVth-century Milan', *Renaissance Quarterly*, 44 (1991) 231ff. Such prescriptions were dictated by the theory of *fomites* (see below).

7 This exclusion was not absolute, but the exceptions were limited. Galen admits that there are 'occult virtues' in certain remedies (especially the antidotes) which are efficient *tota substantia* (*De simplicium medicamentorum facultatibus*, ed. Kühn, XI, 823; *Methodus medendi*, XII, 356–8); these remedies, required in case of plague (see note 4 above) were identified *a posteriori* by the observation of their effects. Arab medics had been very active and inventive in this field. On the evolution of Galenism and on Arabic medicine, see O. Temkin, *Galenism: The Rise and Decline of a Medical Philosophy* (Ithaca, NY: Cornell University Press, 1973); and D. Jacquart and M. Micheau, *La Médecine arabe et l'Occident médiéval* (Paris: Maisonneuve et Larose, 1990). On the importance of alchemical remedies with occult properties, see C. Crisciani, 'Oro potabile ra alchimia e medicina. Due testi in tempo di peste', in F. Calascibetta ed., *Storia e fondamenti della chimica. Atti del VII Convegno Nazionale* (Roma: Accademia Nazionale delle Scienze, 1997), pp. 83–93; C. Crisciani and M. Pereira, 'Black Death and Golden Remedies: Some Remarks on Alchemy and the Plague', in A. Paravicini Bagliani and F. Santi, eds, *The Regulation of Evil: Social and Cultural Attitudes to Epidemics in the Late Middle Ages* (Firenze: Società Internazionale per lo Studio del Medioevo Latino, 1998) Micrologus' Library, 2, pp. 7–39.

8 'Car souvent apperent bonnes urines et bonnes digestions au pacient quand nonobstant ce il tent a la mort', Jasme, *Traicté de la peste*, ch. 2. See also Paré, *Traité de la peste*, ch. 24.

9 See E. Grant, 'Medieval and Renaissance Scholastic Conceptions of the Influence of the Celestial Region on the Terrestrial', *Journal of Medieval and Renaissance Studies*, 17 (1987) 1–23.

10 Avicenna, Canon, l. IV, F. I, tr. IV, c. 1.

11 See B. Goldstein, *Levi ben Gerson's Prognostication for the Conjonction of 1345* (Philadelphia: American Philosophical Association, 1990).

12 Campbell, *The Black Death and Men of Learning*, p. 40 ; L. Thorndike, *A History of Magic and Experimental Sciences* (New York: Columbia University Press, 1935) III, pp. 289–92, 303–9, 326–37; Rudolf Sies, 'Das Pariser Pestgutachten von 1348 in altfranzösischer Fassung', *Untersuchungen zur mittelalterlichen Pestliteratur*, IV (Würzburg, 1977).

13 H. Pruckner, *Studien zu den astrologischen Schriften des Heinrich von Langenstein* (Leipzig and Berlin, 1933); W. Coopland (Liverpool: The University Press, 1952).

14 'Affluxum, qui hasce res tractant, nominarunt syderum unicuique tributam vim peculiarem, ab eo qui summæ rerum præest, qualis est Magnetis cum ferrum trahit, Electri cum aceres. Formam specificam alii vocare malunt. Ea est conveniens aptitudo, ejus quod afficit, cum eo quod afficitur, ut commode et facile alterum suam vim exerat, alterum excipiat.' Raimundus a Vinario, *De peste libri tres*, ed. J. Dalechamps (Lyon, Guil. Roville, 1552), p. 23.

15 *Ibid.*, p. 26.

16 On the astrological explanation of syphilis, see Paola Zambelli, *L'Ambigua natura della magia* (Milano: Il saggiatore, 1991), ch. 4.

17 According to the pseudo-Aristotelian *Problemata* (*ca.* first century CE) phthisis, psoriasis and ophthalmia were transmitted by putrid particles contained in breath and sudation (VII, 8, 887A). See Nutton, 'Seeds'.

18 These three types of contagion were described by the school of Salerno in the tenth or eleventh century. See Nutton, 'Seeds'. We find the same distinction in Fracastoro's *De contagione* (I, 2 'De contagionum differentia'); *De sympathia et antipathia rerum liber unus. De contagione et contagiosis morbis et curatione libri III*, (Venice: Giunta, 1546; Lyon: Bacquenoys for Gazeau, 1550 cited here).

19 Fracastoro, *De contagione*, I, 2: 'fomitem appello vestes ligna et ejusmodi, qui incorrupta quidem ipsa existentia conservare nihilominus apta sunt contagionis seminaria prima, et per ipsa afficere' (29r°).

20 *De natura rerum*, VI, 1098 sq.

21 See A. Mundella, *Epistolæ medicinales* (Lyon: Junta, 1556), n° 3, Nov. 1538, quoted by Nutton, 'The Reception of Fracastoro's Theory of Contagion: The Seed that Fell among Thorns?', *Osiris*, 2nd ser. VI (1990) n. 13; and the letter to G.B. Ramusio (10 May 1549), in *Opera* (Cominiania, 1739), section *Quædam fragmenta*, I, pp. 96–7.

22 Other separate editions: Venice: Scoto, 1546; Lyon: Nicolas Bacquenoys, 1550; Lyon: Bacquenoys for Gazeau, 1550; Lyon: De Tournes and Gazeau, 1554. From 1555, the treatise was included in the successive editions of Fracastoro's *Opera* (Venice: Giunta, 1555; 1574; 1584; Lyon: F. Faber, 1591; Genève: S. Crispin, 1621; Genève: J. Stoer, 1637). On the text and its reception, see C. and D. Singer, 'The Scientific Position of G. Fracastor (1478?–1553), with Special Reference to the Source, Character and Influence of his Theory of Infection', *Annals of Medical History*, I (1917) 1–34; Nutton, 'The Reception'.

23 See L. Baumgartner and J.-F. Fulton, *A Bibliography of the Poem Syphilis sive Morbus Gallicus by Girolamo Fracastoro of Verona* (New Haven, CT: Yale University Press, 1935); F. Cairns, 'Fracastoro's *Syphilis*, the Argonautic Tradition and the Ætiology of Syphilis', *Humanistica Lovaniensia*, 43 (1994) 246–61.

24 See F. Pellegrini, Trattato inedito in prosa di G. Fracastoro sulla sifilide (Verona, 1939); *Idem*, 'Frammento inedito di G. Fracastoro riguardante la pestilenza del 1534–1535', *Rivista di Storie delle Scienze Mediche e Naturali*, 4a serie, 26 (1935) 253–9; *Idem*, *Scritti inediti di G. Fracastoro*, (Verona: Valdonega, 1955); *Idem*, *Origini e primi sviluppi della dottrina fracastoriana del 'contagium vivum'* (Verona, 1950). On Fracastoro's medical activities, see also O. Viana, 'L'atto di ammissione del Fracastoro al Collegio medico di Verona', *Rivista di Storia Critica delle Scienze Mediche e Naturali*, 5 (1914) 382–3; F. Pellegrini, 'L'epidemia di *Morbus peculiaris* del 1546–1547 e il medico del Concilio del Trento', *Castalia*, 2 (1946) 271–8.

25 This is the view of C. and D. Singer, and of F. Pellegrini. Nutton, conversely, has stressed the traditional aspects of Fracastoro's conceptions.

26 On this term, see P. Michaud-Quantin and M. Lemoine, *Etudes sur le vocabulaire philosophique du Moyen Age* (Roma: Edizioni del Ateneo, 1970), pp. 113–50.

27 Alkindi's *De radiis* (before 866), inspired by Plotinus, achieves a synthesis of neo-Platonic metaphysics and Greco-Arabic optical theory.

28 On this conception, completely theorized by Alhazen, and on its transmission, see D. C. Lindberg, 'Alhazen's Theory of Vision and its Reception in the West', *Isis*, 58 (1967) 321–41; idem, *Theories of Vision from Al-Kindi to Kepler* (Chicago: University of Chicago Press, 1976), ch. 4; G. F. Vescovini, *Studi sulla prospettiva medievale* (Turin: Giappichelli, 1965), ch. 7. Alhazen's *De aspectibus* (or *Perspectiva*) was translated into Latin towards the beginning of the thirteenth century. Roger Bacon was probably the first European philosopher who knew and understood thoroughly Alhazen's great optical treatise *De aspectibus*: his own *Perspectiva* (composed in the 1260s) is the fifth part of his *Opus majus*. Witelo (*Perspectiva*, ca. 1274), and John Pecham (*Perspectiva communis*, ca. 1274–79) were also influenced by Alhazen. See D. C. Lindberg, *John Pecham and the Science of Optics* (Madison: University of Wisconsin Press, 1970).

29 See M.-Th. D'Alverny and F. Hudry, 'Al-Kindi, *De radiis*', *Archives d'histoire doctrinale et littéraire du Moyen-Age*, 41 (1974) 224. On the geometrization of this conception, see the preceding note.

30 See, for example, Robert Grosseteste, *De lineis, angulis, et figuris*, in E. Grant, *A Source Book in Medieval Science* (Cambridge: Cambridge University Press, 1974), pp. 385–6; and Bacon: 'A species is the first effect of an agent; for all judge that through *species* "all" other effects are produced ... the agent sends forth a *species* into the matter of the recipient, so that through the *species* first produced, it can bring forth, out of the potentiality of the matter "of the recipient" the complete effect that it intends.' *De multiplicatione specierum* I, 1, trans. in D. C. Lindberg, *Roger Bacon's Philosophy of Nature: A Critical Edition with English Translation, Introduction, and Notes, of De multiplicatione specierum and De speculis comburentibus* (Oxford: Clarendon Press, 1983), p. 6.

31 See D. C. Lindberg, 'Roger Bacon on Light, Vision, and the Universal Emanation of Force', in J. Hackett, ed., *Roger Bacon and the Sciences* (Leiden: Brill, 1997), pp. 243–75.On further developments of the theory, see K. Tachau, *Vision and Certitude in the Age of Ockham* (Leiden: Brill, 1988) (centred on optics and cognition); *Eadem*, 'Et maxime visus, cujus species venit ad stellas et ad quem speciem stellarum veniunt.* Perspectiva and Astrologia in Late Medieval Thought', *Micrologus*, 5 (1997) 201–24; C. Gagnon, 'Le statut ontologique des *species in medio* chez Nicole Oresme', *Archives d'Histoire Doctrinale et Littéraire du Moyen Age*, 60 (1993) 195–205. Leen Spruit's monumental thesis (*Species intelligibilis. From Perception to Knowledge*, 2 vols [Leiden, Brill, 1995]) provides a general survey, but unfortunately does not investigate the link between the *species intelligibiles* (in other words, mental concepts) and the *species* in the sensible world.

32 In Augustine *species* designates the incorporeal image of an object in the senses or in the intellect.

33 *De multiplicatione specierum* I, 2, pp. 32–41; *Perspectiva* I, 1, 3; I, 10, 1). On the other kinds of sources, see Lindberg, *Roger Bacon's Philosophy of Nature*, pp. lvii–lviii.

34 The species of a substance is always substantial, etc. (*De multiplicatione* I, 2, pp. 42–3).

35   *De multiplicatione* I, 3, pp. 47–57. Lindberg, *Roger Bacon's Philosophy of Nature*, pp. lviii–lxi.

36   *De multiplicatione* I, 4; the propagation of light is similar (*Perspectiva*, I. 9, 4).

37   On other aspects of Fracastoro's philosophical achievements, see A. Orlandi, 'Malinconia e antropologia nel *De intellectione* e nel *De anima* di Girolamo Fracastoro, in *Psicopatologia e filosofia nella tradizione veronese. Atti del seminario di studi* (Verona: Università di Verona, 1994), pp. 9–17; E. Peruzzi, *La nave di Ermete. La cosmologia di Girolamo Fracastoro* (Firenze: Olschki, 1995); *Idem*, introduction to his edition and translation of Girolamo Fracastoro, *L'Anima* (Firenze: Le Lettere, 1999).

38   E. Peruzzi, 'Antioccultismo e filosofia naturale nel *De Sympathia & Antipathia rerum* di Girolamo Fracastoro', *Atti e Memorie dell'Accademia toscana di Scienze e Lettere. La Colombaria*, 45, nuova serie 31 (1980) 41–131.

39   See *De sympathia*, dedication to Alessandro Farnese, pp. 14–15, and cap. 2: 'quando hic non universalem et primam causam quærimus, sed particularem et propriam, quale esse non potest eorum ullum quæ immateralia sunt: sic enim periisset natura.'

40   *De contagione* I, 6.

41   'ut sint ac conserventur' (*De sympathia*, cap. 2).

42   *Ibid.*, cap. 1, 'De sympathia & antipathia multorum'.

43   *Ibid.*, cap. 2, 'De primo rerum omnium consensus'.

44   *Ibid.*, cap. 5 'De attractio similium ad similia': 'Quoniam igitur nulla actio fieri potest nisi per contactum ...' (ed. Lyon: Gazeau, 1550), pp. 45–6.

45   'necesse est, si applicari invicem debent, demitti aliquid ab uno ad aliud', ed. cit., p. 46.

46   As a scientific poet and as a philosopher, Fracastoro was well acquainted with Lucretius's *De natura rerum*. His friend Andrea Navagero had prepared the edition of the poem printed by Aldus in 1515. See C. Goddard, 'Lucretius and Lucretian Science in the Works of Fracastoro', *Res Publica Litterarum*, 16 (1993) 185–92.

47   *De sympathia*, cap. 5, pp. 46–9. See Peruzzi, *La nave di Ermete*, pp. 103ff.

48   Cf. the case of the magnetic attraction of the pole: it supposes a propagation at an immense distance, and the elements transmitted cannot be material: 'Attractio autem hæc ad Athomos et corpuscula reduci posse non videtur, quoniam corpus nullum in tanta distantia demitti potest ... speciem autem spiritualem nihil demitti prohibet, quod & in lumine manifestum est, cujus speciem a supremo orbe & stellis huc usque demitti constat, qua de causa attractiones similium ad corpora universaliter reduci posse, supra non arbitrati sumus.' *Ibid.*, cap. 7, p. 75.

49   Fracastoro recalls this conception (which he rejects): 'Porro nec videtur quomodo spiritualia hæc movere possint, præsertim trahere, quoniam productio eorum non cum motu locali, sed per quandam magis generationem partis post partem momento factam.' *Ibid.*, p. 50.

50   'Alii dicunt alterari ferrum a magnetis specie, et ita alteratum per se moveri ad ipsam.' *Ibid.*, p. 45.

51   *Ibid.*, p. 50.

52   'Recipiendum autem est, ut multis placet, spirituales species ejusdem rationis esse cum formis illis quarum sunt species nec differre ab iis nisi modo subsistendi: eatenus enim materiales sunt et dicuntur, quatenus crassa quadam existentia in materia sunt et certos terminos poscunt.' *Ibid.*, cap. 5, pp. 50–1).

53 'hæ igitur tenues et superficiales formæ aptæ imprimis sunt id quod sunt reprentare, usque ad crassas illas a quibus productæ sunt: propter quod simulacra earum et species sunt appellatæ: propter tenuitatem autem, et quod momento gignuntur spirituales dici consuevere: ejusdem tamen rationis cum iis, quæ et crassæ et materiales dicuntur.' *Ibid.*, cap. 5, pp. 51–2).

54 *Ibid.*, cap. 5, pp. 52–4. Cf. cap. 4, 'De consensu partium in toto'.

55 *Ibid.*, cap. 7, 'De sympathiis & antipathiis mistorum et attractione similium' (65–75). The species of the magnet do not attract iron *fortasse per id quod actu sunt sed par latens aliud in ipsis principium, quod simile ferro est aut ipsi aut principio in eo,* p. 69.

56 Sometimes with some touch of mecanicism, as in ch. 4 of the *De sympathia,* when he describes the movement of the *species* in the medium, using the comparison of waves propagated in water as the result of an *impetus (ibid.,* pp. 36–8).

57 'postremo quum hæc contagio undique, & ad omnem partem "se propagat", imitari quidem videtur spiritualium motum, qui in orbem fit. corpora vero quæ notis qualitatibus constant, unum tantum motum habere sursum aut deorsum', *De contagione,* I, 5, p. 234.

58 *De sympathia,* p. 51.

59 *De contagione,* p. 250.

60 See the text quoted above, note 52.

61 *Ibid.,* I, 6, pp. 236–7.

62 'spiritualia hæc tandiu solum durare consuevere, quandiu præsens est illud, a quo effluxere, nisi forte fuerint in intellectu'. *Ibid.,* I, 6 ('Quod causa contagionis, quæ ad distans fit, non sit reducenda ad occultas proprietates'), pp. 238–9.

63 'at quæ ad distans faciunt contagionem, absente etiam primo perdurant nihilominus, & in fomite, & in aere, quinimo de loco ad locum ferunt trans etiam maria, quod signum est corpus esse...' *Ibid.,* I, 6, p. 239.

64 See *ibid.,* I, 1: 'Quid sit contagio'.

65 'si enim recte definita contagio est, oportet tale in secundo fieri, quale in primo fuit, et idem esse in utroque principium, idemque et in quarto & quinto: & in aliis quæ contagionem recipiunt. tale autem non / potest facere ullum spiritualium per se, per accidens quidem nihil prohibet spiritualia enecare & dissolvere etiam mistionem aliquam fugando quædam contraria, quod & fœtor facere potest, & Catablephæ animalis aspectus (ut dictum in sympathiis fuit), generare autem tale in secundo, quale in primo fuit non possunt spiritualia.' *Ibid.,* I, 6, pp. 239–40.

66 'oportet autem in hisce contagionibus non putrefactionem solum fieri, sed a primis seminariis, et alia quoque gigni et propagari, quæ ipsis similia natura sint, & mistione, non aliter quam spiritus in animali e sanguine solent alios sibi consimiles generare, quod spiritualium nullum efficere per se potest ...' *Ibid.,* I, 6, p. 240.

67 *Ibid.,* I, 12: 'quoniam dictum est eam seminariis inesse vim ut sibi simile propagare, et gignere possint, sicuti et spiritus faciunt.' On *semina,* I have not yet been able to consult H. Hirai's *Le concept de semence dans les théories de la matière à la Renaissance de Marsile Ficin à Pierre Gassendi* (Turnhout: Brepols, 2005).

68 See *ibid.,* II, 3.

2
The Animism of Ambient Air at the End of the Middle Ages

Claude Gagnon

Greek pneumatics will serve as our horizon point. The *pneuma*, or vital breath, has as its source the soul of the world, animating Ptolemy's sky as well as Galen's medicine. In Ptolemy's sky, the stars have a determining influence on earthly mixed bodies, along with the elementary composition of bodies; the sky has great explanatory power for the most important thinkers of Antiquity and the Middle Ages. This primary causal role played by the sky in all natural and human events is totally absent from modern calculations and problematics: astrology has been marginalized in the world of knowledge for the past several centuries. On the contrary, physicians and other scientists of Antiquity and the occidental Middle Ages uphold the existence of the *Astrum*, composed of fluxes originating from each planet and of the influence of the constellations upon the events under examination.

Thus, in his *De Mineralibus* (mid-thirteenth century), Albertus Magnus explains the formation of different metals in different latitudes by the influence of the stars at the moment of their formation and by the composition of elements in that location.[1] This notion of the influence of constellations is a Ptolemaic postulate to which all scholars of the sky adhere. Albert refers explicitly to this in his metallurgic investigation: 'For as Ptolemy says, in no place does any of the elements receive so much of the rays of all the stars as in Earth, because Earth is the invisible center of the whole heavenly sphere.'[2]

It should be noted that Albert also explains the composition of mixed bodies by the combination of sulphur and mercury, a theory that Albert, like all the scholars of his time, borrows from Arab science. To the theory of the stars' influence from a distance in the Ptolemaic sky can thus also be added the physics of sulphur and mercury, which will condition the mental horizon of physicists and doctors well beyond the thirteenth century. The astral origin of substances and notably of metals, as well as the search for remedies originating from sulphur and mercury, thus constitute the scientific dogma that orient scholars' and physicians' research well before the arrival of Paracelsus in the sixteenth century.

The soul of the world

The sky, which has been an influence since Greek Antiquity, is alive. It is identified with the soul of the world in the Platonic tradition and also with a universal vital breath, the *pneuma*, where Galen situates the driving force behind all of his medical theory:

> For him, life is maintained by the perpetual regeneration of the *pneuma*. He recognizes three. The natural *pneuma*, formed in the liver, circulates in the veins, and once in the right ventricle, passes with the blood into the left ventricle, then is transformed into vital *pneuma* that the arteries carry along to the brain. From there, by a mysterious process, it is transformed again into animal *pneuma* that the nerves diffuse thus in the members.[3]

Pneumatics is dedicated especially, but not exclusively, to the study of the human soul, exploring as well the multiple forms of observed or supposed life: 'angels, demons, elemental spirits, disincarnate souls'.[4]

Emblematic of the Neoplatonic tradition, *The Theology of Plato* of Proclus (fifth century) is one of the most influential works in which the sidereal sky is considered to be the soul of the world. As Luc Brisson reminds us, 'in the central myth of the Phaedrus, the perceptible universe is described as an immense sphere, from which troops of gods, of demons and of human souls not yet incarnate emerge in order to go and contemplate intelligible forms'.[5] The substance of the sky is thus identified with the principle of life characterizing the totality of Nature: 'In short, Nature, identified with Fatality, is the Soul of the world, considered not in and of itself, but as the motivating cause of all movements. ... As a soul, Nature is assimilated to a bird endowed with wings.'[6]

This principle necessitates in turn the intervention of a spirit of the world (*spiritus mundi*) 'by the means of which it communicates life to the body of the world, everywhere. ... this spirit of the world is directly produced by the soul of the world itself ...'[7] There is thus a living divine sky which impregnates all things and it is by the *spiritus mundi* that the passage is made between the substance of the sky and the world of the four elements.

The spirit of the world

Distinct from the soul of the world, *spiritus mundi* chiefly develops within the alchemical tradition. Ten centuries after Proclus, thanks to

Marsilio Ficino, it became the major arcanum of the alchemical process.

Sylvain Matton's important study on Ficino aims to resituate him among authentic alchemists.[8] Ficino articulated his 'spirit of the world', which consolidates the link between the sky and the human soul: 'Between the soul of the world and its manifested body is its spirit (*spiritus*), in the virtue of which are the four elements. And by way of our own spirit, we can absorb it.'[9] This spirit of the world is a scientific idea both precise and vast, which Ficino extends all the way to the medical horizon: 'It is indeed upon the good usage and the mastery of the *spiritus mundi* that are founded not only medicine, hygiene and dietetics able to provide man health and long life, but it is also this astral magic that permits one to acquire material goods,' in the view of the Italian philosopher.[10] Intermediary between the divine soul and our bodies, this spirit is the object on which the alchemists, according to Ficino, will work: 'Separating this spirit of gold from fire by a certain sublimation, the diligent physicists will apply it to any metal and will make gold.'[11]

Thus, at the end of the fifteenth century, and before the arrival of Paracelsus (1493–1541) and his successors, the sky, the substance of which is a soul, finds itself placed oddly in the alchemists' alembic: 'Ficino here again explicitly identifies the celestial "fiery and vital spirit", that is to say the *spiritus mundi*, the sky "being everywhere", with the quintessence of the alchemists Arnaldus of Villanova and Raymond Lulle.'[12]

The medical alchemy of Paracelsus

The celebrated Swiss doctor is an authority in the history of modern medicine, and cannot be ignored. With regard to the arcana of the sky and of the world, he rejected Greek theories of the four elements and the four humours of Galenic medicine, preferring the Arab theory of sulphur and of mercury, known to medieval scientists of Christianity, to which he added a third fundamental principle, salt: 'If wood is burnt it will be resolved into its three true components: the flame – its "sulphur", the smoke – its "mercury" and the ash – its "salt".'[13]

In the specific case of contagion, it is necessary to put the infectious agent in contact with the living organ. This paradigm of Aristotelian physics (no action at a distance) is absent from the medieval doctrine of signatures, which triumphs thanks to Paracelsus.[14] The theory of signatures is articulated instead around concepts of similitude and of multiple correspondences that exist between forms or colours of plants or organs and around the observation of the capacity of certain plants to reproduce the symptoms of the patient.[15] This is the foundation of the theory or doctrine of signatures as well as the correspondence between the small world (microcosm) and the big world (macrocosm), a theme which appears as early as the fourth century in the *De natura hominis* of Nemesius of Emesia.[16]

The celebrated itinerant physician applied alchemical postulates explicitly and systematically to his medical theory and in the making of his remedies. As a doctor, he explained illnesses by astral influences, as medieval doctors had done. But he refused to consider the humours as primary agents of illness, following Galenic medicine. He proposed chemical medication which, according to some, makes him the father of chemotherapy.

For the Paracelsians, the concept of *spiritus mundi* continues to be a fundamental theoretical postulate. In 1594, the Paracelsian Gilles Pinot insists:

> Or comme le père universel de toutes choses ici-bas est le ciel, il est nécessaire que toutes choses ici-bas participent de la nature et de la substance du ciel. Et le ciel est ainsi caché intérieurement en toutes choses, lui qui est manifeste extérieurement hors de toutes choses. Et de même que par son corps, il est immensément apparent à la surface du monde, de même par son esprit il est étroitement resserré au centre du monde ... Et comme de par son corps il renferme et recouvre tout, ainsi par son esprit il traverse et remplit toutes choses.

> [As the universal father of all things in this world is the sky, it is necessary that all things in this world participate in the nature and substance of the sky. And the sky is thus hidden inside all things, and is manifest externally outside of all things. And just as by its body, it is immensely apparent on the surface of the world, it is closely tied to the centre of the world by its spirit ... And as with its body it closes and covers all, thus by its spirit it crosses and fills all things.][17]

Paracelsus and his adepts thus treat all kinds of pathologies with the *spiritus mundi*, the quintessence of the soul of the world, for this soul descended from the sky is omnipresent: the soul of the world flows constantly in us, conveyed by the foods that we absorb because it is this same soul that resides in vegetables, animals and man.[18]

Contagion by ambient air

Rejecting elements and humours, Paracelsus none the less welcomes the elemental spirits of the Jewish cabbalistic cosmology, who people the environment and participate in the diffusion of illnesses.[19] Entirely indebted, like all his contemporaries, to Antiquity's theory of the soul of the world and the form of the sky concealed in the body, the physician-healer imagines a new chemistry of the air:

> [Life] comes to us from and through the air – for 'air gives all things their life'. There is nothing corporeal that has not 'a spiritual thing' hidden in itself. ... The world of bodies is a replica of the world of spirits. For there

are as many spirits as there are bodies: celestial, sublunar, human, metal, mineral, salt, herb, wood, flesh, blood, bone.[20]

Like Guillaume Postel and many other learned men and Gnostics of the time, Paracelsus believes that 'these spirits come to us from the *stars*; they are "astral bodies". They carry with them life, function and individual specificity.'[21]

One needs no more in order to imagine the cause of disease as being a specific spirit which enters the body via ambient air: 'disease is not endogenous or constitutional; to engender it, a foreign invader, a "seed", is required.'[22] The atmosphere makes man ill by means of 'injection', that is by the transmission of an astral poison. The general vector of astral emanations is the vaporous 'chaos' that surrounds us: air, which belongs to the *Ens Astrale*. We see the importance Paracelsus attached to the air – the milieu of normal life, in the universe as in man, and at the same time the carrier of morbid agents: 'It is not "astral inclination" that causes disease, but simply transmission of "astral poison" by the air.'[23]

As Lucien Braun says, 'the Sky is that within which we live. We breathe the visible and the invisible, and each star has its correspondent in us. If the air in a room is polluted, says Paracelsus, everyone breathes it, and each will suffer according to his own vulnerability.'[24] This is the essential originality of Paracelsian thinking: the external origin, be it from the sky, of the primary cause of many illnesses that were traditionally explained by humoral imbalance. From this time on, disease has agents in the ambient air: 'Paracelsus visualized diseases as beings ready made in the form of "Semina" which invade man from outside.'[25] 'Disease in Paracelus' doctrine thus assumes the character of an entity in itself which is identical to its "seminal" cause. Diseases are seen as substances rather than something that happens to the body.'[26] None the less, astral poison is not yet liberated from its mythical and theological chains because Paracelsus' Christianity points to human will as the primary cause of disease: 'But first it is man who creates the astral semina of the disease, the *contagium*. This is a physical entity, a body. But it is created by something non-corporeal, the sinful passion and imagination of man.'[27] It is Van Helmont who will liberate the *semina* from the sphere of perversion: 'he subordinates man to the action of the poison which exists independently, whereas in Paracelsus' opinion, the poison is the product of the sinful actions of man.'[28]

The new imaginary of ambient air

Simultaneously with Paracelsian reform, new intuitions arise. As of 1509, the *condottiere* Pietro del Monte, in his speech against alchemy, insists that 'quicksilver, fixed or not, is of a different *species* from all metal'.[29] A century later, in 1607, the doctor Jacques Fontaine disputes the notion 'that there was to be found in the mixes a celestial quintessence, distinct from the four elements'.[30] These long-suppressed doubts concerning the presence of mercury in all metals and the substance of the sky in all things ends up triumphing over the diagnosis of illness by celestial influence and of its healing by the injection of the quintessence of the sky into the sick organism. Ambient air will no longer be the carrier of the soul of the sky on earth and in our humours.

During the seventeenth century, air will be the object of all kinds of physical and chemical experiments that will lead, at the end of the following century, to discoveries and new imaging. But as James Hillman demonstrates, it is not certain that the old theory of pneumatism did not play a role in the modern conception of ambient air and its physical and chemical properties: 'Pneuma or air is the body of the stars, the froth of animal sperm, the seed of the metals.'[31]

Analysing the imagination of the alchemists and the chemists of the first centuries of modernity, the psychoanalyst Hillman does not explain otherwise the appearance of certain discoveries attributed to the multiple experiments on ambient air. At the end of the eighteenth century (1783), the Montgolfier brothers ascended in their first hot air balloon. A decade later, gaslight made its appearance in a foundry in Birmingham, then spread to various cities in England. At the very beginning of the following century (1807) carbonated water destined for commercial consumption appeared. Hillman observes that surrounding these discoveries there are imaginings similar to those of the alchemists. Concerning carbonated water, he writes:

New soda-powders were dispensed to make water fizz. Scientific papers stated the benefits of aerated water. Physicians used 'fixed air' (air of the laboratory) in treating putrid kinds of diseases such as small pox, driving out the putrifactio by means of enemas of fixed air and carbonated waters. Soda-water, like gas-lighting by Accum and ballooning by Franklin, was presented as a public boon, hopes for peace and health, darkness dispelled.[32]

Hillman and Debus reassert the value of the work of the iatro-chemists: 'Not only mass, motion, gravity and direction of Newtonian physics determine the vectors of change in physical bodies and so in our lives, but odours. Chemistry is a science of airs, the determination of the Blas in the world, a truly pneumatic study.'[33] The spiritus mundi, orphan of its celestial father who remained a prisoner of the alembics, will continue to be affirmed in the classical age of the scientific revolution. Pierre Jean Fabre, the king's doctor and a contemporary of Descartes, grants great importance to the spiritus mundi in his philosophy of nature and in his alchemical speculations.[34] The pneumatist philosophy had perhaps, in some respects, triumphed: 'They [the chemists] were as well inspiring the world with new visions of things, that balloons and ventilators, wind charts and air-pumps were carriers of highest aspirations, a pneumatic vision parallel with free-thinking.'[35] More recently, Philip Ball, in his introductory work on the elements, corroborates Hillman's judgement: 'The second half of the eighteenth century was the age of "pneumatic chemistry", when the properties of gases, typically called "airs", were the focus of the discipline';[36] Rutherford, Cavendish and Priestley were all pneumatic chemists.[37]

Creatures of the Chaldaic oracles

There was more than the Paracelsian spirit of the world in the ambient air of early modern Europe. The Platonic soul of the world had engendered numerous other types of creature, of which some survived throughout the Middle Ages. For it is the soul of the world that maintains the links between the three existing worlds: the empyrean, the ethereal and the material tangible world. The doctrine of the soul of the world was largely developed by Proclus in *The Theology of Plato*, which distinguished the souls of the celestial regions from human souls, prisoners of the material world. The first includes 'gods, angels, demons, heroes, disincarnate souls'.[38]

The tradition of the oracles influenced many commentators and thinkers of the medieval occident, among them Michael Psellus in the eleventh century, and especially Gemistus Plethon at the end of the fifteenth century.[39] At the dawn of the modern age these oracles, however archaic, are still part of the landscape of sources when Henry More, with the group of Platonicians of Cambridge, 'continued in the seventeenth century many of the themes and traditions of Renaissance Platonism'.[40]

Unlike the *spiritus mundi*, the oracles did not contribute to the creation of new images to account for disease. They reveal a sky and an atmosphere which have as their structure a ternary subdivision of all celestial regions: the empyrean region, the ethereal region, the material region.[41] The first corresponds to the intellective world and its creatures are pure intelligences: the Iynges. Following in the ethereal region are the Builder-Maintainers, while the Perfectors occupy the material world.[42] Besides these gods, other hyper-cosmic beings are called the chief gods. Then come the seven encosmic gods who preside over the celestial bodies,[43] then the soul of the world, identified with the whole of Nature. The oracles speak little of angels but much of demons, who are associated with the sublunary world, with water and air.[44]

All of these triads of industrious gods come directly from ancient Chaldaea or were transmitted by texts like the Chaldaic oracles, of which only fragments remain. Certain Chaldaic beliefs lasted for millennia as far as scholars are concerned, for example, the belief in the link that unites each man with the stars and each with a precise star.[45] This belief was the object of an astrological directive in all medical prescriptions of Antiquity, the Middle Ages and especially in the work of Paracelsus and his adepts during the sixteenth century.

The celestial and aerial space of the oracles, in addition to postulating a meridian between a man and a star, also suggests action at a distance. In effect, all of these divine figures are so many forces of intercession. The sky, the air, the wind, the rain are mechanics personified by beings very distant from men, but quite close to them by the effects of their actions. Action at a distance to combat an evil spell or to produce a remedy thus has its source in the Mesopotamian sky as imagined by those who, furthermore, invented our calendar.[46]

The demons of the Jewish Cabbala

The corpus of texts of the hermeneutic tradition translated into Latin by Marsilio Ficino in the last third of the fifteenth century constitutes one of the major axes of the Renaissance. At the same time, writes the historian Moshe Idel, Flavius Mithridates translated into Latin a collection of Cabbalistic treatises.[47]

Unlike the translations by Ficino, which were published at the time, the translations by Mithridates remained manuscript. None the less, they would be just as influential, in that they were read by Giovanni Pico della Mirandola, who was in contact with both translators.[48] Thus, at the end of the fifteenth century, Pico became the first Western author to synthesize the two marginal oeuvres and to attempt to harmonize them with Christianity.

The reception of Jewish Cabbalistic literature and its integration into Christian mysticism constitutes an autonomous current that developed continuously during the sixteenth century, influencing different cosmological or theurgical theories of the age. To the quintessential vital breath and to the masters and demons of the Chaldaic oracles were added Satan and the numerous demons of Christian literature. During the sixteenth century, Cabbalistic demons further enlivened the already charged atmosphere inhabited by theologians and demonologists but also by the physicists and physicians of the time. Besides the multiple angels, archangels and demons that influence the life of men, there are other entities, called 'elemental spirits': 'Gnomes, Sylphs, Undines and Salamanders that inhabit respectively the earth, the Air, Water and Fire.'[49]

It is not the addition of a few demons or spirits that characterizes the influence of the Jewish Cabbala on the medicine of the Renaissance; it is rather the maintenance of theurgical or magical actions in diagnosis and prescription. The folklore of the Cabbala legitimized the talismanic practice of the doctors of the Europe that was coming into being. The Cabbala added demons to the air and reinforced the determining effect of the sky upon men and their humours.

Cognitive theories of the philosophers

In addition to the influx of the sky, the gods and demons engendered by Christianity and by the Hebraic Cabbalists, there were other imaginary entities in the ambient air at the end of the Middle Ages. Since Antiquity, two opposing theories posited the way in which the knowing subject appropriates certain qualities of the object to be known. The Pythagoreans imagined a ray (the 'visual ray') leaving the eye and travelling to the object; this is the theory of extramission. Empedocles and the atomists surmised instead a ray leaving all objects and travelling in the direction of the subject; this is the theory of intromission. Galen combined both theories: there is 'a fluid directed from outside towards the eye, and another interior one that, without leaving the eye, sensitizes this organ and renders it able to receive impressions from the first fluid'.[50]

All thinkers thus are investigating the mode of being of these forms or corpuscles that are in the ambient air, water or other intermediary transparent medium between the knowing subject and the known object. The sensory qualities like colour and odour but perhaps also other qualities such as movement, luminosity, similitude, beauty, etc., are emitted by the different objects of our perception, according to John Pecham, celebrated medieval optician of the thirteenth century, who recognized 22 kinds of emanations from objects, towards us (*species intentiones*).[51]

Katherine Tachau's definitive study of the *species* and their medieval history explores both optics and the theories of knowledge.[52] She notes that there are in the ambient air emanations of different qualities from all objects, which impact not only on the senses and on the intelligence, as specified by Roger Bacon, but also on the totality of the sensory world still forming.[53] Bacon explains that this emanation by multiplication takes place without the displacement of flow but instead by a 'corporeal form' that is not at all in the matter of the ambient air.[54] But if it is not in the matter of the air, in what mode does the *species* reside, for it effectively reaches the sensory organ? The medieval physician and psychologist Nicole Oresme, one century later, will suggest an answer.

Nicole Oresme and his air purifier

As David Lindberg explains, the concept of *species* was strongly linked to the concept of emanation of Being, dear to the Platonists. This concept, far from being secondary, over centuries came to furnish the explanation for efficient cause in the sublunar world. Lindberg specifies that the logical link between emanation and the causality of events was developed by Plotinus,[55] confirming the primary importance of the metaphysical role of the *species*, given that so many physicians and theoreticians of knowledge from Greek Antiquity to the classical age had studied them. Historians of knowledge often emphasize William of Ockham's critical work on these *species*, which he considered quite useless in establishing the foundation for tangible or intelligible knowledge. Katherine Tachau sees therein an application of Ockham's 'razor'.[56]

Whatever Ockham's solutions, another medieval thinker a little later in the fourteenth century will propose a theory of the mechanics of *species* in knowledge and their mode of existence in the ambient air and other mediums (water, glass) that separate us from the objects we perceive. We are referring to Nicole Oresme, nicknamed *Doctor anticipator*.

Nicole Oresme's *De Anima* remained unpublished until very recently.[57] Reading this work in light of what he says about these same *species* in other works reveals his solution concerning their very existence.[58] In opposition to Ockham's view, Oresme saves the *species* from pure extinction. In effect, Oresme gives a definition of the *species* in seven points which radically and absolutely differentiates them from primary physical qualities (hot, dry, etc.). This means that their substance is not at all of the tangible order and that it is, on the contrary, purely intelligible, existing solely in the immediate space of the meeting between a subject and an object.

Oresme's removal of the *species* from the materiality of air is as interesting a solution as Ockham's, according to the historian of physics André Goddu: 'So we could say that in Oresme's version we have another reductive interpretation of *species* (as in Ockham), yet one that affirms their existence in an otherwise Aristotelian account of the cognitive abstraction of forms from individual things.'[59]

Oresme's rationality also had its limits. His multiple demonstrations of what he called the incommensurability of celestial movements would have as an indirect consequence the creation of distance, however small, between the sky and human destinies. However, like all the theologians and doctors of his time, he considered demons to be acting on our constitutions and never questions their existence. According to Eugenia Paschetto, who has explored this very question, Oresme would have justified the existence of demons in order to preserve the existence of miracles, for he was a fervent Catholic.[60] While the existence of demons is not demonstrable and the philosopher must consider as improbable the action of demons in our world, their existence must not be questioned according to this precursor of the fourteenth century.[61]

We should make explicit the rational foundation which makes it possible for these demons, who are not composed of matter, to act on our bodies. In his work dedicated to the Jewish alchemists, Raphael Patae summarizes this foundation by showing how the demons, which are not made of elemental matter, can nevertheless be eliminated by the doctors' medications. Moreover, only theology can guarantee the effectiveness of the doctor in the face of an illness provoked by supernatural beings. It is the Dominican and alchemist Raymund de Tarrega, in the fourteenth century, who writes:

How can it be that the demons can be expelled from the bodies, when they have no bodies in which they could receive the impressions of medicine? ... The demons are attached to human bodies because of bad disposition and corrupt humour, or because of melancholic infection which generates evil, black and horrible images in fantasy, and disturbs the intellect, for the demons habitually take on such forms, and generally dwell in obscure and solitary places. When by virtue of the fifth essence and of other things this humour, which is the reason they enter such a body, is expelled from it, then at the same time also the demons vanish at once together with the humour. ... so that their punishment should be greatly multiplied they feel the punishment by order of divine justice.[62]

It is thus divine justice that empowers the medication to banish the demon by acting on the humour.

The historian of medicine Danielle Jacquart reminds us of the seven kinds of natural things that medieval physicians had studied since Galen: this is the classical physiological doctrine of the elements, the humours, the complexes, the membranes, the spirits, the virtues and the operations.[63] With regard to the spirits, 'it is not an exaggeration to say that there are as many descriptions of the spirits as there are authors, in the fourteenth and fifteenth centuries'.[64] Despite a clear definition by Avicenna, the prince of physicians, of the humours as 'a liquid body into which the nutriment is first converted',[65] a thousand and one questions occupied medieval doctors. Generally speaking, the appearance of disease was explained by an imbalance of the humours. But the medical consensus stopped there. The famous fifteenth-century physician Jacques Despars formulated some of the medical questions surrounding the humours:

les humeurs engendrées dans l'estomac, dans le foie, dans les veines, qui répondent au même nom, sont-elles de la même espèce ... les humeurs sont-elles inanimées ... est-ce que le sang nourrit ... les esprits, est-il en substance une vapeur du sang?

[the humours engendered in the stomach, the liver, the veins, and called the by the same name, are they of the same *species* ... are the humours inanimate ... does the blood nourish ... the spirits, is there in substance a vapour of the blood?][66]

The central role of the humours and of temperament in medical diagnosis resisted Paracelsian critique which downplayed their role. Medieval medicine already had in its doctrine a clear idea of contagion, for example in hippiatrics, the care of horses. Like humans, horses were cared for according to Galen's theory of humours. This did not prevent the celebrated medieval veterinarian Ruffo from designating two causes of disease: an imbalance of humours and contagion![67] The phenomenon of contagion, the second cause of disease, is explained thus:

Le cheval souffre de prurit; il se gratte et se mord. Il corrompt ainsi son haleine et l'haleine corrompue agit facilement sur l'homme et les bêtes. La raison de cette action est la suivante: une affection chaude et sèche se transmet facilement à un organisme chaud et humide par l'intermédiaire de l'haleine qui est elle-même chaude et humide.

[The horse suffers from pruritus; he scratches and bites himself. He thus corrupts his breath and the corrupted breath acts easily on man and on beasts. The reason for this action is the following: a hot and dry ailment is easily transmitted to a hot and moist organism by the intermediary of the breath which is itself hot and moist.][68]

These two determining causes of hippic illness, imbalance of humours and contagion, are also influenced by occasional causes, of which some are not, contrary to the humours, internal, but rather external to the sick organism. These causes are, respectively: the meteorological factor (intense cold and heat), the emanations from manure, the penetration of air into the interior of the body, diet and overworking the beast.[69] Thus Galenic pneumatism already contains within itself a conception of air that travels and is a vehicle for disease coming from the exterior.

We see to what extent air is still charged with animism in cosmological and medical theory at the beginning of modern times. During the seventeenth century, pneumatism is redefined in such a way as to take part in several modern discoveries regarding the properties of air. But the sky, in the eighteenth century, has lost its voice; the zodiac was absent from diagnosis and from new pharmaceutical prescriptions which were more and more chemical. The spirit of the world was no longer around us; its image was stealthily displaced from then on into the compact matter of new medications–pills and vitamins![70]

Notes

1 Albertus Magnus, *De Mineralibus* (anastatic rpt; Paris: Manucius-Bibliothèque interuniversitaire de médecine, 2003), p. 145.
2 Albert the Great, *Books of Minerals*, trans. Dorothy Wickoff (Oxford: Clarendon Press, 1967), p. 182.
3 R. Bouissou, *Histoire de la médecine* (Encyclopédie Larousse, 1967), p. 70: 'En physiologie, il est pneumatiste. Pour lui, le maintien de la vie est dû à la

régénération perpétuelle du *pneuma*. Il en reconnaît trois. Le *pneuma* naturel, formé dans le foie, circule dans les veines, et arrivé dans le ventricule droit, passe avec le sang dans le ventricule gauche, puis il se transforme en *pneuma* vital que les artères charrient jusqu'au cerveau. De là, par un processus mystérieux, il se transforme à nouveau en *pneuma* animal que les nerfs diffusent alors dans les membres.'

4   A. Lalande, *Vocabulaire technique et critique de la philosophie* (Paris: PUF, 1968), p. 785: 'anges, démons, esprits élémentaires, âmes désincarnées'.

5   L. Brisson, 'Les oracles chaldaïques dans la *Théologie platonicienne*', in A. P. Segonds and C. Steel, eds, *Proclus et la* Théologie Platonicienne (Leuven-Paris: University Press-Les Belles letters, 2000), p. 130: 'dans le mythe central du Phèdre, l'univers sensible est décrit comme un immense sphère, de laquelle sortent les troupes de dieux, de démons et d'âmes humaines non encore incarnées, pour aller contempler les formes intelligibles.'

6   *Ibid.*, p. 155: 'Bref, la Nature, identifiée à la Fatalité, c'est l'Âme du monde, considérée non pas en soi, mais comme cause motrice de tous les mouvements. ... En tant qu'âme, la Nature se trouve assimilée à un oiseau pourvu d'ailes.'

7   S. Matton, 'Marsilio Ficino et l'alchimie; sa position, son influence', in J.-C. Margolin and S. Matton, eds, *Alchimie et Philosophie à la Renaissance, Actes du colloque international de Tours, 4–7 décembre 1991* (Paris: Vrin, 1993), p. 143: 'au moyen duquel elle communique partout la vie au corps du monde. ...cet esprit du monde est directement produit par l'âme du monde elle-même ...'. This passage summarizes the theory of Porphyrus glossed by Proclus and taken up by Marsilio Ficino.

8   *Ibid.*, pp. 123–92.

9   *Ibid.*, p. 145: 'Entre l'âme du monde et son corps manifeste se trouve son esprit (*spiritus*), dans la vertu duquel sont les quatre éléments. Et nous, au moyen de notre propre esprit, nous le pouvons absorber.'

10   *Ibid.*, p. 144: 'C'est en effet sur le bon usage et la maîtrise du *spiritus mundi* que se fondent non seulement la médecine, l'hygiène et la diététique susceptibles d'assurer à l'homme santé et longue vie, mais c'est aussi cette magie astrale qui permet d'acquérir les biens matériels.'

11   *Ibid.*, p. 145: 'Séparant par certaine sublimation au feu cet esprit de l'or, les diligents physiciens l'appliqueront à n'importe lequel des métaux et feront de l'or.' Matton notes that to speak in a non-pejorative and not ironic manner of the alchemists, Ficino uses the term physicists (*physici*) or 'philosophers of nature', thereby avoiding irony and pejorative connotations (p. 148).

12   'Ficin identifie ici encore expressément le céleste "esprit igné et vital", c'est-à-dire le *spiritus mundi*, le ciel "étant partout", avec la quintessence des alchimistes Arnauld de Villeneuve et Raymond Lulle.' *Ibid.*, p. 149.

13   W. Pagel, *Paracelsus: An Introduction to Philosophical Medicine in the Era of the Renaissance*, 2nd rev. edn (Basel; New York: Karger, 1982), p. 105. The analysis of the opposition of principles and elements in Paracelsus' work by Pagel is definitive; see pp. 82–104.

14   A. G. Debus, *The French Paracelsians* (Cambridge: Cambridge University Press, 1991), p. 12: 'In their opposition to ancient philosophy, the Paracelsians turned to sympathetic action in nature and argued that magnetic action was accomplished not by contact as in Aristotelian physics, but at a distance.'

15   L. Braun, 'Le foisonnement des signes' in *Paracelse* (Lauzanne: Editions René Coeckelberghs, 1988), pp. 65–72; E. Lev, 'The Doctrine of Signatures in the Medieval and Ottoman Levant', *Vesalius*, VIII, 1 (June 2002) 13–22.

16 F. Hudry, 'Le *De secretis nature* du ps.-Apollonius de Tyane' in *Cinq traités alchimiques médiévaux* (Paris-Milan: S.É.H.A-Archè, 2000), p. 8.

17 Cited in D. Kahn, *Alchimie et Paracelsisme en France à la fin de la Renaissance* (Geneva: Droz, forthcoming), ch. 3, section 4, p. 21. I wish to thank the author for sharing his work in press.

18 Pagel, *Paracelsus*, p. 226.

19 Debus, *The French Paracelsians*, p. 27.

20 Pagel, *Paracelsus*, p. 117.

21 *Ibid.*, p. 118.

22 *Ibid.*, p. 140.

23 *Ibid.*, p. 141. Pagel refers to the *Paramirum*.

24 Braun, *Paracelse*, p. 127: 'Le Ciel c'est ce dans quoi nous vivons. Nous respirons le visible et l'invisible, et chaque étoile a son correspondant en nous. Si l'air est vicié dans une pièce, dit Paracelse, tout le monde le respire, et chacun va en souffrir selon sa vulnérabilité propre.'

25 Pagel, *Paracelsus*, p. 216.

26 *Ibid.*, p. 342.

27 *Ibid.*, p. 181.

28 *Ibid.*, p. 187.

29 M.-M. Fontaine, 'Pietro del Monte', in *Documents oubliés sur l'alchimie, la kabbale et Guillaume Postel offerts à François Secret*, ed. S. Matton (Geneva: Droz, 2001), p. 121: ' le vif-argent, fixé ou non, est d'une autre espèce que tous les métaux'.

30 Kahn, *Alchimie et Paracelsisme en France à la fin de la Renaissance*, ch. 4, p. 41: 'qu'il se trouvait dans les mixtes une quintessence céleste, distincte des quatre éléments'.

31 J. Hillman, 'The Imagination of Air and the Collapse of Alchemy", *Eranos Yearbook*, 50 (1981) 273–333.

32 *Ibid.*, p. 181.

33 Hillman, 'The Imagination of Air', p. 309. Hillman specifies that 'The blas originates in the stars. It is the dynamic impulse within natural phenomena such as air-motions, weather, storms, and the subtle dynamic of human beings in their neurovegetative symptoms' (*ibid.*).

34 See F. Grenier, *Pierre-Jean Fabre, L'alchimiste chrétien* (Paris and Milan: Chrysopoeia, 2001). Also consult the works of B. Joly on Fabre.

35 Hillman,'The Imagination of Air', p. 329.

36 P. Ball, *The Ingredients: A Guided Tour of the Elements* (Oxford: Oxford University Press, 2002), p. 32.

37 *Ibid.*, pp. 33–4.

38 *Ibid.*, p. 155: 'dieux, anges, démons, héros, âmes désincarnées'.

39 *Ibid.*, p. 118.

40 F. A. Yates, *Giordano Bruno and the Hermetic Tradition* (1964; rpt Chicago: University of Chicago Press, 1991), p. 423.

41 Brisson, 'Les oracles chaldaiques', p. 133.

42 *Ibid.*, p. 130. Proclus, in his commentary on the oracles, attributes to the 'theologians' 'l'association de l'empyrée avec l'intellectif' (n. 55).

43 *Ibid.*, p. 153.

44 *Ibid.*, p. 155.

45 L. Chochod, *Histoire de la magie et de ses dogmes* (Paris: Payot, 1949), p. 92.

46 *Ibid.*, p. 84.

47 M. Idel, 'Kaballah and Hermeticism in Dame Frances A. Yates's Renaissance', in *Ésotérisme, gnoses et imaginaire symbolique : mélanges offerts à Antoine Faivre*, ed. R. Caron and J. Godwin (Leuven: Peeters, 2001), p. 71.

48 *Ibid.*, p. 72.

49 Chochod, *Histoire de la magie*, p. 69: 'les Gnomes, les Sylphes, les Ondines et les Salamandres qui habitent respectivement la terre, les Airs, l'Eau, et le Feu'.

50 V. Ronshi, *Histoire de la lumière* (Paris: Armand Colin, 1956), p. 33: 'un fluide dirigé de l'extérieur ver l'œil, et un autre intérieur qui, sans sortir de l'œil, sensibilise cet organe et le rend apte à être impressionné par le premier fluide'. On the history of visual and luminous rays in vision, see C. Gagnon, 'Métaphysique de la vision' in *Voir*, L'Agora, July 1999.

51 D. C. Lindberg, *John Pecham and the Science of Optics* (Milwaukee: University of Wisconsin Press, 1970).

52 K. H. Tachau, *Vision and Certitude in the Age of Ockham; Optics, Epistemology and the Foundations of Semantics* (Leiden: Brill, 1988).

53 For a detailed analysis of the ontological status of the Baconian *species*, consult C. A. Ribeiro do Nascimento, 'Une théorie des opérations naturelle fondée sur l'optique: le *De multiplicatione specierum* de Roger Bacon', *Manuscritto; Revista de Filosofia*, V (October 1981), 33–57.

54 Lindberg, *John Pecham*, p. 337: 'nor is it body which is generated there, but a corporeal form; ... and it is not produced by a flow from the luminous body, but by a drawing forth out of the potentiality of the matter of the air.'

55 D. Lindberg, *Theories of Vision from Al-Kindi to Kepler* (Chicago: University of Chicago Press, 1976), p. 335.

56 Tachau, *Vision and Certitude*, pp. 130–4.

57 P. Marshall, *Nicolas Oresme's Questiones super libros Aristotelis de Anima* (Ithaca, NY: Cornell University Press, 1980). Benoît Patar later found the Expositio of Oresme's treatise and has edited it: *Nicolai Oresme expositio et questiones in Aristotelis De Anima*, (Louvain: Peeters, 1995). I myself collaborated in the doctrinal study preceding the editing of the text; cf. C. Gagnon, "Description analytique et commentaire synthétique de l'Expositio", pp. 139–61.

58 C. Gagnon, 'Le statut ontologique des *species* in medio chez Nicole Oresme', *Archives d'histoire doctrinale et littéraire du Moyen Âge*, 60 (1993), 195–205.

59 A. Goddu, review B. Patar, *Nicolai Oresme...*, in *Speculum*, 72 (January 1997) 207.

60 E. Paschetto, *Demoni e Prodigi; Note su alcuni scritti di Witelo e di Oresme* (Turin: G. Giappichelli, 1978), p. 61.

61 *Ibid.*, p. 61: 'Oresme, che ha diachiarato naturaliter indimonstrabile l'esistenza dei demoni, non solo liammette per fede, per non compromettere l'authenticità di quanto narrato le Scritture, ma finisce per assumere la loro esistanza come se fosse un dato razionalmente dimostrato, per convalidare la realità dei miraculi.' See Marianne Closson in this volume.

62 R. Patae, *The Jewish Alchemist* (Princeton, NJ: Princeton University Press, 1994), pp. 201–2.

63 D. Jacquart, *La médecine médiévale dans le cadre parisien, XIV–XVᵉ siècles* (Paris: Fayard, 1998).

64 *Ibid.*, p. 349: 'il n'est pas exagéré de dire qu'il y a autant de descriptions des esprits que d'auteurs, aux XIVᵉ et XVᵉ siècles.'

65 *Ibid.*, p. 331: 'un corps liquide dans lequel le nutriment est converti en premier'.

66 *Ibid.*, pp. 332 and 347.

67 G. Beaujouan, *Médecine humaine et vétérinaire à la fin du Moyen Âge* (Geneva: Droz, 1966), p. 61.

68 *Ibid.*, p. 64.

69 *Ibid.*, p. 66.

70 Note the descriptions of results promised by different vitamin supplements and other optimizers consumed today. For example, the Usana Health Science firm, one of the most credible in the domain, shows the effects of ginkgo biloba and phosphatidylserin with the support of references and clinical tests: 'Des effets mesurables dont une amélioration de la mémoire, de la concentration et de la fonction d'apprentissage ont été constatés. [Le ginkgo] pourra vous aider à préserver une pleine acuité mentale et à prolonger votre joie de vivre' ('Measurable effects included helping memory, attention and learning abilities. [Gingko-PS] can help you maintain full mental acuity and extend the enjoyment of your life for years'), *Product Information*, 2004–5, p. 39. Could this be the ultimate imaginary refuge for the *pneuma*?

3
Windows on Contagion

Donald Beecher

The study of infectious disease ... is a field where argument from analogy is always dangerous and sometimes misleading.

J. A. Boycott

Contagion is a phenomenon of transmission, generally of pathogens that attack the physical organism. *Tangere*, or touch, is the root of the word, suggesting a necessary proximity so that potential parasites may leap from host to host. Nevertheless, it was recognized throughout the second millennium of Western medicine that ideas and images, communicated from afar, retain their power to injure the mind and redound on the body with full psychosomatic potency. Thus it followed that the mechanisms of infection might be extended to include the transmission of those images that were capable of provoking states of disease in the body. No medical condition demonstrated the phenomenon more clearly than lovesickness (*amor hereos, ilischi* or philocaption), which was a form of melancholia or mania due to the unavailability of the beloved object, but more precisely to the retention in the memory of an image of the beloved that had become entirely detached from reality, resulting in desires that could not be socially directed. There was wide agreement that the eye was the point of entry of the malady because it was through the eye that the victim received the *species* of the object deemed beautiful and hence desirable. There was wide agreement, as well, that because this simulacrum of the beloved progressed through the mental faculties in a tyrannously coercive fashion, it must have been imbued with an innate power, such as one attributes to poison, potions, *maleficia* or the evil eye. Such a construction of contagion, for the modern mind, hovers between magic and metaphor, which invariably reopens the discussion concerning what constitutes natural phenomena for the Renaissance mind.

Lovesickness was a prime example of contagion from a distance, and by dint of being a disease it was entitled to an explanation in pathological

32

terms. But given the nature of sight, the definitions imposed on lovesickness by analogy with ostensibly related phenomena contained both occult and material elements which the syncretists could only blend rhetorically by defining each in the terms of the other.

The medical analysis of *amor hereos* as a disease of the imagination was well established by the late Middle Ages. The founding ideas were a legacy of the Arabic physicians Rhazes, Haly Abbas and Avicenna, whose works had been made available to Western physicians through translations and commentaries beginning in the eleventh century. Moreover, poets as early as the thirteenth century began to write meditations on love that reflected the symptoms and causes spelled out in the medical treatises of Gerard of Berry, Peter of Spain and Arnaldus of Villanova; for that reason their works were, in turn, commented on by medical philosophers. This peculiar reification of the two cultures led to diagnostics that were both poetic and scientific, resulting in a *contaminatio* that brought the darts of Cupid ever closer to the best scientific analyses of the physiology of vision. These observers, from the point of an elite culture, had little to say about *maleficia* and the occult. The lady's eye had powers, but they were not demonic and there was no reason to think that in the Petrarchan world order, any more than in the medical, sorcerers or devils had any role in the matter, even though the popular culture of the late Middle Ages was full of lore on amulets, philtres and witchcraft in the promotion of illicit love.[1] Yet the logic of occult causes deemed natural rather than demonic began to make its incursion into the analyses of love produced by the elite and philosophical culture in the fourteenth century, and came to bear particularly on theories of contagion. The question of the aetiology of lovesickness ultimately attracted the attention of one of the watershed minds of that age, that of Marsilio Ficino, whose reflections on the causes of 'vulgar love' would play a principal role in turning medieval *amor heroes*, or erotomania, into an *idée force* among medical philosophers in the sixteenth and early seventeenth centuries.

Ficino's analysis of love as a form of contagion through the transfer of blood vapours comes to light in his Seventh Oration, the last in his celebrated *Commentary on Plato's Symposium on Love* – a work he began as early as 1469, but published only in 1484. In a phrase of his own, his preoccupation was with 'the flying up to divine beauty, aroused by the sight of corporeal beauty'.[2] Hence, his description of lascivious love originating in the frenzy of desire and seeking fulfilment only in the bestial coupling of bodies was a detour into an undesirable opposite. It was an expression of the 'venerean demon' which had run amok. In describing the origins of this diseased passion, however, he resorted to the rays of the eyes in full symmetry with the rays and beneficent demons serving as agents throughout his cosmological system. The final impression is an aetiology of love advanced in purely medical terms that nevertheless functions in perfect symmetry with his world of spiritual operations. All forms of beauty are *species* of reality held in the imagination, just as the image of the beautiful woman is a *species* created in the eye as it is passed to the *virtus estimativa* for evaluation. The creation and modification of these phantasms in the imaginative faculty became a topic of intense speculation on the part of the Neoplatonists, leading them incessantly to those notions of benign or natural magic by which they explained the transfer of spiritual influences. The consequences were a fine muddle because such language tends to 'occultify' natural causes, tempting those who read him to invest their theories of amorous contagion with darker shades of magic. Ficino's analysis of vulgar love remains faithful to the elite medical tradition around *amor hereos* with its methodical, surgical and pharmaceutical cures originating in the *Canon* of Avicenna. Yet the embedding of the received nosology of lovesickness within his greater economy of love resulted in modified notions of infection: that the *species* of the beloved in the corrupted imagination of the ardent lover was the same as the *species* of beauty in general, and that by the same means that man has a burning desire to return to his origins in the Creator, lovers seek each other out to reclaim the blood vapours they had lost in the fatal glances. Ficino, in this, was an original thinker, even though all of the material and occult dimensions of sight, and the principles of reciprocity based on love and beauty which he brings together are not without antecedents in the western tradition.

Returning to axioms, lovesickness as a disease caught through the eyes was a disorder principally of the imagination where the noxious *species* was lodged. Such a *species* was subject to all the agents that influenced the processes of mentation in general, including the conditions that arose from the surfeit of black bile in natural or burned states. Ficino reasoned that as a somatopsychic condition, love required an aetiology that joined the operations of vision with a palpable transfer of pathogenic material, thereby accounting for all the perturbations of the body experienced by lovers, from trembling and flashes of heat to sleeplessness and loss of appetite. Overheating was a literal state of affairs, associated with the combustion of biles and the production of vapours, with amatory fever, and with the sensation of poisoning in the extremities. His answer to these questions was found in the transfer of blood vapours by which the *species* of the object deemed beautiful is conveyed across space. Not only were the vapours credited with the power to retain such precise data as the complete image of the beloved, but as alien materials they were highly toxic to the host, in no way differing from conventional poison or venom. Ficino aligned the effect with several other contagious conditions, which he also understood in terms of poisons or toxins: the 'itch, mange, leprosy, pneumonia, consumption, dysentery, pink-eye, and the plague'.[3] It is an astonishing revelation, which speaks volumes about their understanding of contagion in general. So conceived, it is less surprising that the stare of the beloved should be compared to that of the basilisk, and by extension to the evil eye. But if love is ocular poisoning, then the issue is no longer metaphorical. The blood vapours were not merely an option; they were a necessity. Medical tradition had long held that through the physiology of sight the *species* of the beloved is reduced to a simulacrum in the mirror of the eye before it is delivered, in diminished but fully potent form, to the *virtus estimativa*. Once the object is deemed beautiful, and hence desirable, it is passed as a phantasm to the imaginative faculties where, in dialogue with a memory increasingly devoted to the image, it polarizes the mind in a state of compulsive absorption: the *complexio venerea*.[4] Through the principles of faculty psychology, erotic captivation achieved a full level of medical analysis. Ficino worked a new synthesis of his own, calling on all the contributing systems to philocaption, by reassessing all these operations in terms of the animal spirits. According to received doctrine, the vital and animal spirits were the essential agents in the communion between the body and the soul. Blood, in its ideal condition, is thin, clear, warm and sweet. Such properties are conveyed into the by-products of the concoction of blood such as sperm and the blood vapours. Through heat generated in the heart, quantities of blood were regularly converted into the thinner and more volatile vital spirits. These vital spirits, composed of inhaled air and exhalations of the blood, were subject to further refinement and rarefication in the *retiform plexis* of the brain. The animal spirits generated there occupied the parencephelon, where they performed the operations of the rational soul.[5] Their pneumatic circulation extended to the eyes. Ficino reasoned that radiation was a property of the rarest of spirits, and particularly through the eyes because of their resemblance to windows, and because they were the most luminous part of the body. By extension, the eyes became the most apt instruments for the projection of vapours as rays, still carrying not only their initial properties of the blood, but an encoded identity of the sender, along with all the lifelike contours of her appearance. This densely encoded essence now becomes an inter-body messenger, strangely reifying the rays that elsewhere in Ficinian philosophy are the conveyors of image and idea.

As a syncretist, Ficino is under no obligation to resolve whether these rays are also darts, whether they are shot, or merely absorbed by the viewer, whether they are actively or passively emitted and received, and whether their toxicity is due to a residual poison or merely to the alien status of the material. Ficino seems, rather, to be gathering lore of any useful kind to extend, by analogy, the significance of the transaction. Physicians had already discussed the seat of the disease, whether in the heart, the head, the hypochondries or the sexual organs, and it was through the thinness of the vapours that Ficino could incite all of these contributing centres. Later physicians would continue to adapt the direction of the poisoned blood to those areas of the body associated with the origins of the disease.

To this new pathology Ficino adds a remarkable coda, again attributing to the mechanical force of the vapours precisely what his Platonic habits of reasoning called for in terms of reciprocity. In the *Timaeus*, Plato himself speaks of the vision that sends out its own force to catch the image to be brought into the eye.[6] Beauty is actively pursued by the five senses, even as beauty radiates its own essences. Ficino's world, in the image of Plato's, is a place of dynamic exchange and continual transaction. Thus, even for lovers, what begins as a poisonous stare concludes in mutual toxification, again through properties invested in the vapours. He reasoned that eyebeams are not only an expenditure of spirit dangerous to the sender because their loss causes desiccation and even death, but a property of the self which the sender recalls to its rightful place of origin. But now, in so far as it is natural for the self to seek the self, a 'double bewitchment' occurs when ray joins to ray and vapour to vapour in a state of mutual affliction. For with the return of the rays the sender finds herself tainted by the presence of the alien vapours of the host. A strange reciprocity. In this way, the love that is a disease results in mutual contagion by the same mechanisms.[7] The operative term is 'bewitchment', for the glance of the lover is not only *species* and poison, but a spell, as though cast by some form of ritual magic, or by a philtre. A nexus has been formed around a notion of contagion that now includes a dimension of malign volition or coercively occult forces. Ficino, at this point, does not explain whether the double bewitchment becomes a double enchantment and thus self-cancelling in pathological terms, or whether lovers are caught in a vortex of depleting glances that waste them both away. Vulgar love was a mere antinomy to transcendental love, and his intent may have been more rhetorical than scientific. But he wrote with a cogency that satisfied the critical minds of his age, and thereby left a more potent medical idea than he perhaps realized. There is a sense in which the entire explanation of love was a literalizing of metaphor as the best explanation of a condition requiring material causes, and those metaphors arose with the best analogical thinking whereby donor systems were mapped on target systems.

Ficino provided numerous examples of his methods of reasoning. The semen, by dint of its capacity to replicate features of the parent in the offspring, must of necessity encode those features by being drawn uniformly from all parts of the body. Hence, coitus was a means of conveying the entire organism in micro-form, encoded in the seed, a process of reduction and recomposition entirely analogous to the transmission of vapours whereby the entire image of the donor is replicated out of micro encoding in the mind of the receiver. In this way, blood vapours gain a synonymous power to imprint images which have themselves been imprinted on the blood.[8] At the same time, he allows that love is a frenzy or *furore* caused by the shining of beautiful eyes, in direct relationship to the operations of the basilisk, seizing on the heart as a toxin that spreads throughout the body.[9]

That there is material light in the eye apt to radiate is proved by the eyes of animals that glow in the dark, he tells us,[10] and that Aristotle suggested that menstruating women can by staring at a mirror produce drops of blood on the surface is proof for him that thin blood escapes by the eyes and can hence enter by the eyes and there be congealed again in blood.[11] He then asseverates that young boys are bewitched by the stinking breath of old men[12] which is the authority on which he builds his theory of vulgar love as a bewitchment that is finalized by the attempt of vapour donors to reclaim their own blood spirits. The passage is a classic in the slippage by analogy that propels one set of ideas, through the authority of others, until the best set of similitudes confirms phenomenological fact. Through the conflation of these several operations, he is able not only to link many of the poetic features of love in materially causal terms, but tie his diagnostics to the conventional cures, including phlebotomy, whereby the poisoned blood is expended from the body. Quite sensibly, too, he calls first for a separation of the lovers so that the gazing is interrupted. Amusingly, he takes up the recommendations of Avicenna to have a trusted old crone abuse the desired object in the hope of altering the cherished image, to allow the victim to drink, even to excess, and to expend seed in any way possible. His proposed cures are confirmation that he is thinking in conventional medical terms.

The afterlife of this reconfiguration of eroto-contagion entailed, for the most part, a verbatim adoption on the part of the medical philosophers and writers of the *trattati d'amore*. But that was not the case for Battista Fregoso, Duke of Genoa, in his *Anteros, sive tractatus contra amorem* (Milan in 1496). In this remarkable *omnium gatherum* of lore on love, the author divides his perspectives among the three interlocutors, representing in the character of Fregoso the entire Ficinian analysis. He explains the rays in terms of Cupid's darts and the loss of vital and animal spirits through erotic activity. He agrees with Ficino that eyebeams have the capacity to alter blood and cause infections in the same way that stinking air, such as the strong odours from nostrils and stomachs, causes contagion 'as a thousand examples will confirm'.[13] He too includes an allusion to the evil eye, shifting imperceptibly to the mechanisms of *maleficia* as reinforcement, returning to the transfer of phthisic disorders and dysentery from body to body by the breath as parallel examples of contagion by vapours. Even more revealingly, he concurs with Ficino that such spirits always desire to return to their original bodies, either through coitus because the sperm is rich in animal spirits, or through the reciprocal gaze. But the final word is given to Platina, who speaks for the wisdom of conventional humoral medicine, thereby taking issue with all that Fregoso had argued on the basis of Ficinian analysis. In the manner of the humanist dialogue, the residues of rejected theories and the relativity of final voices leave room for doubts and syncretisms, but the importance for our purposes is the prominence enjoyed by Ficino's speculation on the blood vapours as pathogens and the nature of love as a long-distance infectious disease within the polemic.

Following the trail of Ficino's invention down to its citation in Robert Burton's *Anatomy of Melancholy*,[14] one passes through the case study of the rich merchant of Arles suffering from erotomania recorded by François Valleriola among his *Observationum medicinalium libri sex* (1588),[15] the *Second discours, ou quel est traicté des maladies melancholiques, et du moyen de les guarir* of André Du Laurens, written in the 1590s when he was rector of the medical faculty in Montpellier and published in 1613 in his complete works,[16] *L'antidote d'amour* of the Bordeaux physician Jean Aubery, published in 1599,[17] and *De la maladie d'amour ou melancholie erotique* of Jacques Ferrand, first published in 1610, followed by a revised and amplified version in 1623.[18] For our purposes, these works may be treated as one, having in common their reiteration of the fundamental components of Ficino's final oration. Valleriola offers, in a no-frills résumé of some 40 pages, a complete medical digest of philocaption in which he harmonizes the views of Avicenna, Aristotelian faculty psychology, Galenic humoral medicine, and the poetic culture pertaining to the eyes medicalized by Ficino. In a matter-of-fact way he endorses most of the common ideas: the power of the eye to concentrate images, the transfer of vapours, the fiery radiance of the eye itself, the generation of sun-like heat, the projection of a flame, along with the darts that he now describes as sliding and gliding into the soul of the beholder. He speaks of the tainted blood, its heating and excitation, the burning poison that causes vehement desire, all of which he links to the *fascinatio* or the occult powers of the eye to generate enchantment. Valleriola concentrates on matters of heating and frenzy because his patient was in a state of mania rather than depression over loss; clearly he had been enchanted. Central to this analysis is the vividness of the image held in the imagination and the craving to pour the self into the beloved, in reference to Lucretius and Plutarch, both of whom are cited in similar contexts by Ferrand.

André Du Laurens begins his account of eroto-contagion with the theory of rays, tracing their mode of invasion and their means for targeting the liver and hypochondries, which for him was the seat of the cause of the disease where the combustion of the melancholy vapours took place. He simplifies the aetiological system, in a sense, by provoking the adustion directly through the contaminated blood, before naming the exterior signs of the disease, namely the foolish speech, depraved imagination, sleeplessness and loss of appetite. Du Laurens had little to say about the occult powers of the eye, conscious as he was of separating occult from material causes, but Heinrich Kornmann in his *Linea amoris*, published in 1610, makes the *fascinatio* a basic component of his theory of amorous contagion,[19] while Jean Aubery debates in his scholastic way whether the power of the eyes to enchant is the same as the basilisk's.[20] This and other questions had become fixed *topoi* for rhetoricians such as Aubery and Jean de Veyries,[21] losing in the process something of their clinical urgency. Among these writers, Jacques Ferrand was the most sceptical, but for the sake of inclusiveness he entertained every mechanism that properly pertained to the causes and treatments of love by tradition or hearsay, making his treatise a balanced performance between the medical practitioner and the encyclopaedist. He accepts the visual projection of the vapours and their attack on the hypochondries, disinclined as he would have been to depart from André Du Laurens's teachings. But he attributes to Ficino and to Valleriola by name the notion of enchantment, disinclined to associate love with magic after seeing his first treatise called in and burned by the Inquisition in Toulouse, in part for its tolerance of magical causes. But he does outline the in-bound and out-bound trajectories of the beams, their thinness and subtlety, and their capacity to cause perturbations throughout the body and to generate melancholy blood.[22] By the early seventeenth century, this association of ideas had established its now iterative vocabulary, where we can leave it to its destiny – albeit one that may have lasted longer than expected. Esquirol, in the 1830s, on matters pertaining to erotomania, which he called 'monomanie erotique', was still citing Ferrand as his principal source.[23]

Returning to the origins of the 'meme' as it was passed from physician to physician, we find ourselves confronted not only by the inventiveness of Ficino, massaging new ideas out of mental habits to explain apparent effects by best assessments of their necessary causes, but by a cultural legacy stretching back to the ancients in which nearly every component of Ficino's medical digest is anticipated. All of them are potential 'ideograms' in the history of an idea, namely that love is not only like a disease, but by dint of its symptoms a veritable disease, which by extension must be contracted through the transfer of pathogens from body to body – pathogens that in turn produce results similar to parasitic agents in an unwilling host. In this idea-cluster there are many provisional parts. Poets and philosophers had agreed that desire was first aroused by the effect of beautiful eyes upon the beholder. Even Aristotle confirms in the *Nicomachean Ethics* that 'no one falls in love without being first pleased with the personal appearance of the beloved object'.[24] By the simplest of projections, the sight of the beloved could be invested with energies whereby the resulting emotions are attributed to vectors of the eyes – the extromitted visual rays. Early among those voices was Museus who, in his poem on the fatal love of Hero and Leander, spoke of the eye as the cause of the illness of love. Ficino quotes him on this very score in his oration.[25] Sappho, too, was favoured by the medical philosophers because she described in such vivid terms the many physical manifestations of the inner turmoil of lovers in quasi-medical terms. Ferrand suggested that no physician had surpassed her in the diagnostics of love.[26] Guido Cavalcanti, late in the thirteenth century, still writing in the tradition of the 'fins' amour' of the Provençal poets, played the trick of treating old conceits as new realities. He spoke of hearts wounded by glances,[27] and how foolish eyes permitted to gaze could expect nothing but death.[28] By the eyes of a gentle country girl he found himself wounded in the heart and trembling.[29] But above all, his 'Donna me prega' functioned as an esoteric gloss on the medical theories of his age, providing verification of their facts in the poet's own experience. That circularity of data from philosophy to experience and back to philosophy in the commentaries written by physicians on his poem became a particular form of factification concerning the conceits of the love poets.

Thus it came about that before Ficino, Egidio Colonna, in a commentary based on another by Dino del Garbo, based on an imperfect copy of the poem by Cavalcanti, advanced the doctrine of the eyes as shining rays.[30] Egidio, whose exact identity remains unknown,[31] proposed in the name of Cavalcanti, that beauty recreates itself in the ray-like image conveyed to the beholder, which is the kindling of love, and that the beholder is then set on fire through the concentrated heat. Cavalcanti had said nothing about such rays, but on Egidio's authority, Ficino attributed these ideas to the poet himself. It was a convenient discovery for corroborating his own theory of divine rays, now replicated in the phenomenology of contagion.

That was but one of the 'ideas' invested in the diagnostics of ocular contagion. By analogy, the operations of the evil eye appeared to have a 'natural' part, even while the doctrine of the double enchantment seems to be an opportunist invention derived from the Platonic story of the Androgyne, recounted by Aristophanes in the *Symposium*.[32] Magic and fable join in common cause because their structural operations can be superimposed upon those of 'natural' contagion. Quite simply, if evil eyes and beautiful eyes can cause affliction, they may do so by the same mechanisms, or they may achieve the same results by independent mechanisms, the one occult, the other pertaining to the properties of the animal spirits. Similarly, the myth of the androgyne relates a fundamental truth about erotic craving in so far as the other 'half' sought in the sexual embrace is in reality an original component of the self; in coupling it is the self seeking the self because men, divided by the gods, would naturally seek reintegration with their severed parts. The blood vapours are the vehicles of this physical–metaphysical transaction.

Concerning the occult powers resident in the eye, entire treatises were written. The stare even today retains its bewitching powers, manifested in the calculations incumbent on us all in letting our gaze fall on proximate anatomies. There are but slight discriminations separating the invasive stare from the illegal stare and from the malefic stare that makes the object feel compromised, unclean, withered. It is telling that a word for the evil eye 'occurs in nearly all languages, European and non-European'.[33] It is telling, too, that many of these words carry equivocal connotations such as the word 'glamour', Scots for the evil eye, by attributing to the word the alluring qualities in the object that provoke the stare. In the world of Martin del Rio, the *fascinatio* still retained a capacity to injure by the powers invested in the gaze, a capacity even sought out by the gazer.[34] In associating the eye of the beautiful lady with the basilisk, a quality in the eye of the transfixed beholder has been conflated with the harmful properties of the rays. Only then can the darting of Cupid's arrows be subjoined to the physiology of sight in its pathogenic mode, together with the folk tales of eyes invested with occult powers. The effect, for the sake of reinforcing systems by analogy, is to draw the mechanical operations of the vapours back into the sphere of magic powers.

Now, everywhere one turns, there are parts in kaleidoscopic array. Epicurus, inheritor of the Democritean theory of atoms, as far back as the third century BCE, proposed that every event had a natural cause, and that hence love was to be understood as the substance of sight turned to atoms joined with the semen in order to incite the condition. Plato in the *Phaedrus* discussed the stream of beauty coming through the eyes, how the image begets a special warmth making the soul flourish, how the sender of the stream 'throbs with ferment in every part', and how the beholder feels 'a ferment and a painful irritation'.[35] Already, metaphor is a factor in the diagnostics of love. As the beautiful eyes gaze they emit a 'flood of particles', and once the gazers are parted, both suffer from desiccation. Then the soul 'is stung and goaded into anguish'. The one who sends the beams is also afflicted with madness and robbed of sleep, yearning for the one in whom the image dwells. This is Eros, a condition in which the one afflicted is the only physician for the one who initially sent the infection.[36] Then Plato himself creates a transition from the clinical to the occult, for he claims that lovers seek to make their beloveds as like themselves as they can. This follows from a pouring forth of the soul, leading to capture or enchantment through the infliction of a divine madness. Only then does he go on to speak of heat and the tickling and prickling of desire leading to a carnal assault.[37] The order of his discourse in this celebrated dialogue is doubtless one of Ficino's principal prompts. Plato, too, had set out to teach the avoidance of erotic love, but became absorbed in describing the mechanisms whereby love passes from eye to eye.[38]

Similarly, Alexander of Aphrodisias, in his *Problemata*, Book 98, discusses love as a ray drawn from the eyes, which he links to arrows and the lighting of a fire, the actions of Cupid, and the captivating eyes of a beautiful woman.[39] Apuleius knew the mechanisms well; in Book X of his *Metamorphoses* the wicked stepmother complained of her stepson that he had bewitched her by the rays of his eyes, causing a burning in her heart and an alteration and corruption of her blood.[40] Avicenna explains that debilitation follows from the sexual embrace through the evacuation of vital and animal spirits.[41] Love so conceived is based on an economy of bodily fluids in relation to the desired homeostasis of the body. Rhazes is reported to have pronounced similarly that excessive sexual activity harms the nerves and eyes and diminishes the forces in the same manner as advancing old age.[42] Galen, in his treatise *On Seminal Fluids*, says simply that coitus causes the loss of vital spirits, and that too great a loss will cause death.[43] Pliny, in Book VII, chapter 2 of his *Natural History*, concurs that there are eyes with powers sufficient to cause diseases in others.[44] Guglielmo de' Corvi, in his *Practica*, written around 1275, stated that sight kindled love in the heart just as sunlight through a magnifying glass sets straw on fire.[45] Albert the Great, in his *De animalibus*, confirms that the basilisk shoots poisonous rays from the eyes, possessing the power to kill instantly, and that the principle of contagion involved is identical to the tainted breath by which phthisis (such as pulmonary tuberculosis) or dysentery is spread, or the vapours from excrements or the gummy eye.[46] Toxins remained the clearest markers of contagion, so that by extension their operations came to stand for contagion *tout court*; the case of love was no exception.

Alain de Lille, in the *Salernitan Questions*, discusses the *spiritus* as the medium for the destructive interchange between the body and the soul.[47] Gerard of Berry, in his commentary on the *Viaticum*, describes how the mind of the lover first receives the pleasing impression and then turns the inward gaze on that image, corrupting it in the imagination, thereby drying out the brain and spreading the melancholy disorder throughout the body. Sight as the cause of the disease goes back to Avicenna in his chapter on *amor hereos* in the *Canon*.[48] Arnaldus of Villanova, in his *Liber de amore heroico* (*ca*.1280), understood overheated spirits to be the cause of many adverse conditions. Vapours expelled by excogitations left the brain in a dried state and the eyes hollow. From Peter of Spain, in his commentary on the *Viaticum*, Dino del Garbo learned the principal materials required for his treatise on Cavalcanti.[49] Together, these related sources provided the vocabulary of operations that constituted the Renaissance syntheses and adaptations.

The comparative innocence of Ficino's conflation of medical matters with the lore of the *male d'occhio*, for example, is not to be compared with another work of his age of most pernicious influence, except in the practice of including by analogy causal agents of diverse kinds within single diagnostic analyses. The work in mind is the *Malleus maleficarum*, published in 1486, the work of two Dominican monks in the service of the Inquisition. Their avowed purpose was to reveal precisely how the devil conducted his temptations of the human soul through the very mechanisms endorsed by the most celebrated physicians. Thus it came about that Kramer and Sprenger rehearsed the now familiar description of philocaption as an imprudence of the eyes. They wrote of beautiful eyes, glances, phantasms and the polarized imagination, but with the telling difference that through those very avenues of the senses, directly or through the agency of sorcerers, the devil can stir up images as real as if they were freshly received impressions of the senses. Those with melancholy complexions were particularly prone to such abuses. The devil himself could invest the stare with its occult powers, obviating the need for beams and vapours. It was the malefic power of the eyes that were the agents of contagion. In this way, Sprenger and Kramer adapted the channels of perception, the arrival of the *species* in the *virtus estimativa*, and all that pertained to the erotic disorders described by the physicians to their theory of demonic temptation. Such diseases of the soul as philocaption relied on the devil as the pathogenic carrier. For those suffering from the simple abuses of desire, there were the traditional medical cures which the authors faithfully rehearsed, but for those abused by the devil, only spiritual powers could intervene. The book represents a crux in the development of Western thought, for it manifests an uncritical conflation of material, occult and demonic causes. Contagion and disease are an integral part of their analysis of the sin-sick soul.[50]

The astonishing part of this story is the number of serious medical philosophers who, during the sixteenth century, were lured, as it were, into the devil's camp. Paolo Grillando, in his *Tractatus de sortilegiis*, qualifies love as one of the three forms of magic, even in states entirely divorced from demons or sorcerers, divinations or poisons, philtres or spells.[51] Once love succumbed to the aetiology of theologians, exorcism replaced clinical regimen. That chapter in the history of contagion is invoked here to show how easily the language of natural magic from poetic metaphor to material causation could be gathered into the language of demonic causes. Even as early as the 1440s, Jacques Despars, in his commentary on Avicenna's *Canon*, stated his objections to the belief that witches or demons could produce love mania (Niiir).[52] Ficino presumably would have found the thesis of the inquisitors abhorrent; Satan was not at work through the blood vapours any more than Cavalcanti's 'donna' was an agent of Lucifer. Yet the power of beams to enchant grants them an occult motif susceptible to theological interpretation. No doubt it was Despars' intent to open and close the debate at once. But it was not a topic that would be contained; Johann Bokel in his book on love philtres, published in 1599, was still reading the origins of love in terms of malign forces.[53] As early as 1520, Pietro Pomponazzi likewise sliced the egg over the matter of incantations, labouring on the side of the empirical. Yet even this radical Aristotelian accepted that the human imagination was subject to the *fascinatio* when it functioned according to natural laws. Amatory magic was simply a form of contagion resulting in debilitating desire that was provoked by occult forces at work in nature. Valleriola insisted on the corruption of the mental faculties by a poisoning likened to the basilisk's stare; the blood vapours worked on the beholder by magic powers.[54]

So many disparate examples do not make a history in linear fashion. They constitute representative paradigms for conducting related forms of association by structural analogy, the literalizing of metaphor, and the synthesizing of medical and poetic lore. At the same time, they contain intimations of every component in Ficino's 'Seventh Oration', in a sense suggesting that nothing was truly new in his reconfiguration of axioms and demonstrations concerning love. What changes is the rhetorical ambience. Ficino reiterates the received notions of clinical or pathological love in the context of a treatise on love as a cosmic and aesthetic force, consolidating in his unique way all that he knew of vapours and eyebeams, together with the forces of reciprocity that he associated with a mechanized mutuality of love, building a psychology of love out of deterministic conditions. In the process he created an *idée force*, one that achieved the mass of credibility needed for preservation in a succession of treatises on the diseases of Eros. For a time it achieved the status of a cultural 'meme', having its own capacity to work as a virus in the body of Western culture. The idea had come together from parts that had been simmering in the Western mind from the time of Plato's *Phaedrus*. Ficino's 'idea' of vulgar love is indicative of methodologies in scientific matters closely linked to the operations of metaphor whereby one property is both the analogue and the equivalent of the other. In the collapse of likeness into 'isness', a correspondence becomes natural fact. The mind is given to identifying the remote in terms of its closest familiar parallel, thereby including the less known into the circle of understanding by having the attributes of the known imposed on it according to degrees of similarity in properties, systems, appearances, ends or means. Love was understood among the Ancients as possessing mysterious qualities that were coercive, disorienting, that produce extreme emotions, that alter mental functions in ways beyond rational explanation and that link contemplation to bodily appetites and sexual cravings. That the symptoms of love were so much like the symptoms of disease developed into a full nosological entity, including theories of contagion synonymous with those generally current. Lovesickness was a communicable disease.

Ficino employed current theories of infection to explain the reciprocities of Platonic love expressed in the *Phaedrus* and the *Timaeus*; that was his invention. Fortified with the rays of Egidio Colonna, the lore of the basilisk, faculty psychology and the conduct of the visual *species* among the faculties of the brain, and the received definitions of *amor hereos*, Ficino constructed, according to the reasoning habits of the Neoplatonists wherein magic was a necessary ancillary to the natural world, his own digest of lovesickness. Not only is it an example of Renaissance contagion theory, but it is exemplary of the Renaissance mind confronted by the question of pathogens, particularly in relation to remote victims. Central to every consideration of the transmission of disease was the matter of occult causes on a scale extending from the merely obscure to *maleficia* and the manner in which these agents joined forces in the spreading of disease.

Notes

1 M. F. Wack, 'From Mental Faculties to Magical Philters: The Entry of Magic into Academic Medical Writings on Lovesickness, 13th–17th Centuries', in *Eros and Anteros: The Medical Traditions of Love in the Renaissance*, ed. D. Beecher and M. Ciavolella (Ottawa: Dovehouse, 1992), pp. 9–31.
2 M. Ficino, *Commentary on Plato's Symposium on Love*, ed. S. Jayne (Dallas: Spring Publications, 1985), p. 172.
3 *Ibid.*, p. 162.
4 M. Ciavolella, 'Eros and the Phantasms of Hereos', in *Eros and Anteros*, p. 80.
5 See the introduction to J. Ferrand, *A Treatise on Lovesickness*, ed. D. Beecher and M. Ciavolella (Syracuse: Syracuse University Press, 1990), pp. 79–80.
6 *Timaeus*, trans. R. G. Bury (Loeb Classics, 1966), 45c.
7 Ficino, *Commentary on Plato's Symposium on Love*, p. 161.
8 *Ibid.*, p. 165.
9 *Ibid.*, p. 167.
10 *Ibid.*, p. 159.
11 *Ibid.*, p. 160.
12 *Ibid.*, p. 161.
13 B. Fregoso, *Contramours: Anteros ou Contramour de Messire Baptiste Fulgose, jadis Duc de Gennes* (Paris: Martin le Jeune, 1581), p. 144.
14 R. Burton, *The Anatomy of Melancholy*, ed. T. C. Faulkner, N. K. Kiessling and R. L. Blair (Oxford: Clarendon Press, 1989–97).
15 F Valleriola, *Observationum medicinalium libri sex* (Lugduni: apud Antonium Candidum, 1588).
16 A. Du Laurens, *Second discours, au quel ests traicté des maladies melancholiques et du moyen de les guarir* in *Toutes les œuvres*, trans. T. Gelée (Paris: Chez P. Mettayer, 1613).
17 J. Aubery, *L'antidote d'amour. Avec un ample discours, contenant la nature des causes d'iceluy, ensemble les remedes les plus singuliers pour se preserver et guerir des passions amoureuses* (Paris: Chez Claude Chappelet, 1599).
18 See note 3 above.
19 H. Kornmann, *Linea amoris, sive commentarius in versiculum glossae visus* (Francofurti: typis M. Beckeri, 1610).
20 J. Aubery, *L'antidote d'amour* (Paris: Chez Claude Chappelet, 1599).
21 J. de Veyries, *La genealogie de l'amour divisée en deux livres* (Paris: Chex Abel l'Angelier, 1609).
22 Ferrand, *A Treatise on Lovesickness*.
23 J. E. D. Esquirol, *Mental Maladies: A Treatise on Insanity*, intro. R. de Saussure (New York: Hafner Publishing, 1965).
24 *The Nicomachean Ethics*, trans. D. P. Chase (London: J. M. Dent and Sons, 1949) 1167a.
25 Ficino, *Commentary*, VII, 10, p. 166.
26 Ferrand, *A Treatise on Lovesickness*, p. 272.
27 G. Cavalcanti, *The Complete Poems*, trans. M. Cirigliano (New York: Italica Press, 1992), p. 15.
28 *Ibid.*, p. 13.

29   *Ibid.*, p. 77.
30   J. E. Shaw, *Guido Cavalcanti's Theory of Love: The 'Canzone d' Amore' and Other Related Problems* (Toronto: University of Toronto Press, 1949), p. 149. See also D. del Garbo, *Scriptum super cantilena Guidonis de Cavalcantibus* in *Rime di Guido Cavalcanti*, ed. G. Favata (Milan: Marzorati, 1957).
31   J. C. Nelson, *Renaissance Theory of Love: The Context of Giordano Bruno's 'Eroici furori'* (New York: Columbia University Press, 1963), p. 270.
32   Plato, *Symposium*, trans. W. R. M. Lamb (Loeb Classics, 1977).
33   R. H. Robbins, *Encyclopedia of Witchcraft and Demonology* (New York: Crown Publications, 1959), p. 193.
34   M. Del Rio, *Investigations into Magic*, trans. and ed. P. G. Maxwell-Stuart (1608; Manchester: Manchester University Press, 2000).
35   *Phaedrus*, trans. R. Hackforth (New York: The Liberal Arts Library, 1952), 251c, p. 96.
36   *Ibid.*, 252b, p. 97.
37   *Ibid.*, 253e.
38   *Ibid.*, 255d.
39   *Problemata*, ed. I. L. Ideler, 2 vols; *Physici et medici graeci minores* (Amsterdam: Adolf Hakkert, 1963), pp. 474–5.
40   L. Apuleius, *Metamorphoses*, trans. W. Adlington (Cambridge, MA: Harvard University Press).
41   *Canon*, Book III, fen 1, tract 5, ch. 24.
42   Fregoso, *Contramours*, p. 88.
43   Galen of Pergamon, *On Seminal Fluids* in *Opera quae extant*, vol. 4., ed. C. G. Kühn (Leipzig: Teubner, 1821–23).
44   Pliny the Elder, *Natural History*, trans. H. Rackham, T.E. Page *et al.* (Cambridge, MA: Harvard University Press, 1961).
45   G. de' Corvi, *Practica*, 23v.
46   Albert the Great, *De animalibus* jn *Opera omnia*, vol. 29 (Monasterii Westfalorum: in aedibus Aschendorff, 1972–82).
47   M. F. Wack, *Lovesickness in the Middle Ages: The 'Viaticum' and its Commentaries* (Philadelphia: University of Pennsylvania Press, 1990), p. 58.
48   Book III, fen 1, tract 5, ch. 24.
49   D. del Garbo, *Scriptum super cantilena Guidonis de Cavalcantibus* in *Rime di Guido Cavalcanti*.
50   H. Kramer and J. Sprenger, *The Malleus Maleficarum*, trans. M. Summers (London: John Rodker, 1928).
51   Introduction to Ferrand, *A Treatise on Lovesickness*, p. 86.
52   J. Despars, *Avicennae Canon (liber III) cum Jacobus de Partibus* (Lyon: J. Trechsel, 1498).
53   Introduction to Ferrand, *A Treatise on Lovesickness*, p. 88.
54   *Ibid.*, p. 107.

4
Contagions of Love: Textual Transmission

Nancy Frelick

Pathologizing love is not new to this age of modern psychology and self-help books. Indeed, much was written about 'lovesickness' as a serious medical condition in the early modern period and many cures were suggested for those afflicted with this potentially deadly disease, which is described not only as a physical illness, but also as a disorder of the mind or imagination that afflicts the soul.[1] Yet, one of the most curious aspects of European lovesickness is that it is subtly fostered, if not promoted, by the very texts that decry and pathologize it. As Wack explains: 'The growing body of medical discourse on love made it possible for the literary representations of erotic passion to be interpreted mimetically or realistically, as reflections of real life. The cultural authority of medicine may have in part enabled the poetic fantasies of the troubadours to become the social realities of the late Middle Ages and early modernity.'[2] The discourses providing descriptions of causes and cures for the disease can be said to be its transmitters, the vehicles through which it spread like wildfire across Europe throughout the early modern period, creating and fanning the flames of this contagion. The creation and sanctioning of love as a disease – through the ambivalent language of medical, philosophical, religious and literary discourses – thus makes it possible for individuals not only to identify with the discursive models but also to fashion themselves and their behaviours after texts and, ultimately, to shape new realities: life imitates art.

Part of the difficulty encountered in writing(s) on love is akin to the problem of contagion, and love itself: it is an issue of containment. Not only does it elude logical linear discourse, but it often seems so amorphous and all-encompassing that it is impossible to describe without falling into circular arguments, and definitions rarely appear without contradictions. Thus, love is often simultaneously pathologized (or demonized) and idealized. Indeed, this ambivalence, which is already present in the Greco-Roman tradition, is an important vector in the development of the early modern Christian imagination. Although love is a difficult topic to contain textually, writings on love do contribute to shaping their Protean object and part of what they transmit is a miscellany of cultural constructs, often introjected and projected quite unconsciously. The *topos* of contagion at a distance from the eyes (and other senses) into the imagination[3] thus acquires another meaning, as readers are contaminated by the texts they read, texts that seem to be propagating precisely what they purport to prevent or cure.

Although it is tempting to distinguish medical texts from literary ones with respect to lovesickness, seeing the first as texts of containment and the second as texts of dissemination, it appears that both were discourses of contagion and transmission in the early modern period. Moreover, they interpenetrated each other. Literary examples were used in medical texts and medical formulations affected literary expression. Together, they helped to propagate what they sought to describe or circumscribe, perpetuating received ideas, generating variations on established themes, and opening up spaces for subjects to (re-)create themselves according to the images proffered in these textual mirrors.

While it is impossible here to give a complete accounting of the factors contributing to the spread of lovesickness in the early modern Christian imagination, one can trace a few suggestive lines of thought that contributed to the formation of the Western tradition of lovesickness in ways that influence us still, even though some of the theoretical underpinnings (such as the humoral system) may have fallen by the wayside.

Early modern ideas about love in Europe are rife with paradoxes and contradictions, in large part because they conflate various philosophical and medical traditions. The preference for rationalism over observation and experimentation, the desire to reconcile ideas gleaned from a number of divergent authorities, including stories and myths, the literalization of metaphor, and the reification of thoughts and feelings all contribute to the confusion.[4] For instance, Aristotelian and Platonic ideas are combined with Hippocratic and Galenic theories, often via the interpretations of Arabic medical writers such as Avicenna (Ibn Sina), translated by Gerard of Cremona, or Ibn al-Jazzar, translated by Constantinus Africanus and Johannes Afflacius,[5] and glossed and recombined by clerics eager to reconcile them with Christian doctrine.[6] As recent scholarship has shown, not only did the Greco-Roman notions around *eros* combine with Islamic ideas (such the semantic range of *'ishk*, assimilated with both the desire to possess the beloved and with universal love[7]) to influence these texts, but modifications, neologisms and errors of translation themselves helped redefine lovesickness, as slippages occurred and introduced new terms to the mix. Thus, in some ways, it seems that *amor hereos* is disease of translation, of textual mutation and transmission. *Amor hereos* emerges as a lovesickness reserved for the heroic class or nobility (*erus*) that is rooted in humoral theory and Avicennan faculty psychology, incorporating

Greco-Roman ideas around *eros* as well as the Arabic '*ishk*.[8] It appears to have surfaced as the result of a corruption or proliferation of terms around *eros*, producing neologisms such as *eriosus*[9] and slippages from *eros* to *heros* and *hereos*, to *heroicus* linked through false etymology to *herus* or *erus* (by Arnald of Villanova and others), and later described as *heroical* in texts such Ferrand's *Treatise on Lovesickness*[10] or Burton's *Anatomy of Melancholy*.[11]

As Wack suggests, the idea of a noble love was also supported by the connection between love and melancholy,[12] which became a 'disease of heroes' and exceptional men in ancient Greece, through mythology (Ajax, Heracles, Bellerophon) and through its association with Platonic frenzy, inspiration and spiritual exaltation in the *Phaedrus*, and more explicitly in the (Pseudo-)Aristotelian *Problemata* 30, which linked melancholy with heroism and genius.[13] While it is difficult to say with absolute certainty whether the idea of a noble love or the term came first, there can be little doubt that once *amor hereos* was recognized as a kind of love melancholy, it mediated much of the thinking and literature on the subject, generating new ways of seeing things and, as Wack suggests, encouraging *fin'amor*, or courtly love, to influence social reality.

Paradoxically, whereas lovesickness was increasingly viewed as a disease available only to the nobility, it was also described in terms that suggested it was a subversion of the social order. This is true not only of the lover's 'overestimation' of the beloved described in medical treatises[14] and of the social inversion in courtly poetry in which the lady is elevated to the position of a lord (*Domna* or *Midons*) to be worshipped by a servile poet, but also in some of the hierarchical metaphors used to illustrate the imaginary internal geographies created to explain the etiology of the disease.[15] According to Wack, some descriptions of the psychophysiology of love use the figural language of nobility as a model for the functioning of mental faculties, suggesting that health is the result of these faculties maintaining their natural(-ized) hierarchies (modelled on the feudal order), whereas pathology implies subversion:

Gerard [of Berry in his gloss on the *Viaticum*] describes the interplay of mental faculties responsible for the lover's affliction in hierarchical metaphors. Closely following Avicenna's *De anima*, he retains the language of male governance that the 'prince of physicians' uses to describe the functioning of the mental faculties. After receiving a powerfully pleasing sense impression, the estimative faculty, the *noblest* of the perceptual faculties, *orders* the imaginative faculty to fix its gaze on the mental image of the beloved. The imaginative faculty in turn *orders* the concupiscible faculty to desire that person alone. The concupiscible obeys the imaginative, which obeys the estimative, at whose *rule (imperium)* the other faculties *are inclined toward* the desired person, even though she may not be desirable in reality. The inner world of perception and desire is thus structured like the outer system of hierarchical rule: those at the top rule and give orders, those below bow and obey. The pathological moment thus subverts the 'noble' faculty of estimation from below.[16]

Ferrand also describes 'the generation of love melancholy' in subversive terms, using martial metaphors comparing the invasion of the mind, body and spirit by love to the conquest of a fortified city, overthrow of its good government and enslavement of its inhabitants by a devious and deceptive tyrant.[17]

These kinds of texts may have encouraged early modern subjects to (re-)imagine themselves both internally (in terms of fragmented inner spaces) and externally (in relation to others) in ways that opened up sites for both the perpetuation and destabilization of dominant ideologies. According to Easthope, courtly love (as expressed in the poetry of Bernart de Ventadorn and Petrarch, for example) both uses and contravenes the feudal order: on the one hand, it perpetuates concepts and obligations related to vassalage along with Christian values and language, while on the other, it promotes the rise of individualism by presenting a proto-bourgeois subject with private feelings and aspirations that break with both the Christian and feudal order with respect to sexuality, gender roles, marital laws and social hierarchy, for example.[18]

One could say that although courtly love and Christian love are described as having different objects and natures[19] and often conflicting ideologies, they also share some commonalities, including the promotion of feelings and desires that are at odds with the feudal order, such as the idea that nobility can be derived from moral virtue, as Andreas Capellanus explains in the *De Amore*.[20]

The association between love and nobility allowed for other kinds of ambivalence with respect to the dominant order. While attaching lovesickness or *amor hereos* to the heroic or upper classes could lead to critiques of the more effete ruling elite by moralizing Christians, it could also make love seem ennobling; what was originally seen as a rather effeminate disease affecting men of leisure[21] was made to seem desirable, as if contracting it would somehow ennoble the desiring subject, giving him attributes normally reserved for the higher classes.

Indeed, morbid love was also ennobled and enabled in other ways. As Wack explains, the way in which Christianity cultivated images of love as suffering in the worship of Christ on the cross also contributed to the propagation of the disease: 'Veneration of the Cross and identification with Christ's Passion – *imitatio christi* – forged a bond, in medieval consciousness, between love and suffering. Suffering, moreover, was not weakness but the ultimate act of love. Since Christ's Passion was taken as a model for affective experience, this strain of affective piety deliberately cultivated "lovesickness".'[22] She goes on to explain that the 'the potential for confusion between *morbus eros* and *passio caritatis* was an uncertainty ripe for exploration by vernacular poets of love from Gottfried to Chaucer'.[23]

According to Marsilio Ficino, love is both a divine creator (*Amor est auctor omnium et seruator*) and the worst of contagious diseases. In Book III of the *Commentarium*, it is described as a beneficent force governing the universe and all earthly creations, including the arts, and a source of goodness and beauty in which men can find beatitude. Yet, in book VII, it is viewed as a sudden infection causing a illness graver than the worst kinds of pestilence and epidemics that plagued early modern people:

Certainly [the violence, vehemence and perniciousness of the corruption caused by such a fine ray, such a light spirit, so little blood] will not astonish you if you consider the other diseases born from contagion, such as the itch, mange, leprosy, pleurisy, consumption, dysentery, ophthalmia, plague. Amatory contagion occurs easily and proves worse than all these infectious diseases.[24]

It is a literary commonplace that lovesickness occurs as a result of the lover's first glance of the beloved, which is often described as a kind of wounding expressed metaphorically as darts, arrows or even poison entering the eye at the moment of the *innamoramento*. Ficino explains that this is owing to the transmission of a sanguine vapour or *spiritus* containing contaminated blood that can travel through rays emitted by the eyes, thus infecting the beholder and resulting in a kind of bewitchment or *fascinatio* (VII, 4) wherein the image of the beloved is imprinted or reflected, as in a mirror, in the heart and soul of the lover (VII, 8). The alteration of the blood leads to an imbalance of the humours, to melancholy and obsession (VII, 7). The gravity of the illness depends on factors such as humoral and astrological dispositions (VII, 9–11) and various traditional cures are suggested, including the passage of time, avoidance of the object of love and of idleness, reminders of the faults of the beloved, various purges, bloodletting, sweating, drinking wine and even coïtus (VII, 11). This unhealthy 'vulgar love' is described as a kind of madness (*insania* or *furor*) that torments lovers, first through the burning of bile and then through melancholy adust (burnt black bile), ultimately turning them into beasts (VII, 12).[25]

Celestial love, which is celebrated by Ficino, is distinguished from this lower form, but it is still described as a *furor*, though it leads to God. Ficino names four types of divine *furor*: the first, poetic *furor*, is linked to the Muses; the second, having to do with mysteries, is associated with Dionysus; the third, divination, with Apollo; and love with Venus (VII, 14). Using a pastoral image reminiscent of biblical teachings linking the true lover to a shepherd protecting his flock from wolves (or false lovers), Ficino implicitly compares Socrates to Christ, describing 'the wisest of Greeks' as the exemplary charitable lover who inspires his young pupils to lead virtuous lives (VII, 16).

The opposition Ficino presents between vulgar and celestial love (associated with Venus Urania and Venus Pandemos, presented by Pausanias in the *Symposium*) occasionally seems straightforward enough, when taken in isolation within certain chapters, but becomes more problematic when examined in light of the rest of the work. For example, in reality, how is one to distinguish between good and bad kinds of melancholy or *furor*, the kind that dehumanizes and debases men, from the more angelic kind that elevates them to genius or to the divine? The contradictions inherent in the text are no doubt partly owing to the composition of the work, which took place over a considerable period, as well as to attempts to reconcile divergent views. It is set up as a series of discourses by various discussants after a banquet commemorating the anniversary of the death and birth of Plato. Presumably, Ficino's *Convivium* was meant to reflect the dialogic nature of Plato's works (while attempting to reconcile them with medical theory and Christian doctrine) but its syncretism elides much of Plato's true plurivocity, which remains only in contradictory traces, along with fragments of other myths and traditions.

The ambivalence of discourses around love – the fact that it is presented both as the gravest illness and the greatest good, that it must be avoided as the cause of unbearable pain and suffering and sought after as the source of supreme pleasure – is not new to Ficino. According to Wack the conflation of pleasure and pain already exists in Constantine, who attributes lovesickness to *eros*, which he also associates with extreme pleasure. As she points out, his 'definition of love as a pleasurable disease (*morbus*) suggests the paradoxical inversion of the concept of health in "courtly" love in medieval Islam and Christian Europe (that is, it is salutary to be sick with love)'.[26]

In many Christian texts *eros* or *concupiscentia* – generally associated with sensual love and desire or unhealthy attachment – is opposed to *agape* or *caritas*–a redemptive spiritual love based on compassion. This opposition can be linked with the two Venuses (vulgar and celestial) of Renaissance Neoplatonists such as Ficino. Yet, it is often difficult to make a clear distinction between them in texts like Ficino's except to say that the baser sort involves sensuality and leads to torment (although sexuality is deemed acceptable or desirable for therapeutic and reproductive purposes), whereas the nobler sort leads to creativity and ultimately to the divine. Indeed, even though one might be idealized and the other disparaged (either as an illness or a kind of bewitchment), in practice it may be very difficult to distinguish between vulgar and celestial forms, especially since definitions seem to collapse into each other and since they are generally described as manifesting through the senses, through love of beauty, or love for a particular individual. Moreover, since love is characterized as deceptive – from the medical perspective of faculty psychology it is seen as a disorder of the 'estimative faculty' leading to a disease of the imagination[27] which is over-whelmed by the *phantasma* of the desired object whose image or *species*[28] is imprinted in the imagination and memory resulting in 'compulsive cogita-tion'[29] – one wonders how anyone in love can be expected to have any powers of discernment. Wack points out that ambiguity already exists in the *Viaticum*: 'As Constantine presents it, it is difficult to decide whether *amor eros* is more akin to Neoplatonic striving for beauty or to an attack of sexual need.'[30] Wack also points to a kind of determinism implicit in Constantine's aetiology (a shift from the Galenic view) that contributes to the problem: 'The adventitious sight of a beautiful object and the presence of excessive or corrupt humors involve chance and the material composi-tion of the body; free will or domination of matter through one's philo-sophical attitude are passed over in silence. The ambiguity of the *Viaticum's* causality enabled lovesickness to be idealized in poetry, to be compared with mystical love, and to be condemned as cupidity).'[31]

The ambiguity and determinism of early writers on lovesickness, as well as the cures they propose, may thus have contributed both to the paradox-ical ambivalences in the erotic tradition and to a more fatalistic image of 'falling in love'. The *innamoramento* presented in love lyric for example, generally implies a passivity, helplessness, and powerlessness of the subject caught off guard by what some call a *coup de foudre*, an unsettling moment of rapture at the first sight of the beloved that leads to a desire for recogni-tion from and union with the object of love and an attendant fear and anxiety that these will be denied. This kind of romantic love is associated with lack and is often accompanied by the belief, or fantasy, that comple-tion can only be found through union with the beloved (see Aristophane's myth of the soul's search for its other half in Plato's *Symposium*, for example).[32] All too often the only cure envisaged in courtly poetry is a con-summation of love leading to *joie*. This idea is no doubt reinforced by some of the cures proposed in medical texts, which only seem to encourage love-making: along with drinking wine, bathing, pleasant conversation, music and poetry, sexual intercourse is proposed as a therapy.[33] The Neoplatonic idea (put forth by writers like Ficino) of the transmigration of souls during the *innamoramento* can only enhance this fantasy. According to Ficino, a lover is dead unto himself unless his love is reciprocated because his soul goes to live in the beloved. In mutual love, each lover's soul lives in the other. In reciprocal love both lovers are alive, each living in the other; whereas in unrequited love, the lover is dead, having lost his soul to another. Ficino even goes so far as to vilify an object of love that does not respond to the lover's advances, saying that such a beloved is guilty of theft, homicide and sacrilege and therefore deserving of the worst punish-ments if unwilling to submit to the laws of love (II, 8).[34] This is a dangerous logic because it encourages those who may have unhealthy attachments to divest themselves of responsibility and to project blame onto others for their own inner states. Such thinking has been very detrimental to women, in particular, who have often been victimized either by individual men who used this kind of logic to rationalize their desires by either taking whatever they wanted from them (see the stories of rape in Marguerite de Navarre's *Heptameron* for literary examples[35]) or punishing them for not giving men what they want (see Boccaccio's story of Nastagio in the *Decameron* (V, 8), recounted Flore's *Comptes amoureux*, and Panurge's treatment of the 'haute dame de Paris' in chapters 21–22 of Rabelais's *Pantagruel*, for example) or through more widespread persecution by a society that wanted to other the female sex – to demonize it, make it responsible for men's irrational feelings and behavior, and control it (see Kramer and Sprenger's *Malleus Maleficarum*).[36]

The power of words for good or ill was recognized[37] and, indeed, there is a double discourse in early modern texts regarding the contagion of love through reading/listening to amorous stories/poems/songs.[38] In discussing cures to lovesickness proposed by various authorities, Wack states: '"Recitation of verses" was ... supposed to alleviate the patient's obsession with a particular woman. As Avicenna recognized, sometimes this strategy served only to reinforce the lover's preoccupation, especially when the songs were about unsuccessful love. Constantine, in contrast, though giving the cure only the briefest mention, introduces no doubts as to its efficacy. This particular cure provided a fertile point of convergence between the medical tradition of lovesickness and literary representations of passionate love. From the twelfth century onwards, fictional characters or lyric personae claim to compose or sing in order to relieve love-sorrow, and in the hands of Dante, Boccaccio, and Chaucer, the lovesick poet or poetic lover was variously reinterpreted according to medical ideas about lovesickness.'[39]

While writing could be construed as therapeutic and reading or listening to verses were sometimes proposed as cures, stories of love were at times also portrayed as discourses of contagion, not only in the sense that they speak of the lovesickness of the characters (whether caught by magical means through love potions, in Tristan and Iseult, through the arrows of Love in the *Roman de la Rose*, or through the gaze of the Ladies' basilisk eye in Scève's poetry, for example) but that they themselves seem to have the capacity to engender the disease, presumably through something like the power of suggestion.[40] A classic example of this is the story of Francesca and Paolo (who appear in Dante's *Inferno* V) wherein the moment of the *innamoramento* is tied to reading the story of Lancelot and Guenevere's adulterous love, implying that reading romance leads to lovesickness. This theme is developed in many ways over the centuries both comically as can be seen in Cervantes's *Don Quixote*, for instance, and tragically, even in much later works, such as Flaubert's *Madame Bovary*, in what René Girard terms triangular desire in *Mensonge romantique, vérité Romanesque*, wherein historical or fictional characters and texts are seen as mediators of desire.[41]

Texts about love are often seen as dangerous reading for girls in early modern educational tracts. Females are generally seen as the weaker sex and conservative writers like Juan Luis Vives believe that women are like children and it is dangerous to put ideas into their impressionable minds. While Vives describes reading as an important part of *The Education of a Christian Woman* (in chapter 4), he makes it clear in chapter 5 that love and war are unacceptable subjects, comparing the intake of such material to drinking poison or exposing oneself to fatal infection: 'A woman who contemplates these things drinks poison into her breast, of which such interest and such words are symptoms. This is a deadly disease, which it is not only my duty to expose but to crush and suppress, lest it offend others by its odor and infect them with its contagion.'[42] He also warns his readership that 'women become addicted to vice through reading' and describes the authors of romances (his list contains many of the stories we now study in university curricula) as 'idle, unoccupied, ignorant men, the slaves of vice and filth' whose stories are 'full of lies and stupidity'. He pronounces 'What madness it is to be drawn and fascinated by these tales!' warning that the only clever thing to be found in such books is advice for lovers on how to seduce their lady.[43]

After this diatribe, Vives goes on to praise a number of ancient poets – 'Callimachus, Philetas, Anacreon, Sappho, Tibullus, Propertius, Cornelius Galus' and Ovid – but suggests, following Ovid in the *Remedy of Love*, that 'they must be repudiated by the chaste' (5.33) because they speak of love. In brief, a woman should read only Christian works and her 'whole motivation for learning should be to live a more upright life' (5.35), which means remaining chaste and silent (4.28).

Of course, it can be argued that there is also an opposing tradition according to which stories of love can be read as negative *exempla* instructing women to avoid the dangers of love (see the epistle to the reader in Hélisenne de Crenne's *Angoysses douloureuses qui procedent d'amour*, as well as Boccaccio's *Elegy of Lady Fiammetta*) and those men who feign lovesickness in order to seduce them. The *topos* is illustrated in different ways in the work of Alain Chartier[44] or Christine de Pizan,[45] in many *nouvelles* of Marguerite de Navarre's *Heptaméron* or even Madame de La Fayette's *Princesse de Clèves*, not to mention the many stories of Don Juan, and perhaps most notoriously Laclos's *Liaisons dangereuses*.

At the heart of these texts is a problem common to love and to literary analysis: How are we to read their signs? How do we interpret them? Do they have prophylactic or curative properties? Are they to be read as curative *exempla* or as dangerous discourses of contagion, which are themselves capable of transmitting the dreaded/desired disease either through the power of suggestion, by deliberate abuse or unconscious misuse? Do negative *exempla* keep readers from becoming victims of love? Or do love stories engender what they portray? How can one decide where the signs in a text (or a lover) will lead? Or whether they can be trusted? Part of the difficulty and indeed the pleasure of the literary texts on love is their indeterminacy. To a certain extent, this is owing to the conflation of various, often conflicting notions with respect to love: it is an undesirable disease on the one hand, yet it is somehow ennobling, both because it is the disease of the nobility and also because suffering through love makes one more Christ-like; it is celestial and/or bestial, infernal and heavenly (Ficino's two Venuses); it gives life and death; it is a disease and a cure. Indeed, many texts about love employ what me might call a kind of poetics of indeterminacy, a paradoxical logic, expressed thematically and structurally, that explores and exemplifies the contradictory definitions of love (see Plato's *Symposium*, Andreas Capellanus *De Amore*, Erasmus's *Praise of Folly*, etc.). In some cases, this can be seen not only in the ways the works are constructed, but also in the wide-ranging disagreements among readers in their reception.

Hélisenne's *Angoysses douloureuses*, often described as the first French sentimental novel, is an interesting case in point as the authorial persona makes opposing claims for her book, the writing of which is variously described as a harmless pastime inspired by literary examples rather than personal experience (*Epîtres familières et invectives*), a personal story of grief resulting from lovesickness meant to warn women not to fall prey to love (*Angoysses*), a kind of writing cure,[46] a forbidden pleasure she hides from her jealous husband, a direct means of communication to her lover who is to read her words, and a way to prolong and cultivate the wretched disease. At the outset she states that she wants pity and compassion from her female readers for the torments she endures, while at the same time giving them a negative *exemplum* encouraging them to avoid the snares of love through useful occupation. Yet, the reader's faith in the effectiveness of literary *exempla* is quickly undermined. In the second chapter, where she describes the moment of the *innamoramento*, as a kind of psychological wounding that occurs when she first gazes upon the beautiful youth by whom she is smitten, and whose image she internalizes (in true Ficinan fashion), she states that reason prompted her to think of stories of destructive love in order to resist it. The negative examples include literary characters from classical and early modern times: Helen of Troy, Medea and Jason, Lucretia and Eurial, Lancelot and Guenevere, and Tristan and Iseult.

Even though she (very) briefly resolves to resist love, she quickly dismisses her fears of misfortune along with the negative *exempla* cited and rationalizes her sensual desires. Indeed, the speed with which she sets aside these caveats makes one wonder both whether these tales of love have not encouraged (rather than discouraged) her to pursue her own amorous adventure, and also whether her narrative is not set down in order to encourage others to follow her example and thus make her own behaviour and circumstances more acceptable, as misery and sin love company. As the authorial persona puts it: 'One thing comforts me, and that is that a person who sins along with many others is not worthy of such great blame.'[47]

These words may caution us not to trust our unreliable narrator, or perhaps any of the writers on love who offer us textual mirrors through which to examine ourselves.[48] The dangers are great. Perhaps the ubiquitous images of Narcissus in discourses on love are there to remind us that readers are as vulnerable to contamination through these discursive mirrors as he was to the image of his othered self. Reading, writing and love all comprise specular dimensions. They involve projection, introjection and transference. As such they can be profoundly ambivalent, as they may contain both therapeutic and contaminating elements. Moreover, like any medicine, they can be curative or poisonous, depending on how they are taken. They can transmit knowledge, wisdom, truth or nonsense, if not fear, paranoia, and hatred, depending on the discrimination and insight of the reader. Indeed, how can we examine ourselves separately from the images and words in these discursive mirrors, when they engage us with captivating reflections that have already subtly penetrated us, changed us, 'othered' us, perhaps beyond our conscious perception; when even those characters who consciously (re)fashion themselves according to texts on lovesickness with an eye to using them for their own ends find that they too have changed, that they have been deceived, or rather have deceived themselves through the seductive refractions and reflections in the endless hall of mirrors constructed by the discourses of love (see Laclos, for instance)? To put it another way, how can we trust our 'estimative faculties' when the very language for examining or even constructing ourselves, others and the world around us has already been (re)shaped by the logic and rhetoric of the discourses we seek to investigate? How do we distinguish ourselves from the images and words that are supposed to represent us, when we have already been subtly influenced if not refashioned by them? How can we be certain that other subtle ideas have not left an impression in our imagination, shaping our reality at some unconscious level? How do we fashion ourselves, if not in part through the readings (or other media) that offer themselves to us as mirrors, proffering images of our inner and outer worlds? Are these mirrors true? Or are they filled with distortions? What is lost (or added) in the translation? What the literary tradition seems to suggest is that we must strive to be good readers, not only of books, but of ourselves and others, which, like love, seems to require striking a balance between wisdom and ignorance, between our desire for the inaccessible and the endless pleasures of the open text.

Notes

1 The list of symptoms gleaned by Constantine from the Greco-Arabic synthesis 'include the alteration of appearance (sunken eyes, jaundiced color) and of behavior (insomnia, anorexia, depressed thoughts [*profundatio cogitationum*])'; M. F. Wack, *Lovesickness in the Middle Ages: The* Viaticum *and its Commentaries* (Philadelphia: University of Pennsylvania Press, 1990), p. 40. See also M. Ciavolella , *La 'Mallattia d'amore' dall'Antiquità al Medievo* (Rome: Bulzoni, 1976).

2 *Lovesickness in the Middle Ages*, p. 50.

3 See Gagnon's, Beecher's, Pantin's and Shuttleton's contributions to this volume.

4 See Beecher in this volume.

5 See M. F. Wack, 'The *Liber de heros morbo* of Johannes Afflacius and Its Implications for Medieval Love Conventions', *Speculum*, 62, 2 (1987) 324–44.

6 See *ibid.*; Ciavolella La *'Mallattia d'amore' dall'Antiquità al Medievo* and Beecher also offer excellent overviews.

7 See Wack, *Lovesickness*, 35–8. Wack specifies that Ibn al-Jazzar 'fuses a Platonic conception of frenzy with a medical view of madness. This Platonic theme in the chapter on *'ishk* offered a point of contact in the west, through Constantine's translation, between medical views of love and those of other disciplines in influenced by Platonism, such as philosophy and literature' (p. 36).

8 'Commentators on Avicenna, whose word for the disease (*al'-ishq*) was transliterated into Latin as *ilisci* or *ylisci*, equated it with *amor hereos* and declared it a hazard of noble life' (*ibid.*, p. 150).

9 A neologism introduced by Constantine in the *Viaticum* meaning 'sufferer of lovesickness' that is consistently replaced by *heroicus* in the *Liber de heros morbo* (Wack, 'The *Liber de heros morbo*', p. 337).

10 In chapter 3, Ferrand explains: 'Avicenna, along with the whole Arabian clan, calls this sickness in his language *alhasch* and *iliscus*, Arnald of Villanova, Bernard of Gordon, and their contemporaries call it heroical or lordly love, either because the ancient heroes or demi-gods were often afflicted with this ill according to the mythical recitations of poets, or because the great lords and ladies were more inclined to this malady than the common people, or finally because love rules and dominates the hearts of lovers.' J. Ferrand, *A Treatise on Lovesickness*, trans. and ed. D. Beecher and M. Ciavolella (Syracuse: Syracuse University Press, 1990), p. 232.

11 R. Burton, *The Anatomy of Melancholy*, eds. N. K. Kiessling, T. C. Faulkner and R. L. Blair (Oxford: Clarendon, 1989–97). See the groundbreaking article by J. L. Lowes, 'The Loveres Maladye of Hereos,' *Modern Philology*, 11:4 (1914) 491–546; also Wack on 'The *Liber de heros morbo*, and Beecher and Ciavolella's edition of Ferrand's treatise. Much of the research on the subject of *amor hereos*, spawned by Lowes's findings, has centred on Chaucer's use of the term. See C. F. Heffernan, *The Melancholy Muse: Chaucer, Shakespeare and Early Medicine* (Pittsburgh: Duquesnes University Press, 1995), Wack, etc. There also seems to be quite a bit of research on the topic in Spanish literature; see R. Folger, *Images in Mind: Lovesickness, Spanish Sentimental Fiction and Don Quijote* (Chapel Hill: North Carolina Studies in the Romance Languages and Literatures, 2002), for example.

12 The link exists in Avicenna. Wack explains that from the fourteenth-century medical writers classify *amor hereos* as a subspecies of melancholy and it certainly is identified as such in Burton's seventeenth-century *Anatomy of Melancholy* (*Lovesickness*, p. 101).

13 See R. Klibansky, E. Panofsky and F. Saxl, *Saturn and Melancholy* (New York: Nelson, 1964), pp. 16–17. *Saturn and Melancholy* contains a side-by-side translation of 'Problem 30'. The notion that outstanding men are melancholic finds widespread artistic and literary expression, notably in Dürer, Erasmus, Ficino, Montaigne, Cervantes, Shakespeare and Burton; In addition to Klibansky *et al.*, see L. Babb, *The Elizabethan Malady: A Study of Melancholia in English Literature from 1580 to 1642* (East Lansing: Michigan State University Press, 1951), W. Schleiner, *Melancholy, Genius, and Utopia in the Renaissance* (Weisbaden: Harrassowitz, 1991), M. A. Screech, 'Good Madness in Christendom', in *Essays in the History of Psychiatry*, vol. 1. ed. W. F. Bynum, R. Porter and M. Shepherd (London: Tavistock, 1985); *idem, Montaigne and Melancholy: The Wisdom of the Essays* (London: Penguin, 1991) and Heffernan, *The Melancholy Muse*, for example).

14 As Wack explains, when discussing Gerard of Berry's gloss of Avicenna's theories, for example (*Lovesickness*, p. 56). Lacan likens the overestimation of the love object to transference; see especially his seventh, eighth and twentieth seminars: J. Lacan, *Le Séminaire VII: L'éthique de la psychanalyse (1959–60)*, ed. J.-A. Miller (Paris: Seuil, 1986); *idem., Le Séminaire VIII: Le transfert (1960-61)*, ed. J.-A. Miller (Paris: Seuil, 1991). See also my own work on the subject: *Délie as Other: Toward a Poetics of Desire in Scève's* Délie (Lexington: French Forum, 1994); 'Friendship, Transference, and Voluntary Servitude: Montaigne and La Boétie', in *Le Visage changeant de Montaigne/The Changing Faces of Montaigne*, ed. K. Cameron and L. Willett (Paris: Champion, 2003), pp. 195–206; '"J'ouïs-sens": Thaumaste dans le *Pantagruel* de Rabelais et le "sujet-supposé-savoir"', *Etudes rabelaisiennes*, 30 (1995) 81–97; 'Lacan, Courtly Love and Anamorphosis', in *The Court Reconvenes. Courtly Literature across the Disciplines. Selected Papers from the Ninth Triennial Congress of the International Courtly Literature Society, University of British Columbia 25–31 July, 1998*, ed. B. K. Altmann and C. W. Carroll (Woodbridge, Suffolk and Rochester, NY: D. S. Brewer 2003), pp. 107–14; 'Poétique du transfert et objets a: l'exemple de la *Délie*', in *Poétiques de l'objet. L'objet dans la poésie française du Moyen Age au XXe siècle. Actes du Colloque international de Queen's University (mai 1999)*, ed. F. Rouget and J. Stout (Paris: Champion, 2001), pp. 73–82; 'Sex, Lies, and Anamorphosis: Love as Transference in Scève's *Délie*', *Romanic Review*, 90:3 (1999) 301–16.

15 Avicennan faculty psychology generally divided the brain into three ventricles responsible for different mental abilities. See Wack's discussion of Gerard of Berry, Urso of Calabria, and 'Master Alain' (*Lovesickness*, pp. 56–61).

16 Wack, *Lovesickness*, pp. 58–9.

17 Introduction to Ferrand, *A Treatise on Lovesickness*, p. 252. It is interesting to note that whereas the subversion of the body politic is used as a metaphor for the diseased body and mind in these texts on lovesickness, contagion and disease are used as metaphors for perceived threats to the body politic in political and religious tracts (see Greenspan's contribution to this volume).

18 A. Easthope, *Poetry and Phantasy* (Cambridge: Cambridge University Press, 1989), pp. 73–5.

19 *Ibid.*, p. 73.

20 *On Love*, trans. P. G. Walsh (London: Duckworth, 1982), 1.6.13–15. Wack mentions this passage of the *De Amore* in *Lovesickness* (p. 62).

21 Even though it is seen as a womanly disease, Wack points out that it is not available to women in the same way in the medical tradition: 'For passionate love to be "womanly love" meant that ancient and medieval culture saw excessive love

as characteristic of women, yet women are with few exceptions absent from medical accounts of love. "Womanly love" befell men, who were the subjects of medical discourse' (*Lovesickness*, p. 12); Wack discusses gender issues in more detail in chapter 8 and in 'The Measure of Pleasure: Peter of Spain on Men, Women, and Lovesickness', *Viator*, 17 (1986) 173–96.

22   Wack, *Lovesickness*, p. 25.

23   *Ibid.*

24   *Commentaire sur le* Banquet *de Platon*, trans. and ed. P. Laurens (Paris: Belles Lettres, 2002), VII, 5, my translation.

25   According to Wack, in the medieval medical tradition, humoral imbalance could either be the cause or the result of lovesickness, which was sometimes considered a precursor to and sometimes as a type of melancholy. Either way, this link forged a powerful image in the early modern imagination: 'the kinship between love and melancholy in the Constantinian corpus reinforced the notion that love could become a disease; it provided an authoritative medical background for the figure of the "melancholy lover" in European literature; and it suggests that the psychodynamics of melancholy may illuminate the genesis of lovesickness' (*Lovesickness*, p. 40).

26   *Ibid.*, pp. 38–9.

27   Wack, discussing the two versions of Peter of Spain's glosses of the *Viaticum*, states that in version A, lovesickness is a disease of the imaginative faculty whereas in version B it is a disease of the estimative; *ibid.*, pp. 90–3.

28   See Wack's 'From Mental Faculties to Magical Philters: The Entry of Magic into Academic Medical Writings on Lovesickness, 13–17th Centuries', in *Eros & Anteros: The Medical Traditions of Love in the Renaissance*, ed. D. A. Beecher and M. Ciavolella (Ottawa: Dovehouse, 1992), pp. 9–31, as well as Pantin, Gagnon and Beecher's contributions to this volume for more on *species*, which Wack defines as 'likenesses of objects that mediated between the material world and the mind' (p. 15).

29   See M. Ciavolella, 'Saturn and Venus' in *Saturn from Antiquity to the Renaissance*, ed. M. Ciavolella and A. A. Iannuci (Ottawa: Dovehouse, 1992), pp. 176-9, and introduction to Ferrand, *A Treatise on Lovesickness*, p. 81. See also Ciavolella's 'Eros and the Phantasms of *Hereos*', and L. Bolzoni, 'The Art of Memory and the Erotic image in 16th and 17th Century Europe: The Example of Giovan Battista Della Porta', in *Eros & Anteros*, pp. 9–31 for a discussion of the relationship between the estimation, imagination and memory in various texts.

30   *Lovesickness*, p. 39.

31   *Ibid.*, p. 40.

32   Psychoanalysts, such as Freud and Lacan, link it with transference and the overvaluation of the object of love. For Lacan (whose work on transference includes analyses of Plato's *Symposium* and the courtly tradition) love is always ambivalent and fundamentally deceptive and narcissistic; it is an imaginary projection, 'a specular mirage', a desire for the impossible that is often expressed in aggressive terms as a demand for love. Clearly, early modern thinkers were aware of this dimension, if we are to judge by the ubiquitous image of Narcissus in texts and images about love.

33   *Ibid.*, p. 41. For some writers, the best cure is giving the lover the object of his desire, preferably through legitimate marriage (see Burton, *The Anatomy of Melancholy*, 3, 2, 5). This is the subject of *nouvelle* 9 of Marguerite de Navarre's *Heptameron*, ed. M. François (1943; rpt Paris: Garnier, 1991), wherein the girl's

mother promises to give her to the ailing lover at his deathbed and compels her to kiss his corpse-like body. However, the cure arrives too late (he kept his noble love silent for too long); the lover dies in her embrace, which renders her inconsolable and incapable of ever finding true happiness.

34 This kind of suggestion exists in the *De Amore* of Andreas Capellanus wherein women who refuse love suffer many torments (I, 245–6), an idea that is reified in misogynist texts such as Jeanne's Flore's *Comptes amoureux*, a compilation of unknown authorship, which includes a series of stories in which women who refuse love are vilified and/or punished. See my 'Attribuer un sexe à Jeanne Flore', in *Actualité de Jeanne Flore*, eds D. Desrosiers-Bonin, E. Viennot and R. Renolds-Cornell (Paris: Champion, 2004), pp. 239–50.

35 See M. J. Baker, 'Rape, Attempted Rape, and Seduction in the *Heptaméron*', *Romance Quarterly*, 39, 3 (1992) 271–81, and P. R. Cholakian, *Rape and Writing in the HeptamÈron of Marguerite de Navarre* (Carbondale: Southern Illinois University Press, 1991).

36 *The Malleus Maleficarum*, trans. M. Summers (1928; rpt New York: Dover, 1971). For more on the demonization of women and witch hunts, see Closson's contribution to this volume.

37 Words and images were also sometimes seen to have magico-religious powers. For belief in the power of poetry as incantation, see Ferrand on 'Homerical remedies' (pp. 345–6). See San Juan's contribution to this volume on the powers of votive images.

38 The curative power of texts was sometimes seen as all encompassing. As N. Siraisi points out, Jean Bodin asserts that reading history could 'cure all illnesses of the body and the mind'; 'Anatomizing the Past: Physicians and History in Renaissance Culture', *Renaissance Quarterly*, 53, 1 (2000) 1–30. Rabelais's narrator Acofribas playfully advertises his book as wonder drug in the Prologue to *Pantagruel* and Burton suggests his writings be taken as a 'gilded pills'. See introduction to Ferrand, *A Treatise on Lovesickness*, p. 170 n. 28. At the beginning of his *Anatomy of Melancholy* Burton also states: 'I write of Melancholy, by being busie to avoid Melancholy. There is no greater cause of Melancholy then idleness, *no better cure than business*, as Rhasis holds' (I, 6–7) and he speaks of writing as a kind of 'evacuation' (I, 7) and comfort as well as to help others. His image of textual borrowings is also medicinal: 'As Apothecaries we make new mixtures every day, poure out of one Vessell into another' (I, 9). Burton's book can be seen as a distraction from idleness, a therapy for melancholy and a symptom of his malady; see J. Pigeaud, 'Reflections on Love-Melancholy in Robert Burton', in *Eros & Anteros*, p. 229.

39 Wack, *Lovesickness*, p. 46.

40 As Shuttleton suggests, smallpox was also considered to be transmissible through a similar dynamic (see his contribution to this volume).

41 R. Girard, *Mensonge romantique et vérité romanesque* (Paris: Grasset, 1961). See Fournier's contribution to this volume for more on 'the novel as an agent of contagion'.

42 J. L. Vives, *The Education of a Christian Woman: A Sixteenth-Century Manual*, ed. C. Fantazzi (Chicago: University of Chicago Press, 2000), 5.30. See Cazes's contribution to this volume for more on odour.

43 *Ibid.*, 5.31–2.

44 See Chartier's *Belle Dame sans mercy The Poetical Works of Alain Chartier*, ed. J. C. Laidlaw (Cambridge: Cambridge University Press, 1974).

45 See B. Altmann *The Love Debate Poems of Christine de Pizan* (Gainseville: University Presses of Florida, 1998).

46 For more on reading and writing as therapy in Hélisenne's works, see Nash's '"Si oncq' lettres ou paroles eurent rigueur et puissance de pouvoir prester salut": Writing as Therapy in Hélisenne de Crenne', in *Parcours et rencontres: Mélange de langue, d'histoire et de la littérature offerts à Enea Balmas*, t. I, Moyen-Âge-XVIIe siecle (Paris: Klincksieck, 1993), pp. 519–25. Consult the edition of *Les Angoysses douloureuses qui procedent d'amours (1538). Première partie*, ed. P. Demats (Paris: Belles Lettres, 1968).

47 *The Torments of Love*, trans. L. Neal and S. Rendal (Minneapolis: University of Minnesota Press, 1996), p. 12.

48 One might suggest that just as the lover's body and soul are translated into images and texts to be read for their symptoms or signs in medical treatises describing lovesickness, the literary explorations of love themselves play with the early modern idea that books are mirrors for the readers' self-examination.

5

The Devil's Curses: The Demonic Origin of Disease in the Sixteenth and Seventeenth Centuries

Marianne Closson

The witch hunts at the beginning of the early modern era greatly broaden the question of the demonic origin of certain diseases as attested by the Bible, which at several points shows a demon capable of acting, by divine permission, on bodies and spirits. Until that time, beneficial or evil spells cast by witches on men or animals had a mysterious origin, and their effectiveness was not questioned. Beginning in the fifteenth century, these magical practices, which we find in all traditional societies, became extremely suspect: they could not but come from a pact with Satan; how else could the sorcerers provoke storms, kill people and animals, spread disease? The proliferation of Satan's henchmen thus represents an immense threat. Vying in evil, during the sabbath sorcerers prepare powders and unguents and receive the power to make the one they designate as their victim fall violently ill by a single gesture or word. They are also able to send demons into the bodies of the possessed. All direct contact with them – true agents of contagion – runs the risk of bewitchment.

The belief in demonic witchcraft inspired a heated polemic among doctors, demonologists and theologians. In the public eye, especially at the time of the many cases of possession in convents in France in the seventeenth century, this debate helps to found modern science because it results in the exclusion of diabolical and supernatural hypotheses in the origin of certain illnesses at the time the witch hunts were coming to a close.[1] Despite their apparent diversity – epileptic fits, violent skin eruptions, paralysis, sudden blindness, vomit and excrement containing objects foreign to the human body, seemingly impossible contortions – the illnesses generally attributed to a demon constitute what we today would call the psychosomatic field.[2] The arguments of the age about the devil's action in these pathologies were a way of exploring, and conquering, those marginal places in medicine where the possibility of scientific knowledge was fighting to impose itself; indeed, these troubles appeared all the more challenging to medical discourse as rituals of exorcism, the condemnation of sorcerers or the discovery of spells sometimes brought an end to them.

Relying on biblical texts – no physical or mental disease is in fact excluded from the long list of divine curses we find in Deuteronomy XXVIII, and the ulcers of Job, like the melancholy of Saul, are endlessly recalled in other texts – many authors at the beginning of the witch hunts affirm, like the inquisitors Kramer and Sprenger in the *Malleus maleficarum* (1486), that 'there is no bodily infirmity, not even leprosy or epilepsy, which cannot be caused by witches, with God's permission'.[3] This radical position, which would lead to suspecting systematically the presence of an evil influence the moment that there is illness, will be blurred in later texts. Certainly, the anecdotes related by Kramer and Sprenger will be retold in numerous treatises of demonology,[4] and Henry Boguet in his *Discours exécrable des sorciers* writes in 1606 that witches 'afflict people with all kinds of ills of the stomach and the head and the feet, with cholic, paralysis, apoplexy, leprosy, epilepsy, dropsy, strangury, etc.',[5] but demonologists will henceforth take into account medical discourse, all of which affirms that for there to be evil, the illness must be 'prodigious'.

In witchcraft trials, verbal threats followed by an effect, or the fact of finding spells in the house of the ill person, constituted undeniable proofs of bewitchment, but doctors are more dubious: if they do not generally contest the existence of spells, attested by so many biblical and mythological examples, they do question the criteria which permit verification of an illness truly sent by the devil.[6]

In *Des monstres et des prodiges* (1573), Ambroise Paré thus repeats the anecdote taken from the celebrated work *De abditis rerum causis libri duo*[7] – which appeared for the first time in 1548 and was republished more than 30 times – of the doctor and mathematician Jean Fernel, who recounts the care that he gives in vain to a young gentleman who appears epileptic until

le troisieme mois suyvant, on descouvrit que c'estoit un diable qui estoit autheur de ce mal, lequel se declara luy mesme parlant par la bouche du malade du Grec et du Latin à foison, encores que ledit malade ne sceust rien en Grec. Il descouvroit le secret de ceux qui estoyent presents, et principalement des medecins se moquant d'eux, pour ce qu'avec grand danger il les avait circonvenus et qu'avecques des medecines inutiles ils avoyent presque fait mourir le malade. ... Ce Demon, contraint par les ceremonies et exorcismes, disoit qu'il estoit un esprit et qu'il n'éstoit point damné pour aucun forfait. Estant interrogué quel il estoit, ou par quel moyen et par la puissance de qui il tourmentoit ainsi ce gentilhomme, il respondit qu'il y avoit beaucoup de domiciles au-dedans où il se cachoit, et qu'au temps qu'il laissoit reposer le malade, il en alloit tourmenter d'autres; au reste qu'il avoit esté jetté au corps de ce gentilhomme par un quidam qu'il ne vouloit nommer et qu'il y avoit entré par les pieds, se rampant jusques au cerveau et qu'il sortiroit par les pieds, quand le jour pactionné entre eux seroit venu.

[three months later, it was discovered that it was a devil who was author of this sickness, and he declared himself, speaking Greek and Latin copiously through the mouth of the patient, although said patient knew nothing about Greek. He uncovered the secret of those who were present, and especially of the Physicians who were present, making mock of them because with great peril he had entrapped them and because with their useless medicines they had almost caused the patient to die. ... This Demon, constrained by ceremonies and exorcisms, said that he was a spirit and that he was not damned for any crime. Being questioned as to what he was or by what means and through the power of whom he was thus tormenting this gentleman, he answered that there were a lot of domiciles in which he hid and that during the time when he let the patient rest, he went off to torment others. [He said] besides that he had been cast into the body of this gentleman by an individual whom he didn't wish to name and that he had entered into that body through the feet, creeping clear to the brain, and that he would go out through the feet when the day agreed upon by the two of them had come.][8]

This devil that moves about inside the bodies of the sick and travels from one body to another will inspire a profusion of commentary. In fact, for some, the signs of diabolical presence were already there, even before the sick gentleman began to speak Greek: thus an anonymous doctor, commenting in 1648 on the possession of the nuns of Louviers, writes that Fernel, in diagnosing this young man whom he treated as epileptic, 'did not notice that the convulsion appeared sometimes in one finger only, other times in the whole hand, first one side, then the other; sometimes also in the whole body, and among all these movements, the mind remained stable, which are *things naturally impossible*'; thus, it is in the disease itself that the devil was already furnishing 'this mark of possession'.[9]

In a word, the fact that the symptoms surpass the natural order signals the presence of the demon, and we are hardly surprized that the archetype of the strangest demonic illness of all is, as Father Martin Del Rio writes,

lors que le malade jette tantost par la bouche, et tantost par le bas des espines, des os, des brins de bois, des pierres des morceaux de verre, des aiguilles, des cousteaux ...

[when the sick person ejects first from the mouth, then from below thorns, bones, splinters of wood, stones, pieces of glass, needles, knives ...][10]

On this subject, the same anecdotes circulate from text to text; for example, a certain Madeleine de Constance believed she was pregnant by the devil, and gave birth as Ambroise Paré writes, to iron nails, small pieces of wood and glass, bones, stones and hair, tow, 'and several other fantastic and strange things, which the devil through his trickery had put there'[11] or Catherine Gautier of Louvain, who in 1571, in the eighth month of her illness, produced a live eel among her excrement and then vomited for fifteen days – without drinking more than usual – barrels of water similar to urine, and then tufts of fur, bones, and so on. [12]

These two texts from medical works are not at all contested, but explanations for the surprising phenomena are sought; is it a demonic illusion that blurs the spectators' vision – and thus the vision of the doctor himself – as claimed by Johan Weyer? He devotes an entire chapter of his *De praestigiis daemonum* (first edition 1563) to 'Monstrous Objects ejected from the mouth. Many arguments to demonstrate that these objects had not been in the body'.[13] Or must one concur with Del Rio (who does not deny that it might be a hallucination) in thinking that the objects truly come out of the patient's body? In effect, the devil could secretly have operated on the victim in order to place in his body the object that will appear, or he could reduce it to powder and small pieces, then make it come out whole: are these not in fact probable and plausible means, as Del Rio suggests?

One of the important debates surrounding the powers of the devil is to 'know if all the spells are naught but illusion and magic, or if there is something real', writes Pierre de Lancre, who devotes a large part of *L'incredulité et Mescreance du sortilege plainement convaincue* (1622) to this extremely complex question, because it amounts to asking whether or not the devil has the power to change the order of divine creation.[14] Without entering fully into these debates, it must be noted that much greater powers were attributed to the imagination than is the case today: it was well known that for the majority of doctors of the age, the images seen by the mother over the course of her pregnancy give shape to the unborn child, or that one can fall ill oneself by seeing a sick person. In other words, the imagination can act on reality; moreover, the belief in miracles and supernatural apparitions is an object of faith. Thus, as Pierre Le Loyer writes in the *IIII Livres des spectres ou apparitions et visions d'esprits*, demonic illusion falls into the category of 'ghosts', defined as the 'imagination of a substance without a Bodie, the which presenteth itselfe sensibly unto men'.[15] This means that the 'prestige' of the devil is not simply a trick of the mind; there is truly the *presence* of the supernatural. Even if diabolical illusion – as distinguished from divine apparitions – is 'false' and comes from the blurring of sight and of spirit, it belongs to the world of the senses; whether or not we believe that the objects that came from the body of the sick person are real or imaginary, we can not only see them, but feel them and smell them; as for the sick person, his or her suffering is not fictitious. In a word, whatever the debates on the 'reality' of this illusion, there is a point on which all are in agreement; such phenomena, whether they are real or imaginary, evidence the intervention of the devil since they are against the natural order, the only *sure* sign of a demonic illness.

Numerous quotations attest to this principle, like that of the well-known doctor Michel Marescot who establishes that, in 1599, Marthe Brossier was not possessed by the devil at all, because 'nothing ought to be attributed to the Devill, that hath not something extraordinarie beyond the lawes of Nature'.[16] This restrictive position, which limits the field of illnesses attributed to the work of demon, played as we know a large role in cases of possession,[17] where people confronted one another on the subject of manifestations of the supernatural (speaking an unknown language, revealing hidden things, knowing an unlearned science, rising above the ground, and so on) which alone permitted one to distinguish someone demonically possessed from people who were 'crazy' or otherwise ill.

This simple, almost universally accepted principle was nevertheless subject to discussion. Thus speaking a language that one did not know was not always recognized as a sure sign of the supernatural. The doctor Levinus Lemnius, in his work *Les occultes merveilles et secrets de nature*, translated into French from the Latin as of 1566, upholds the Platonic thesis that our knowledge being nothing more than collective memory, the human mind is 'imbued with the arts before it learns and practices them': if drunkenness and fever at times make vulgar men eloquent, and if approaching death gives the gift of prophecy, these are 'sparks' that come from 'natural faculties', just like speaking unknown languages: it is thus not the evil spirit that explains the phenomenon, but 'only the strength of the disease, & the violence of the humours by which, like some ardent flame, the soul of man is set ablaze'. The proof is 'that if this was done by evil spirits, such illnesses would not be cured by laxative medicines, nor would they disappear through the use of sedatives'.[18]

Placing himself in the medical tradition which from Hippocrates[19] to Avicenna seeks to combat the idea that certain illnesses, such as epilepsy, have a supernatural origin, Lemnius affirms with force that it is the humours and not evil spirits that cause our illnesses. He none the less recognizes that demons, especially 'aerial spirits', have a role because they 'insinuate themselves among the humours, but also incite human spirits to great wickedness'. Saul's melancholy can thus be traced back to natural causes for 'such furor is appeased by the gentle sounds of the harp' and it is thus in a body *already* ill that the demons had received divine permission to interfere: 'we read that Satan had *worsened* the melancholy of Saul, and had incited him to murders and betrayals and several miserable things.' The devil in fact 'looks for all occasions and means by which he can surprise us feeble and frail beings', in order to tempt us to evil.[20] In other words, while the humours are the principal cause of illness, evil spirits, the stars, air quality and other external things affect us by chance. The secondary role of demons, in all meanings of the term, is confirmed by the efficaciousness of the doctor's treatments, as much in troubles of the spirit as in those of the body.

Lemnius, whose theory is in part inspired by Pomponazzi,[21] stages a direct confrontation between the 'naturalists',[22] as they were called, and the demonologists, all the more pointed since not believing in evil spells meant questioning the very foundation of the witch hunts. Accused of rejecting the evidence of the Holy Scripture and of having a 'depraved and corrupt' conception of nature as Pierre Le Loyer states it,[23] the naturalists are often put on the same level as 'Epicurian infidels'.[24] That said, the idea that there must be favourable terrain for the intrusion of demons is more widely acknowledged, and melancholy – or more precisely melancholic madness – will find itself at the centre of investigations of the devil's action.[25]

Certainly, for a long time black humour – the *'balneum diaboli'* of Saint Jerome – seems propitious for the action of the demon, but it was not conceived of in the resolutely organic form that it takes in diverse treatises at the end of the sixteenth century. Taking up the theories of the Spanish physician Francescus Valesius (Francisco de Vallés), who distinguishes 'external' causes (the origin of the illness is external to the body, 'pestilential breath' for example) from 'internal' causes, where the devil, a true pathogenic agent, disrupts the balance of humours, Del Rio writes:

> This is how he starts melancholic diseases. First he stirs up the black bile which is within the body, and drives black specks into the brain and the cells of the internal sense-organs. Then he increases the amount of black bile by moving it to agents which dry it up with excessive heat, or he retains it so that it cannot be expelled. He also causes epilepsy, paralysis, and similar injuries by bringing down fluid which is somewhat too thick, which he does by blocking one of the ventricles of the brain, or by obstructing the roots of the nerves. He is the cause of blindness or deafness ...[26]

Indeed, the devil's action would be analogous to the deregulation of humours which, as we know, provokes most illnesses. This sort of initiating role of melancholy, the disease of 'the wounded imagination', in unleashing troubles that are as much psychic as physical marks its importance in medical thought at the end of the sixteenth century,[27] but the demonologist Del Rio insists that it is the devil himself who provokes melancholy. This is not the dominant opinion: for many the devil acts upon the humours only when the door is already open to him, as Pierre Le Loyer affirms:

> si le diable void que le cerveau soit offensé des maladies qui lui sont particulieres, comme l'Epilepsie, ou mal caduc, la Manie, la Melancholie, les

Fureurs lunatiques, & autres passions semblables, il prend l'occasion de le tourmenter davantage, & s'emparant du cerveau par la permission de Dieu, brouïlle les humeurs, dissipe les sens, occupe la fantaisie, offusque l'âme, & parlant par les organes propres de l'homme, & se coulant en iceux, se decele & montre ce qu'il est, prononce divers langages, raconte les choses qui adviennent par le monde, prophétise l'advenir combien qu'il se trouve le plus souvent menteur, & fait des merveilles qu'on ne peut croire venir d'un corps humain naturellement.

[If the devil sees that the brain is afflicted with ailments, such as epilepsy, mania, lunatic furors, and other such passions, he seizes the occasion to torment it and, by overpowering the brain with God's permission, he disturbs the humours, dissipates the senses, occupies fantasies, irritates the soul, thus expressing himself through man's own organs, and mingling with them, he appears and shows what he is, speaks several languages, relates things happening throughout the world, prophesizes about the future, even if he most often turns out a liar, and accomplishes marvels which cannot be thought of as coming of a human naturally.][28]

In a word, the criterion of diabolical intervention remains the extraordinary event, even if the disease has a natural origin. Moreover, this explains the effectiveness of the remedies, because when one is purged of 'one's superfluities and refuse', the devil who lodged there also departs. Le Loyer hesitates, however, to think that medicine has the power to chase away devils: certainly, there are 'natural means', 'herbs, simples, minerals, animals', and also music that entertains and soothes the patients, but the supreme remedy remains the name of God and of Jesus Christ and the Holy Trinity, in other words, exorcism and prayer. [29]

When confronted with this problem of the effectiveness of natural remedies against spells and curses, Del Rio adopts a position that aims to re-establish the supernatural dimension of healing, and thus of disease itself. In fact, he allows that these treatments can be effective against the devil, but they are nothing more than an indirect and secondary force, because in order to act against a powerful devil, there must be a supernatural cause, meaning the angel or God himself.[30]

It is unlikely that such an argument was able to convince a doctor who was giving hellebore to one possessed! And here we clearly see the line of demarcation drawn between theologians and doctors: on one side, there is an attempt to reduce all symptoms to natural causes – whose mechanisms are yet unknown, but that one day will be unveiled – and on the other, while using terminology borrowed from medicine, the will to return the supernatural to its rightful place, in other words to a form of sacralization of illness; one of the increasingly open issues of this conflict[31] during the seventeenth century was the medicalization of witchcraft and possession, which had until then belonged to the field of religion and had been integrated into its rituals and meaning.[32]

Weyer considered witches ordinary elderly women, weak of mind, into whose fantasies the devil slips easily, to the point of making them believe that they cause all misfortune, calamity and death.[33] He reminds us that the female sex is 'inconstant, credulous, wicked, uncontrolled in spirit and ... melancholic; [the Devil] especially seduces stupid, worn out, unstable old women'.[34] Clearly, melancholy in the understanding of the age, as we have seen, does not in the least exclude the presence of the devil, who uses the black humour to claim minds and to 'variously agitate and corrupt the thoughts and the imagination ... lulling or stirring ... the bodily humours and spirits ...'.[35] But the sorceress is thereafter in the eyes of Weyer bewitched and must be considered a victim of the devil. The physician thus appropriates the comments of Cardano when noting the resemblance between witches and the possessed:

> They are misshapen, pale, and somewhat gloomy-looking; one can see that they have an excess of black bile just by looking at them. They are taciturn and mentally infirm, and they differ little from those who are thought to be possessed by a demon.[36]

Weyer questions at great length the different cases of collective possession in convents, remarking that the most learned doctors are often tricked by the demoniacal and that many are merely tormented by melancholy.[37] He writes that

> The Devil takes great delight in immersing himself in this humour, as being the proper moisture for himself and his activities by virtue of its analogous properties; with its assistance he induces wondrous phantasms and rare imaginings.[38]

To illustrate his point, the author evokes the oft-cited example of the young girl who claims to have a knife in her body following an encounter with a black dog, and is believed to be insane until, at the end of one year, people see the knife exit the body, proof that she was demonic: again, we come back to the border between the natural and the supernatural, but the question of melancholy remains extremely problematic.[39]

Indeed, the argument of melancholic folly was often used in the debates of the age in order to demonstrate that witches or those possessed were not what they *believed* to be. Witch hunters and the supporters of possession alike were quick to respond: Jean Bodin, in his *Refutation des Opinions de Jean Wier*, affirmed thus that witches could not be melancholic because a woman's humour, cold and moist, was directly contrary to black bile, which was hot and dry.[40] A half-century later we find the same argument in the *Traitté de la melancholie, sçavoir si elle est la cause des effets que l'on remarque dans les possédées de Loudun* (1635) by La Mesnardière, for whom the 'illnesses of those possessed are more imaginary than from the imagination',[41] thus we can absolve these well-born girls from folly, even though their extraordinary contortions, levitations and other marvels attest to the presence of the devil. [42]

But to the extent that melancholy is the preferred disease of the demon, this clear opposition does not always apply: a certain ambiguity is already present in the New Testament where possession is most often linked with other illnesses, like paralysis, blindness, deafness or epilepsy.[43] To give an idea of the confusion that reigns, we can quote Jean Taxil who, in his *Traicté de l'epilepsie* (1603), wonders why the demonic are epileptic, in the manner of the Sibyls, who 'convulsed, fell, foamed and tormented themselves when they were bedevilled'; Taxil attributes this phenomenon to the melancholic humour.[44] For this physician, as for Weyer and so many others, there is thus no incompatibility (far from it) between melancholy and demonic possession.

Going even further, Robert Burton in his celebrated work, *The Anatomy of Melancholy* which appeared for the first time in 1621, writes:

> The last kinde of madnesse or melancholy is that demoniacall (if I may so call it) obsession or possession of divells which *Platerus* and others would have to bee praeternaturall: stupend things are said of them, their actions, gestures, *contortions*, fasting, prophecying, speaking languages they were never taught &c. many strange stories are related of them which because some will not allowe ... I voluntarily omit.[45]

Possession is given explicitly as one of the forms of melancholic folly, but Burton prudently refuses to follow this proposition to its logical conclusion because 'supernatural' demons might indeed be active. In any case, faithful to sacred texts, he continues to think that the devil – and more so sorcerers and magicians – are at the origin of melancholy.[46]

But in 1643, young doctor Yvelin, examining the possession of the nuns of Louviers and their extraordinary contortions, develops an argument that thereafter seems definitively to exclude the devil from human disease. Staunchly naturalist because it is 'natural heat' which 'creates the miracles of the prophets, natural ecstasies, divination by dreams, makes the ignorant speak revelations and other similar effects, that for being rare, seem supernatural'.[47] Yvelin sees in those possessed the effects of 'the wounded imagination':

> Ne se peut-il faire par folie et erreur d'imagination, elles se croient possédées ne l'estant pas, je crois cette pensée aussi facile comme à d'autres de s'estimer Saints ou Rois, ou autres ainsi que l'on voit tous les jours.

[Can it not be through madness and error of imagination, they believe themselves possessed when they are not, I believe this idea as easy as for others to believe themselves Saints and Kings, or still other things, as one sees every day.][48]

But the characteristic of the melancholic humour which 'stagnates in the hypophyses' but also in the wombs of these nuns, is that it acts upon the body:

Il s'en eslèvent des vapeurs et des vents de qualité assez maligne pour produire tous ces effets qui semblent si étranges et si extraordinaires, pour ce que la chaleur qui travaille pour les dompter, esmet non seulement les humeurs, les vapeurs et les vents en diverses sortes, et ces meslanges produisent encore des effets tous differents selon les parties qu'ils attaquent. Et que ne feront-ils pas si la vapeur d'une semence pourrie dans la matrice vient se joindre à cet autre humeur, il n'y a presque aucun mouvement bigearre dans la nature qui ne puisse estre fait apres ce meslange des deus matieres.

[Therefrom arise vapours and winds malignant enough to produce all of these effects that seem so strange and so extraordinary, for the heat that works to overcome them, emits not only the humours, vapours and winds of diverse kinds, and these mixes produce still more effects all different according to the parts they attack. And what will they not do if the vapour of semen rotten in the womb comes to join with this other humour, there is almost no strange movement in nature that could not be done after this mixing of two matters.][49]

The 'strange', the 'extraordinary', the 'bizarre' – and we know just how spectacular those somatic manifestations called hysteria[50] in the nineteenth century can be – receive a purely organic explanation. Indeed, Yvelin makes melancholy an illness that can be provoked by suggestion and imitation:

un confesseur leur voyant dire et faire des choses étranges pourraient par ignorance et simplicité, croire qu'elles seraient possédées, ou ensorcelées, et ensuite leur persuader par le pouvoir qu'il a sur les esprits. Or si telles pensées saisissent une fois les esprits de deux ou trois d'entre elles, elles s'estendent soudain et se communiquent à toutes les autres, car les pauvres filles adjoustent beaucoup de foy à ce que disent leur compagnes et n'osent révoquer en doute ce que disent leurs supérieurs.

[a confessor hearing them say or do strange things could in ignorance and simplicity, believe that they are possessed, or bewitched, and then persuade them by the power that he has over minds. So if such thoughts once seize the minds of two or three among them, they spread suddenly and are communicated to all of the others, for the poor girls place much faith in what their companions say and do not dare to cast doubt upon what their superiors say.][51]

In short, it is no longer the devil as the word of authority and the mimetism inspired by this word that is at the origin of the demonic possessions in convents. In the beginning, then, we would not find the devil, but a representative of power whose interpretation of slightly strange symptoms that he sees, or believes he sees, allows in a way an epidemic as much of the symptoms themselves as of the belief that it is a matter of spells.

Extended to all of society, Yvelin's intuition provides an analysis of the witch hunts as an historical phenomenon: the discourse of authority of early modern clerics on the powers of the demon fed the *concrete* manifestations of this power, particularly where the human body was concerned. Thus it is not surprising that their silence put an end to this belief. As a physician of the mid-seventeenth century wrote:

Quand mesme l'on serait bien asseuré que veritablement il possederait quelque personne, il faudrait toutefois le dissimuler, traicter cette personne-là comme une folle et luy faire exercer les plus abjectes functions que l'on pourrait s'inventer. ... Il ne faudrait pas qu'un honneste homme prit la peine de parler à elle; et si l'on exorcisoit, il seroit bon que ce fust en la présence de peu de personnes qui n'escouteraient pas son babil et qui ne luy tiendraient point d'autres discours que ceux que l'église a préparé pour cet effet. Sans doubte il seroit bientost ennuyé de sa possession et rechercherait luy mesme les moiens de la quitter et de s'enfuir.

[Even when one would be sure that some person were truly possessed, it would be necessary nonetheless to conceal it, to treat that person as mad and to make him perform the most abject functions that one could invent. ... It would be forbidden for an upstanding man to take the trouble to speak to him; and if he were exorcised, it would be good that this be in the presence of few people who would not listen to his babble and would tell him nothing but what the church has prepared for this purpose. No doubt he would soon be bored with his possession and would seek out for himself the means of leaving it and escaping.][52]

The devil himself is reduced to silence: a new era begins, one that puts an end to the very ancient idea that relates all mental illness to the action of a supernatural force, divine or demonic ...

Notes

1 See R. Mandrou, *Magistrats et sorciers en France au XVIIe siècle* (Paris: Seuil, 1980).

2 In an article fundamental to our subject, Jean Céard reminds us that 'les "maladies mentales" s'enracinent dans un désordre physiologique, comme toutes les autres maladies' ['"mental illnesses" are rooted in physiological disorder, like all other illnesses'] and therefore there is not a discipline of 'mental alienation' among the medical disciplines of the sixteenth century; 'Folie et démonologie au XVIe siècle', in *Folie et déraison à la Renaissance* (Édition de l'Université de Bruxelles, 1976), pp. 131–2.

3 H. Kramer and J. Sprenger, *The Malleus Maleficarum*, trans. and ed. M. Summers (1486; New York: Dover Publicatons, 1971), Part II, ch. XI, p. 134.

4 See J. Bodin, *De la demonomanie des sorciers* (1580), (Gutenberg Reprint, 1979); *On the Demon-Mania of Witches*, trans. R. A. Scott, ed. R. A. Scott and J. L. Pearl (Toronto: Centre for Reformation and Renaissance Studies, 1995).

5 *An Examen of Witches*, trans. E. A. Ashwin, ed. M. Summers (Bungay, Suffolk: John Rodker, 1929), pp. 89–90; *Discours exécrable des sorciers* (Marseille: Lafitte Reprints, 1979), p. 168: 'affligent de toutes sortes de maladies comme d'estomach, de teste, de pied, de cholique, d'apoplexie, de lepre, d'epilepsie, d'enfleure, de retention d'urine, etc.' In some trials we also find witches accused of being sources of the plague. In Geneva in 1545, for example, 43 witches were accused of having brought the plague and 29 were executed.

6 See J. Céard, 'Le monde obscur. L'occulte et le démoniaque', in *La Nature et les prodiges*, 2nd edn (Genève: Droz, 1996), ch. XIV.

7 'Et morbos, & remedia quaedam trans naturam esse', II, 16 (Lyon: Bartholomée Vincent, 1605).

8 A. Paré, *Des monstres et des prodiges*, ed. J. Céard (Geneva: Droz, 1971), p. 95; *On Monsters and Marvels*, trans. and ed. J. L. Pallister (Chicago: The University of Chicago Press, 1982), p. 100.

9 'Lettre d'un médecin anonyme à M. Philibert de la Marre' in R. Mandrou, ed., *Possession et sorcellerie au XVIIe siècle, textes inédits* (Paris: Fayard, 1979), p. 217: 'ne remarquait pas que la convulsion paroissoit quelquefois en un doibt seulement, d'autrefois en toute la main, tantost d'un costé, tantost de l'autre; quelques fois aussi en tout le corps, et parmi tous ces mouvements l'esprit demeroit entier, qui sont des *choses impossibles naturellement.*'

10 *Les controverses et recherches magiques*, trans. from the Latin by A. Du Chesne (Paris: J. Petit-pas, 1611), p. 407; *Investigations into Magic*, trans. from the Latin and ed. P. G. Maxwell-Stuart (Manchester: Manchester University Press, 2000), pp. 257–8, provides the English translation of Del Rio's observations, but does not quote the entirety of Del Rio's argument.

11 Paré, *Des monstres et des prodiges*, p. 94; French original, p. 89: 'plusieurs choses fantastiques et estranges' que 'le diable par son artifice y avoient appliquées'. This anecdote is borrowed from the doctor Jacob Rueff, whose first edition of the *De conceptu et generatione homini* appeared in 1554.

12 This story comes from C. Gemma, *De naturae divinis characterimis* (Anvers: Plantin, 1575).

13 Johan Weyer, *Witches, Devils, and Doctors in the Renaissance*, trans. J. Shea (Binghamton, NY: Medieval & Renaissance Texts & Studies, 1991), pp. 286–90.

14 (Paris: Nicolas Buon, 1622): 'sçavoir si tout le sortilège n'est qu'illusion et prestige, ou bien s'il y a quelque chose de réel.' Lycanthropy, like the flight of

witches on the sabbath, has inspired the same questions. See on this subject my *L'Imaginaire démoniaque en France, 1550–1650* (Geneva: Droz, 2000), pp. 24–36.

15 P. Le Loyer, *A Treatise of Spectors or straunge Sights, Visions and Apparitions appearing sensibly unto men* (London: Mathew Lownes, 1606), n.p.; Angers: George Nepveu, 1586: 'imagination d'une substance sans corps qui se présente sensiblement aux hommes'. Republished in 1605 under the title *Discours et histoires de spectres* (Paris: Nicolas Buon, 1605).

16 *A True Discourse, Upon the Matter of Martha Brossier of Romarantin, pretended to be possessed by a Devill*, trans. A. Hartwel (London: J. Wolfe, 1599), p. 11; *Discours veritable sur le faict de Marthe Brossier de Romarantin, pretenduë demoniaque* (Paris: M. Patisson, 1599), p. 15: 'rien ne doit estre attribué au démon qui n'ait quelque chose d'extraordinaire par-dessus les loix de la nature.'

17 See R. Mandrou, *Magistrats et sorciers*; and Michel de Certeau, *La Possession de Loudun*, 2nd edn (Paris: Gallimard-Julliard, 1990); *The Possession at Loudun*, trans. M. B. Smith (Chicago: University of Chicago Press, 2000).

18 Levinus Lemnius, *Les occultes merveilles* (Paris: Pierre de Pré, 1567), pp. 215–17: 'imbu des arts avant qu'il ne les apprenne et les pratique'; 'la seule force de la maladie, & la violence des humeurs, par laquelle comme par quelque flambeau ardent, l'âme de l'homme s'embrase'; 'que si cela se faisoit par les malings espris, telles maladies point ne guériroient par medecines laxatives, ny ne s'en iroient à force de dormitoires'.

19 The Hippocrates of *De sacro morbo* in which the father of medicine objected to the notion that epilepsy might have supernatural causes.

20 Lemnius, *Les occultes merveilles*, pp. 210–12: 'ils se meslent parmy les humeurs, mais aussi incitent les esprits humains à toute meschancetez'; 'telle fureur s'appoisoit aux doux sons de la harpe'; 'nous lisons que Sathan avoit *aigry* la melancholie de Saul, et l'avoit incité à meurtres et trahisons et plusieurs choses malheureuses'; 'cerche toutes occasions & moyens comme il nous pourra surprendre faibles & debiles'. Note that Weyer, who here is very close to Lemnius, saw witches as melancholics who could be treated.

21 On Pomponazzi, see Céard, *La nature et les prodiges*, pp. 96–105.

22 For Céard, Lemnius, in removing all distinctive character from the supernatural, actually destroys 'naturalism', because God and nature are completely confused. On Lemnius, see *La nature et les prodiges*, pp. 345–9.

23 *Discours et histoires*, pp. 124ff.

24 As does Jean Taxil, who judges the gift of languages 'supernatural and extraordinary' and blames the naturalists for denying demonic possession in these cases; *Traicté de l'epilepsie* (Lyon: Robert Renaud, 1603), p. 150.

25 André Du Laurens distinguishes the 'healthy' melancholic, 'fit to undertake matters of weightie charge and high attempt', from the unhealthy one whose sick mind is troubled by 'a thousand fantasticall inventions and objects'; *A Discourse on the Preservation of the Sight: of Melancholike Diseases; of Rheumes, and of Old Age* (1599), trans. R. Surphlet (Oxford: Humphrey Milford, Oxford University Press, 1938), pp. 85–7; originally *Discours de la conservation de la veue: des maladies melancoliques, des catarrhes et de la vieillesse* (Tours: Jamet Mettayer, 1594), f° 113b. In other words, there is a difference between a melancholic complexion and mental illness. In any case, the link between melancholy and creative genius, so often formulated during the Renaissance, is naturally excluded from the analysis here.

26 Del Rio, *Investigations into Magic*, p. 127.

27   See Céard, 'Folie et démonologie au XVIe siècle'.
28   Le Loyer, *Discours et histoires des spectres*, p. 146.
29   *Ibid.*, p. 150.
30   Del Rio, *Investigations into Magic*, pp. 261–2.
31   In his article 'Folie et démonologie au XVIe siècle", J. Céard dated the end of the close collaboration between doctors and clerics to the end of the sixteenth century, with Marescot's firm conclusions in the Marthe Brossier affaire as a turning point.
32   See M. Foucault, 'Médecins, juges et sorciers au XVIIe siècle', in *Dits et écrits I, 1954–1975* (Paris: Gallimard, 2001), pp. 781–95.
33   Weyer, *Witches, Devils, and Doctors*, p. 176.
34   *Ibid.*, p. 181.
35   *Ibid.*
36   *Ibid.*, p. 510.
37   *Ibid.*, pp. 304–14.
38   *Ibid.*, p. 315.
39   *Ibid.*, pp. 314–15.
40   In *La Demonomanie des sorciers*, fo 247–8.
41   La Flèche: Martin Guyot and Gervais Laboé, 1635, p. 21: 'Les maladies des possédées sont plustost imaginaires que de l'imagination'.
42   *Ibid.*, pp. 130–1.
43   Thus, among other examples, we find in Mark IX. 4–29, Luke IX. 38–43, Mat. XVII. 14, an epileptic child tormented by a blind and deaf demon.
44   Taxil, *Traicté de l'epilepsie*, pp. 149ff: 'convulsaient, tombaient, escumoient, & se tourmentoient lors qu'elles estoient endiablées'.
45   *The Anatomy of Melancholy*, ed. T. C. Faulkner, N. K. Kiessling and R. L. Blair (Oxford: Clarendon Press, 1989), vol. I, Part I, Sect. I, Memb. 1, Subs. 4, pp. 135–6.
46   *Ibid.*; see Part I, Sect. 2, Memb. 1, Subs. 2 'A Digression of the Nature of Spirits, bad Angels or Divels, and how they cause Melancholy', and Subs. 3, 'Of Witches and Magitians, how they cause Melancholy' (pp. 174–99).
47   *Response à l'examen de la possession des religieuses de Louviers* in *Les Possédées de Louviers*, ed. R. Dubos (Condé-sur-Noireau: C. Corlet, 1990), p. 196: 'fait les merveilles des prophètes, les extases naturelles, les divinations par les songes, fait dire des choses révélées aux plus ignorants et d'autres semblables effets qui pour estre rares semblent surnaturels.'
48   *Ibid.*, p. 203.
49   *Ibid.*, p. 195.
50   Hysteria has existed naturally since Antiquity and is not at all absent from medical treatises of the sixteenth and seventeenth centuries: it is the illness of 'virgins, nuns and widows', as Burton writes, but, like melancholy – and more generally all mental illnesses whose nosology is constantly renewed and contested, because they cannot be conceived of outside of a given historical and cultural context – its definition was not that which was given to it *a posteriori* by Dr Charcot at Salpétrière. Let us recall that we owe the classification among hysterics of the 'démonopathes' (those possessed or witches) and the great mystics like Teresa d'Avila to the staunchly rationalist medicine of the nineteenth century.
51   *Response*, p. 204.
52   Mandrou, *Possession et sorcellerie*, p. 218.

Part II
Practice

6
Apples and Moustaches: Montaigne's Grin in the Face of Infection

Hélène Cazes

All smell is disease.

Edward Chadwick[1]

There could be no more intense and threatening conception for the transmission of disease than a theory of some ubiquitous, potent and inescapable infection whose origins are untraceable and whose contact is unavoidable. Contagion in such a conception would be the unstoppable and universal spreading of death, recognized by the symptoms of the malady and the span of its action; impossible to counter by the isolation of its physical source, the outbreak of epidemics would leave its potential victims with quarantine as their only defence: that is to say, separation from the community of either the patients stricken with the disease or the healthy, fleeing the 'infected areas' to the countryside or some more distant destination. Moreover, this conception of fatal contagion would be all the more appalling if the theory of infection relied on implicit, imprecise and unperceivable principles, such as was the case before germs could actually be observed and accounted for. In other words, implicit theories belong to the imaginary representations of the world, and the place of man in it; before microorganisms were first seen and understood, the descriptions and prevention of epidemics pertained to evasive systems of thought. Medicine, folklore, literature or religion may then be analysed, not as such, but as discourses where a collective conception is at work. Now, when physicians, patients and citizens must turn to imaginary and unquestioned assumptions for describing and understanding the transmission of maladies, when these assumptions are unanimously accepted and used so that they define an undebatable vision of the world, and when the various theories are based on a common but unsaid substratum, the very idea of contagion seems to invade the discourse itself: widespread, recognizable, potent, it has no definite origin and lacks a definition, but it occupies medical and non-medical speech. Unassignable to germs, if not within an abstract

79

construction without pragmatic effects, the imaginary conception of contagion spreads outside the medical domain and serves as a model for neighbouring or distant topics such as love, politics, laughter, religious life, magic. Imprecision and consensus, the specific qualities of the collective imaginary, thus lead to the crossing of boundaries and the theory itself enacts the expansion of its object: in its indetermination, the idea of contagion infects representations of civility, relationships, beliefs and behaviours.[2]

Evil smells in the air

The notion of contagion during the Renaissance becomes achingly crucial, not only because of the innumerable attacks of plague, typhus, cholera and other maladies, but also because the very notion of infection threatens the new sociability and the new trend towards a secular and personal conception of humanity. The spectre of infection is located precisely at the crossing of the self and the group, the city and the community, the just and the unaccountable: not surprisingly, physicians are numerous who try to pry into the mysteries of the onset and spread of illness. At stake is, in fact, the possibility of leading a personal life. The sixteenth century, for all its daring innovations (or revolutions), often thinks of itself as an age it proclaims 'worthless'. There, in the denied inheritance, in the implicit use of categories, lies the foundation for the study of imaginary construction.

In the age of humanism, ancient medical theories prevail in France, even strengthened by the surge of humanism and the pious admiration of Greek and Roman models; according to the two luminaries of scholastic medicine, Hippocrates and Galen, diseases are the result of a lack of balance in the climatic and environmental elements that provokes, or meets, an imbalance in the human body; contagion, then, strikes the individuals who do not internally possess the equal temperament that would fend off the disruption. Maladies appear from a disruption of the environment and are caught by breath: as the historian of medicine B. Gordon summarizes this long tradition, the theory 'in the air' about contagion was, precisely that infection came and went with ... the air.

The physicians of the Middle Ages had an idea that certain diseases are contagious but the mechanism of transmission was not understood. The teaching of Hippocrates that disease in general may arise from the food we eat or the air we breathe generally prevailed throughout the medieval period and even later. Therefore, when a disease gripped a large number of persons of various ages and both sexes at one time, it was considered that its origin must have been the air. This theory was accepted by Galen, who added that climatic factors also contribute to outbreaks: miasmas, given off by water in certain communities, 'taint the air and occasion disease'.[3]

These theories were considered as commonplace: their banality dispensed from quotation, from criticism, and also from thinking. In the 'air', the Hippocratic narrative of the birth of epidemics, relayed by Galen's commentaries and treatises, was unanimously accepted and endlessly repeated.

The climatic doctrine of Hippocratic treatises and its development in the Galenic tradition leave no room for consciousness or prevention: disease is inhaled by its victim as surely as air is breathed. But the modification of the air, a natural and harmless element, into a deadly poison bears a mark: its smell. Odour is described in the very terms, vague and pervasive, which were used for describing the infection of air: born from a disorder, bad odour indicates danger. Repeatedly, and with emphasis, medical treatises insist on the necessity of avoiding foul-smelling places such as swamps, open graves, volcanoes or sickrooms. When physicists and physicians try to assign a material cause to the outbreak of an infectious malady, as Lucretius and the Epicurists attempted to achieve with the theory of 'atoms' and 'seeds of disease', odour remains the only perceivable sign of disease.[4] The symmetric of foul ambient air, bad breath, indicates a bodily disorder and, similarly, is perilous to inhale. The doctor is thus advised to keep at a distance from his patients.

But the distinction between the medium of the smell (the air itself and respiratory organs) and the external source of the smell is difficult, if not impossible, to establish: even if Galen, in his treatise on *The Organ of Smell*, evokes a difference in density and humidity between air – a dry, pure element – and vapours that would alter the composition of the air and possess an essential percentage of water, the definition and the identification of the vapour remains ambiguous and elusive. When Galen tries to account for smells, he turns to a theory of particles, but the very source of odours appears as elusive as the source of infection.[5] A volatile and transient composite element that travels through air and carries infection, vapour is supposed to emanate from bodies and elements and to reach the brain through the olfactory canals, along with the *pneuma*, but it is never observed or characterized. Indeed, the main attribute of smell seems to be its ability to propagate and reach the vital parts of the body. Unassignable because of its fugitive nature, odour possesses composite qualities: humid as a vapour, it mixes with air because of its dryness; and in the same way, it enters the body through the nose but is recognized through the tongue. Thus, Galen–and following him, medieval and early modern medicine – describes the nose as consubstantially linked to the humidity of vapours (and therefore able to perceive them):

[Nature] made the visual organ most light-like, for it alone senses brilliance and light, and air-like the organ of hearing ... So also Nature formed the organ distinguishing tastes, the tongue, from the wetter elements in the body. Between air and moisture and fire, is the matter sensed by smell, being neither as rarefied as air nor as dense as moisture. For whatever streams off the surface of a body is the substance of the odorous matter. One recognizes this quite definitely with roses and with things similarly simple, which swiftly become smaller and drier than they were, showing unambiguously that parts have escaped from their substance.[6]

If smell is to be considered a sign, the perceptible aspect of a larger and hidden object, the latter being the transmission of an infectious disease, then, in a tragic reversal of the usual order of perception and interpretation of the signs, it bears also the meaning that danger has already crossed the boundaries of one's self, has invaded the body and is working its frightful way through the organs. The swiftness of the process, situating the event before its perception, ties together the interpretation of the sign and the diagnosis of infection. The nature itself of this inescapable danger, sensed when it is too late to counter its threat, evades definition: the theorist fails to assign it to a single particular organ, as well as to a single and particular structure; from the substantial intricacies of matters, properties and qualities, the odour appears to be as unnoticeable as air, as swift as water.

A bad smell, rather than the sign, was believed to be the actual means of propagation, especially for the plague. Prior to the dissemination of Pasteur's research on microorganisms (around 1900), unpleasant smells were viewed as having a direct effect on health and life. Nauseating exhalations from sewers, charnel houses, cemeteries, cesspools and marshes were believed to cause many fatal diseases. Does the double meaning of the French word *empester*, establishing as it does an absolute equivalent between stench and death, mean that the plague was actually believed to *be* an odour?[7]

Air, apples and oranges

Smell is sensed as particularly dangerous because it is an invisible transmission (either by the smallness of the 'atoms' or by the nearly immaterial substance it propagates) and it is not possible to isolate the smell from the air. By way of reciprocity, pure and healthy air is necessarily odourless. Two ways of attaining clean air are proposed: filtering the smells or countering them. If odours are in effect a vapour mixed with the air, and if the nasal cavity, according to Galen, could block off the largest elements contained in the air, then a proper and elaborate filter could keep the stench away. Similarly, more potent smells could prevent the weaker ones from reaching the olfactory nerves: by besieging the nose and expelling rival odours, they too would act as filters. The distinction between seeds of disease and sources of odour being so elusive, 'aromatic prevention' would protect from infection itself by protecting against its smell. According to this doctrine, there was more to be gained in perfumes and spice than a pleasing effect on the senses: the prescriptions given by physicians for windows to be opened, incense to be burnt, perfumes to be worn, as well as the 'apple of aromas' held by the doctor when he approached the sick, were believed to be efficient protections.

The procedures of medical visits, lengthily described in similar terms, emphasize the need for the physician not to smell anything 'infected'. The treatises of remedies against the plague are full of recipes for fragrances, perfumed oils and unguents, creams and powders that would keep the stench away. To the same effect, a sponge steeped in vinegar was supposed to 'stop the smell'. Of course, it was of the first importance to keep away from foul air, and thus to stay as much as possible outside sick houses and to ventilate rooms; but once the places were treated by proper airing, good odours could, as well, counterbalance bad smells.

Thus, in a treatise published in 1562,[8] the Toulouse physician Auger Ferrier gives recommendations to his colleagues and to healthy people who wish to keep well. The main principle of his advice is to keep away from bad smells. Here is the 'safe day' he recommends for times of plague:

On arising from bed:
As soon as you are awake, you will have all the windows in your room opened, particularly those looking South and West, taking good care that no infection may enter. Then, ask for your clothes to be prepared, which you will have aired all night long. And it would be good to change them often, so that you would not put on today the outfit you wore yesterday. Afterwards, ask for a warmer full of glowing coals and throw into it some incense, myrrh, benzoin, laudanum, stirax, roses, leaves of myrtle, rosemary, lavender, basil, satura, mother-of-thyme, marjoram, aloe wood, squinanti, macis, cardamom, small pieces of pinewood, cloves, pieces of cypress and juniper and other odoriferous things, the smoke of which you will put on your clothes so that they retain the smell. And then, you will dress as usual.[9]

Starting with the airing of the room and the clothes, the day continues with a walk outside the house: once more, the odoriferous herbs and spices allow the walker not to smell anything other than his own pomander or scented sponge. In Latin, the doctor gives the recipe for these preventative objects, to be prepared by an apothecary; the composition varies according to the season, which reflects the aforementioned link between odour and ambient air.

When you want to go out of the house
Do not go out of your house earlier than two hours after sunrise and, on cloudy days, do not move. And to be safe, do not go out before you have eaten. When you want to go out, take your scented apple. Here is the

recipe, for the hot season: R. *Sandalorum citrinorum, macis, corticum citri, rosarum, foliorum myrti, an. drachmas duas: benjoin, ladani, stiracis, an. drach. et feneis: cardomomi, violarum, croci, an. scrupulos duos: camphorae et ambrae an. scrupulum unum: algaliae, et musci an. grana duorum aqua rosarum infusionis formetur pomum ...*

If you do not have a scented apple, you will choose an apple of Capendu,[10] lemons, limes, oranges and any odoriferous fruit, or a bunch of herbs and flowers of agreeable perfume.

At the least, you should take a sponge and dip it in a mixture of vinegar and rose water, combined together with cinnamon, cloves, macis, safran, or anise, some grains of camphor, amber, civet, according to each one's purse.[11]

For additional security, the reader is advised to eat strong-tasting food, and the chewing of lemons or roses ensuring an internal exhalation of 'good' smell is also recommended. Thus armed, protected from outside and inside odours, the wise man will hold a piece of burning wood in front of him.

You will take as well, and carry with you, a piece of marzipan, where you will have secured lemon seeds; or some rose preserve or agradelle[12] preserve: or the pulp of oranges, cut into little pieces, on which you will throw lots of ground sugar; this recipe being for the summer. ... In addition, if you are a physician, an apothecary, or someone else who visits the pestiferous, you should have handy numerous pieces of juniper wood, or, failing this, a good torch as well as incense.[13]

The figure of this prudent fellow, hidden behind smoke and objects, does not relate to the external world, and even less to his fellow man:

Strolling through town
Put before your nose your scented apple or your bunch of flowers, or the dipped sponge as it was said previously, or some other odoriferous thing. Take from your marzipan a lemon seed, or some bit of the other things inside, and chew on it as you go, then swallow it. Stay away from people's breath and from infected streets, as much as possible.[14]

The visiting physician is even more remote, the apple literally keeping the doctor away:

When the physician, or others, visit the pestiferous
When you approach the pestiferous room, send someone ahead who will have all the doors of the house opened as well as the windows of the room where the patient lies; meanwhile, stay put a little in the street. And ask that a good fire be lit in the patient's room. Then, you

will have brought a warmer full of glowing coals, with incense, roses, myrrh, benzoin, laudanum, styrax, cloves and other similar smells for fumigation. Then, have your piece of juniper wood lit on fire and thus fearlessly enter, having sent ahead of you the aforesaid person with the aforesaid fumigation of incense, myrrh etc.; and follow him, holding in one hand your burning wood and, in the other, pressed to your nose, your scented apple, or your bunch of herbs, or the aforesaid sponge.

In this manner, you will walk to the bedroom, where you will have installed the aforesaid warmer with the aforesaid incense, so that the smell diffuses throughout the room. Thus, holding in your mouth your marzipan, holding one hand before your nose with the aforesaid odours, having in the other hand the aforesaid piece of burning juniper, you will look at your patient from some distance and you will interrogate him about his ailment and his accidents.[15]

Holding odoriferous substances in front of the nose, advising that incenses and essences be burnt in sick rooms, opening windows to north winds and preaching aeration, the physician encounters the stricken people without falling victim to their smell. Thus, the exposed person devises ways of breathing without inhaling the deadly stench: aromas and safe distance act as filters of the infected air. The image of the doctor with his face hidden by a mask in the form of a long beak, filled with odoriferous substances, wearing gloves and a long robe offers the representation of the individual citadel built against contagion.[16] But, not so far from us, the traditional gifts that one brings nowadays to confined people strangely resemble the gear of the plague doctor: oranges for prisoners[17] and flowers for the sick may well be the superstitious continuation of marzipan and bunches of herbs.

The mask and the moustache

The mask is the emblematic object of these protective devices, which complement the filtering functions of the nose itself. Held as a barrier between the self and the contagion, the mask in its various forms of sponge, apple, smokescreen or piece of cloth conceals the person it shelters: there is no communication between the well-equipped stroller described by Ferrier and the possible transmitters of disease. This is, precisely, the exact opposite of the protection that would suit the inquisitive, open-minded and friendly author of the *Essays*. It is no surprise that the writer who devotes the last years of his life and his books to an enquiry into himself would confront the notion of contagion: in asking what the self is, where it stands in the relationships woven with others, how friendship or admiration blur the frontiers between individuals, Montaigne encounters the question of infection as a subsidiary definition of his own ways of being. As early as the first version of the *Essays* in 1580 and continuously throughout the subsequent additions and corrections, Montaigne included a chapter entitled *The Odours* (I, 55).[18] In choosing such a topic, he describes his dealings with his fellow men as well as his reaction to threats of infection. As often with Montaigne's *Essays*, the argument about smells seems contradictory and rambling before the reader settles in and follows the meanderings of thought, corrections and ambiguity: thus, Montaigne seems to disapprove of perfumes, but to advise their use by clerics, and he claims to be safe from infection thanks to his moustache, that is 'full' and 'retains fragrances'. A detailed reading of this short chapter shows an internal consistency in the argument: at the threshold of the self and the world, the moustache is a natural filter against invading diseases that does not preclude communication or even closeness with another. Thus, the paradox and the ambiguity serve a stance of both independence and kinship.

Beginning the exploration of smells with the paradox that some great people are said to emit a good body odour,[19] Montaigne devotes the first part of his reflection to the merits of odourless breath and [odourless?] complexions. The first and last statements of the chapter, composed before 1580 and before 1588 respectively, emphasize the danger of inhaling odours: the good breath is the unnoticeable one, and the good places to live in are the odourless ones.

Thus, the first development counters a rapid evocation of perfumed breaths:

> But the ordinary constitution of human bodies is quite otherwise, and their best and chiefest excellency is to be exempt from smell. Nay, the sweetness even of the purest breath has nothing in it of greater perfection than to be without any offensive smell, like those of healthful children ... [20]

Similarly, in the last paragraph, added in the 1588 edition, the absence of odours is deemed the best criterion for selecting lodgings:

> My chiefest care in choosing my lodgings is always to avoid a thick and stinking air; and those beautiful cities, Venice and Paris, very much lessen the kindness I have for them, the one by the offensive smell of her marshes, and the other of her dirt.[21]

Quoting three Latin authors for their mistrust of body odours and of perfumes, the author argues that the 'foreign' fragrance, applied by deceptive women or barbaric seducers, is a suspicious habit in itself: is the artificial odour not meant to hide and disguise unpleasant, and thus unhealthy smells? The defence of natural body odours sounds indeed like distrust of every kind of smell that would counteract the true, spontaneous ones. And

such as make use of fine exotic perfumes are with good reason to be suspected of some natural imperfection which they endeavour by these odours
to conceal. Even to smell sweet is to stink.[22]

An addition of 1588, omitted by Cotton in his translation because of its
reference to a French idiom, explains the excellence of the absence of
smells: 'comme on dict que la meilleure odeur de ses actions c'est qu'elles
soyent insensibles et sourdes'.[23]

But the second detour of Montaigne's account for odours is a series of
additions in the second edition, that is to say the 'most personal inspiration';[24] this time, smells are shown as pleasurable; moreover, a keen sense
of smell is presented as an asset. The author defines himself as exception-
ally perceptive of smells:

> I am nevertheless a great lover of good smells, and as much abominate
> the ill ones, which also I scent at a greater distance, I think, than other
> men. ... 'Tis not to be believed how strangely all sorts of odours cleave to
> me, and how apt my skin is to imbibe them. He that complains of
> nature that she has not furnished mankind with a vehicle to convey
> smells to the nose had no reason; for they will do it themselves, espe-
> cially to me.[25]

Here is the point where the moustache enters the scene:

> my very mustachios, which are full, perform that office; for if I stroke
> them but with my gloves or handkerchief, the smell will not out a whole
> day; they manifest where I have been, and the close, luscious, devour-
> ing, viscid melting kisses of youthful ardour in my wanton age left a
> sweetness upon my lips for several hours after. And yet I have ever
> found myself little subject to epidemic diseases, which are caught, either
> by conversing with the sick or bred by the contagion of the air, and have
> escaped from those of my time, of which there have been several sorts in
> our cities and armies. We read of Socrates, that though he never
> departed from Athens during the frequent plagues that infested the city,
> he only was never infected.[26]

Part of the face and of the 'natural' person, the moustache is the ideal filter
in that it does not mask or suppress the odours but allows them to be
retained: thus, the aromatic strategy of the odour-sensitive essayist is not to
drown the stench under wafts of perfumes but to choose among his own
repertory of fragrances. A continuation of the face, a natural 'ornament',
the moustache is also part of the costume – and as such is associated
with the gloves and the handkerchief. By his hairy appendage, the person
situates himself both in the public and the personal spheres: without
hiding, he keeps to himself. The extraordinary faculty of the moustache for

olfactory retention is its first advantage; it proves to be a salvation in as much as it provides a library of personal fragrances against the invasion of exterior ones. Thus, the defence of the person against contagion is precisely to evoke and inhale his own treasure of sensations.

By mastering odours, Montaigne maintains his control on relationships and on risks of infection: without renouncing 'conversation', nor social life – even in times of epidemics – he remains faithful to himself by keeping to his own odours. But remarkably the examples given by Montaigne are memories of intimate relations, involving close encounters with others and blurring the frontier between public discourse and privacy: the lyrical remembrance of 'close, luscious, devouring, viscid, melting kisses' whose perfume is stored for long hours intrude on the argument developed by linking the self with his former lovers and by bringing in the inner life that eventually proves to be more potent than the social masquerade. A marker of boundaries, the moustache recalls identity without isolating the subject in quarantine.

Therefore, in a final paradox, praise of nature and unnoticeability is contrasted in 1588 with the aromatic strategies of physicians, then evoked in the same breath as the use of incense in churches: perfumes are for crowds and for persuasion, whereas intimate odours speak of individuals and personal truths.

> Physicians might, I believe, extract greater utility from odours than they do, for I have often observed that they cause an alteration in me and work upon my spirits according to their several virtues; which makes me approve of what is said, that the use of incense and perfumes in churches, so ancient and so universally received in all nations and religions, was intended to cheer us, and to rouse and purify the senses, the better to fit us for contemplation.[27]

As far as I know, the moustache strategy is unique to Montaigne. A close witness and a survivor of the plague of 1585, which he describes at length in the final chapter of the last book of the *Essays*,[28] he does not boast directly of his personal technique when he evokes the terrors and cruelties of epidemics. In passing, though, in the lighter chapter devoted to the odours, he does disdain the allusion: through his enigmatic double Socrates, he shows that the independent and wise person may remain safe from 'popular disasters'. The essayist who tells of approaching sick and infectious people, returning to his homeland when it is stricken with plague, shares the invulnerability of the philosopher. Moreover, he tells his reader about his exceptional survival: in a way comparable to Odysseus encountering the Sirens without deafening himself, Montaigne experiences the disease and common affliction without being destroyed nor kept at distance by an array of aromas and burning woods. Establishing first his sensitivity to smells, and then his unexpected resistance to disease,

Montaigne attributes to his moustache the magical power of filtering the evil seeds without forbidding contact.

The moustache, though, sounds like an anticlimax, compared to the magnificent incenses and ceremonies mentioned above: this very modesty is the last and not least of the subtle paradoxes of Montaigne's confidence. Close to the nose, it obviously fulfils the filter function evoked by Galen and enters into the accepted narrative of aerial transmission. But it quickly departs from this common conception: the peculiar *regimen sanitatis* advocated by Montaigne does not resort to astringent, acidic or corrective odours, all strategic devices having been rejected as artificial. Instead, the author turns to the uniqueness of his face. As he emphasizes sensation, memory or feelings, he ostentatiously avoids the didactic position of the physician: the anecdotal tone forbids any transfer or generalization and sends the reader to his own experience and invention. Paradoxes then serve to centre on the person and situate discourse on the breach of boundaries: always in contradiction – with accepted opinions or even with itself – always in movement – ceaselessly progressing by corrections and objections, the reasoning forces the reader to reconsider his interpretation and to reconstruct the rational path: the rejection of general principles meets here the alert writing of ambiguity.

In the *Essays*, plague is not the only threat to individual integrity: literary influence, mimetic repetition of classics or authorities are another trial for the modern writer. The manner that Montaigne claims as his own is not to evade the encounter with tradition, models and other books: it is to face the authors of his library and converse with them, admiration not limiting freedom. Thus, as the writer can borrow materials from his library without being infected by the other authors, he can encounter smell without paying with his life. He who can breathe foul air without contracting diseases can converse without being absorbed by the other; he can read and quote without ceasing to be an original author himself. Remarkably, it is conversation in the chapter devoted to odours that provides the occasion for contagion: disease is second in naming the risks pertaining to one's integrity. So doing, Montaigne does not use contagion as a metaphor for human sociability: he unveils the underlying risk of any approach of another and claims a new and unique manner of maintaining frontiers when engaged with the community.

The modest and picturesque moustache, both boundary and signature, acts as an imaginary mask. Entailing a representation of the world, this idiosyncratic protection builds on commonplaces and common knowledge; the essay weaves together personal experience and received beliefs; thus, it provides a response that conciliates sensibility with safety, contact with survival. To the collective danger embodied by contagious illness, the writer answers with personal invention, borrowing implicit theories that linger in the air and modelling them to his own purpose.

Notes

1   Cited in C. Classen, D. Howes and A. Synnott, *Aroma: The Cultural History of Smell* (New York: Routledge, 1995), p. 83.

2   The present volume demonstrates this 'infection' of the concept, even if the metaphorical uses of notions such as contagion or illness were removed from its topics.

3   B. L. Gordon, *Medieval and Renaissance Medicine* (London: Peter Owen, 1959), p. 457.

4   Annick le Guérer, *Les pouvoirs de l'odeur* (Paris: F. Bourin, 1988); *Scent, the Mysterious and Essential Powers of Smell*, trans. R. Miller (New York: Turtle Bay Books,1992), pp. 41–2: 'Lucretius ... attributes the origin of all infection to germs of disease and death. Thought of as "atoms", such germs can foul the sky should they happen to combine. And when we breathe in such contaminated air, "we are also allowing these pernicious principles to penetrate our bodies". Stench, an indication of rottenness and poison – two terms that tend to be interchangeable in both Greek and Latin – can be deadly. Seneca believed lightning contained a pestilential and venomous element, a smell of "naturally poisonous" sulphur that spoils whatever it touches and gives a nauseating smell to unguents and perfumes.' See also A. Corbin, *Le Miasme et la jonquille* (Paris : Aubier Montaigne, 1986).

5   See R. E. Siegel, *Galen on sense perception* (Basel and New York: S. Karger, 1970), p. 142: 'We know today that only the uppermost part of the nasal mucosa, close to the base of the skull, contains cells sensitive to odors. ... Galen, however, did not consider any part of the nasal cavity as sensitive to odor, since he regarded nasal channels and the pores in their roof only as passages leading to a sense organ at a higher level. Galen's idea was that the odors travel as very fine particles from the nasal cavity through openings in the lamina cribrosa of the ethmoid and the adjacent pores of the nasal and meningeal membranes into the olfactory area of the brain.'

6   Galen, *De instrumento odoratus*, ed. and trans. J. Kollesh, *Galen über das Riechorgan* in *Corpus Medicorum Graecorum*, Supplementband 5 (Berlin: Akademie Verlag, 1964), pp. 2, 10–12, 38, 24–40; English trans. B. S. Eastwood, 'Galen on the Elements of Olfactory Sensation', in *Rheinisches Museum für Philologie, Herausgeben von Hans Hetter* (Franfurt am Main: J. D. Sauerländer Verlag, 1981), pp. 272–2.

7   Le Guérer, *Les pouvoirs de l'odeur*, p. 38.

8   *Remedes preservatifs et curatifs de peste*, Nouvellement composez par Maistre Oger Ferrier Medecin, natif de Tolose (A Paris: chez Guillaume Julien ..., 1562). Jehan Guidon (*Traicte et remedes contre la peste*, Paris, 1545), Ambroise Paré (*Briefve Institution pour preserver et guerir de la peste*, Paris, 1545) and many others could here be evoked, the following developments being medical commonplaces of the time. The work of Ferrier, though, is the most striking and detailed for our topic.

9   Ferrier, *Remedes preservatifs et curatifs de peste*, pp. 30–1: '*Au lever du lict*: Desincontinent que vous serez esveillé, vous ferez ouvrir toutes les fenestres de vostre chambre: principalement celles qui ont regard vers le Septentrion et l'Occident, vous donnant garde qu'aucune infection n'entre dedens. Tandis faites apprester vos habillemens, lesquels aurez laissées toute la nuict en l'air. Et seroit bon les changer souvent, tellement que ne vestissiez aujourd'hui les

Robbes qu'auriez hyer portées. Puis faites apporter une eschauffette pleine de charbons ardens, et jettez dedens de l'encens, myrrhe, benjoin, ladanum, stirax, roses, fueilles de myrte, romarins, lavende, basilic, saturee, serpoulet, marjolaine, boys d'aloës, squinanti, macis, cardamomi, petites pieces de pin, clouz de girofle, pieces de cypres et de genevrier, et autres choses odoriferantes, sur la fumee desquelles mettrez vos habillemens pour les faire participans de l'odeur: et quant et quant, vous vestirez comme de coutume.' (Trans. H. C.)

10 Apple of Capendu, also called 'Courtpendu' and today known as the short-stemmed apple of Mandeure.

11 Ferrier, *Remedes preservatifs et curatifs de peste*, pp. 48–50: '*Quand vous voudrez sortir hors de la maison*. Ne sortez hors de vostre maison, sinon deux heures apres le Soleil leviés et des jours nubigeux ne bougez. Et pour bien faire ne partez, que vous n'ayez premierement disné.

Quand vous voudrez sortir, prenez vostre pomme de senteur, faite en la forme suyvante, pour temps des chaleurs: R. *Sandalorum citrinorum, macis, corticum citri, rosarum, foliorum myrti, an. drachmas duas: benjoin, ladani, stiracis, an. drach. et feneis: cardomomi, violarum, croci, an. scrupulos duos: camphorae et ambrae an. scrupulum unum: algaliae, et musci an. grana duorum aqua rosarum infusionis formetur pomum ...*

Si vous n'avez encore lesdites pommes de senteur, vous prendrez de pommes de capendu, de citrous, limonnes, orenges, et autres fruits odoriferans, ou bouquet d'herbes et de fleurs de bonne odeur.

A tout le moins il convient prendre une esponge, et la tremper en vinaigre et eau rose meslés ensemble avec un peu de cannelle, girofle, macis, saffran, ou anec, quelque grain de canfre, ambre, civette, selon la capacité de la bourse d'un chacun.'

12 'Agradelle' is a word that does not appear in either French or Latin dictionaries of the time, thus it has been included in the form found in the French text.

13 *Ibid.*: 'Vous prendrez aussi, et porterez avec vous un Massapan, dans lequel tiendrez semences de citrons: ou conserve de roses, ou d'agradelle: ou de la chair des orenges coupee en petites pieces, sur lesquelles ietterez force sucre pulverise: et ce pour l'été. ... Outre cecy, si vous estes medecin, apothicaire, ou chirurgien, ou autre, qui allez visiter les pestiferés, il vous convient avoir prestes force pieces du boys de genevrier: ou en default d'iceluy, une bonne torche et de l'encens aussi.'

14 *Ibid.*, p. 52: '*En cheminant par la ville*. Mettez au devant du neds vostre pomme de senteurs ou vostre bouquet de fleurs: ou l'esponge trempee comme dessus est dit ou quelque autre chose odoriferante. Prenez de votre Massapan une semence de citron, ou quelque morceau des choses qui seront dedens, et le maschez par les chemins, puis l'avallez. Gardiez-vous de l'haleine des gens, et des rues infectes, tant qu'il sera possible.'

15 *Ibid.*, pp. 53–4: '*Quand le medecin, ou autre va voir les malades de peste*. Quand vous serez pres de la maison du pestiféré: envoyez quelcun devant, qui fasse ouvrir toutes les portes de la maison, et les fenestres de la chambre ou gist le patient: et tandis arrestez vous un peu à la rue. Et commandez que l'on allume bon feu à la chambre du malade. Puis ferez descendre une eschauffette pleine de charbons ardens, avec de l'encens, roses, myrrhe, benjoin, ladanum, styrax, clou de girofle, et semblables odeurs pour en faire fumigation. Et quant et quant faictes allumer vostre piece de boys de genevrier: et ainsi entrez hardiment, faisant passer devant vous ledict personnage avec ladite fumigation d'encens, myrrhe etc. et le suyverez tenant à une main ledit boys allumé: et à l'autre vostre

pomme de senteurs, ou vostre bouquet, ou ladicte esponge, l'appliquant au nez. Et en ceste façon marcherez jusques dans la chambre, là ou ferez mettre en ladicte eschauffette avec ledit encens, à celle fin que l'odeur s'espande par toute la chambre. Ainsi, tenant dans la bouche quelque chose de votre massapan, et tenant l'une main aupres du nez avec lesdites odeurs, et ayant en l'autre ladicte piece de genevrier allumee: vous regarderez d'un peu loing vostre patient, et l'interroguerez de son mal, et de ses accidens.'

16 See J.-C. Sournia, *The Illustrated History of Medicine* (London: Harold Starke, 1992); *Histoire Illustrée de la Médecine* (Paris: Larousse, 1991).

17 It used to be customary in France to bring oranges to prisoners. The expression 'porter des oranges', meant as a warning and a promise 'to remain faithful to someone who was caught by the law', still testifies the traditional gift of visitors.

18 As confirmed by the last editions (1588), and the manuscript additions. M. de Montaigne, *Les Essais de Montaigne*, ed. P. Villey (Paris: Presses Universitaire de France, 1965) I: 314–16, a chapter devoted to smells, and … the author's moustache; *Essays of Michel de Montaigne*, trans. C. Cotton (1685), ed. W. C. Hazlitt, Project Gutenberg Release 3600: http://onlinebooks.library.upenn.edu/webbin/gutbook/lookup?num=3600, consulted and downloaded 1 October 2004.

19 The ancient belief concerning Alexander the Great finds a Christian echo in the 'odour of sanctity' given off by the bodies of saints. The current French expression 'être en odeur de sainteté' has maintained this conception throughout centuries.

20 Cotton, trans. *Essays of Michel de Montaigne, ad locum*; ed. Villey, I: 314: 'Mais la commune façon des corps est au contraire: et la meilleure condition qu'ils ayent, c'est d'estre exempts de senteur. La douceur mesme des haleines plus pures n'a rien de plus parfaict, que d'estre sans aucune odeur, qui nous offence: comme sont celles des enfans biens sains.'

21 Cotton, *Essays of Michel de Montaigne*; ed. Villey, I: 316: 'Le principal soing que j'aye à me loger, c'est de fuir l'air puant et pesant. Ces belles villes, Venise et Paris, alterent la faveur que je leur porte, par l'aigre senteur, l'une de son maraits, l'autre de sa boue.'

22 Cotton, *Essays of Michel de Montaigne*, ed. Villey, *ibid.*: 'Et les bonnes senteurs estrangieres, on a raison de les tenir pour suspectes à ceux qui s'en servent, et d'estimer qu'elles soyent employées pour couvrir quelque defaut naturel de ce costé-là. D'où naissent ces rencontres des Poëtes anciens, c'est puïr, que sentir bon.'

23 'As it is said that the best odour of an action is to be unperceivable and unheard.'

24 Villey, pp. xxviii–xxx.

25 Cotton, *Essays of Michel de Montaigne*, ed. Villey, I: 314–15: 'J'ayme pourtant bien fort à estre entretenu de bonnes senteurs, et hay outre mesure les mauvaises, que je tire de plus loing que toute autre … Quelque odeur que ce soit, c'est merveille combien elle s'attache à moy, et combien j'ay la peau propre à s'en abreuver. Celuy qui se plaint de nature dequoy elle a laissé l'homme sans instrument à porter les senteurs au nez, a tort: car elles se portent elles mesmes.'

26 Cotton, *Essays of Michel de Montaigne*, ed. Villey, I: 315: 'Mais à moy particulierement, les moustaches que j'ay pleines, m'en servent: si j'en approche mes gans, ou mon mouchoir, l'odeur y tiendra tout un jour: elles accusent le lieu d'où je viens. Les estroits baisers de la jeunesse, savoureux, gloutons et gluans, s'y colloient autrefois, et s'y tenoient plusieurs heures apres. Et si pourtant je me trouve

peu subject aux maladies populaires, qui se chargent par la conversation, et qui naissent de la contagion de l'air; et me suis sauvé de celles de mon temps, dequoy il y en a eu plusieurs sortes en nos villes, et en noz armées. [C] On lit de Socrates, que n'estant jamais party d'Athenes pendant plusieurs recheutes de peste, qui la tourmenterent tant de fois, luy seul ne s'en trouva jamais plus mal.'

27 *Ibid*.: 'Les medecins pourroient (ce crois-je) tirer des odeurs, plus d'usage qu'ils ne font: car j'ay souvent apperçeu qu'elles me changent, et agissent en mes esprits, selon qu'elles sont: Qui me fait approuver ce qu'on dit, que l'invention des encens et parfuns aux Eglises, si ancienne et espandue en toutes nations et religions, regarde à cela, de nous resjouir, esveiller et purifier le sens, pour nous rendre plus propres à la contemplation.'

28 Villey ed., Essay III, 12, especially pp. 1039–50 in II.

7
Contagion, Honour and Urban Life in Early Modern Germany

Mitchell Lewis Hammond

In his famous treatise *On Assistance to the Poor* (*De Subventione Pauperum*), printed in Bruges in 1526, Juan Luis Vives waxed indignant over the sick poor who risked infecting others by begging in public places. 'What sort of situation is this,' he asked, 'when in every church – especially at the solemn and most heavily attended feasts – one is obliged to enter into the church proper between two rows of the sick, the vomiting, the ulcerous, the diseased with ills whose names are unmentionable ...?' What was more, the presence of the sick was a risk to the young, the old and pregnant women, 'especially since ulcers of this sort are not only forced upon the eyes but upon the nose as well, the mouth, and almost on the hands and body as they pass through. How shameless such begging!'[1] Vives' words resonated with particular force because an outbreak of a highly contagious illness, the so-called 'French pox', had erupted across Europe only three decades earlier, bringing with it severe pain and swollen pustules that terrified people who encountered the infected. When city officials confronted such crises, Vives believed, they should act 'in the same manner as the medical profession who cannot eradicate diseases completely from the population but bend every effort to cure them'.[2] City councils and urban residents across Europe shared these concerns, especially since the sixteenth century was a period of growing urban poverty as well as a tumultuous era of epidemic outbreaks. Even after the publication of Fracastoro's *De contagione* in 1530 and its eventual dissemination, the term 'contagion' itself was mostly restricted to learned discussions and seldom used among the general populace. But everyone knew and feared the prospect of illness that could contaminate by moving, whether by goods, air or people, from one location to another.

For Vives, who had a physician's training, this was a challenge of public health that belonged to the broader category of civic policy, that which in the German lands was discussed in countless ordinances as *gute Polizei*, or effective administration. However, many recent scholars have suggested that it was religious beliefs about sin and divine punishment that defined popular

94

attitudes towards contagious diseases, casting a moralizing shadow over the afflicted and attempts to treat them. Anna Foa presents a widely held view when she observes that, in the sixteenth century, '[i]llness was a punishment from God for the sins of humanity and could strike either the individual sinner or entire communities in order to make them expiate the sins of the world'.[3] Many documents, especially those addressed to a wide general audience, seem to confirm this view, for few authors shied away from invoking God as the primary cause of disease. Civic plague regulations routinely prefaced their instructions with references to the community's sinful disposition, and individual preachers often did the same for their congregations. Martin Luther's Catholic arch-rival, Johann Eck, had the pox in mind when, in 1530, he announced from the pulpit that 'lust is a swollen pustule'; and while some of Luther's own advice about medicines for plague-stricken cities were disarmingly pragmatic, he too once observed that 'if God permits the pestilence to take you, this is a chastisement'.[4] Such pronouncements also influenced the language of the thousands of petitions written to civic charity institutions on behalf of the sick poor. The notaries who drafted these requests routinely prefaced pleas for help with a pious admission of guilt. In the stark words of one petition, submitted to the Alms Office in Augsburg in 1552: 'thus has God the Almighty gravely afflicted me with His punishment, the severe illness of the French [that is, the French pox].'[5]

We cannot completely dismiss the sentiments expressed in these statements, especially since Christian teachings and institutions were vital and dynamic forces at every level of European society. If we wish, however, to understand the everyday circumstances faced by the majority of Europe's city dwellers, it is helpful to differentiate between the collective social impact of an epidemic event, and individual attitudes toward diseases believed to be contagious.[6] This is important for at least two reasons. First of all, not all outbreaks of contagious illness were sufficiently severe to elicit the baleful rhetoric and communal policing measures that we associate today with, for example, plague outbreaks that carried off hundreds or thousands of souls. More typically, city dwellers feared and encountered isolated cases of illness, forces that threatened their immediate social circle with ill-health or defilement but which did not inspire massive public campaigns. Second, it is essential to distinguish between moralizing *topoi* of disease, sin and punishment, conventional in the public discourse of a broadly Christian culture, and perceptions of contagious disease among individual sufferers and medical practitioners. We should not assume that ordinary people ignored the declarations of the religious elite, but their own dealings with contagious illness often foregrounded more pragmatic concerns about their status as neighbours, co-workers and citizens. In short, we must investigate what people did when they felt personally threatened, either by the experience of illness or by the accusations of others who considered them dangerous.

For this reason, alongside ordinances, sermons and other documents composed for calculated rhetorical effect, it is instructive to examine other sources that have received far less scholarly attention: the records of city councils, poor relief offices and medical examiners. Such documents are hardly transparent, but because many of them were drafted to address personal circumstances and were intended to achieve particular results, they indicate the problems that were uppermost in the minds of city residents and the officials they consulted. In some regions such documents are scant for the sixteenth century, but the cities and towns of the Holy Roman Empire offer a particularly good case study of the relationship between everyday fears of contamination, and efforts by medical practitioners to identify and explain bodily conditions. Under the broad aegis of the Emperor, imperial cities such as Augsburg, Nuremberg and Cologne almost completely controlled their public health measures, as did many other smaller cities that were subordinate to dukes or other overlords. Civic institutions varied considerably in these communities, as did each city's position in the widening schism between Catholics and Protestant. However, there were significant patterns in the way the residents of these communities assessed and treated contagious diseases.

One important trend was that, by the early seventeenth century, barber-surgeons and physicians had emerged as the most important judges of individual and collective health, consulted by city officials and less influential residents alike. For barber-surgeons, years of apprenticeship to a master qualified them to administer bloodletting regimens, to bind wounds and to relieve ailments of the body's surface. While most were not members of the higher echelons of guild society, they frequently served as city officials who treated the sick and furnished opinions about the health of individuals and the possible usefulness of medical assistance. Physicians, by contrast, reserved for themselves the identification and treatment of disorders within the body. They were university-educated, often at institutions in France or Italy, which enjoyed greater prestige than German schools of more recent vintage such as Heidelberg, Tübingen, or Wittenberg. Like barber-surgeons, physicians had long been organized in guild-like associations. From the early 1580s onward, in many of the Empire's largest cities they formalized their associations by founding colleges of medicine.[7] These corporate bodies advised city governments on public health matters such as the pricing of apothecary goods, the authorization of travelling healers and herb salesmen, or the civic response to epidemics. The most important tools of the physicians' craft were systems of explanation, largely derived from the ancient authority of Galen, that linked the bodily signs of an illness to an explanation of its origins and its future course.[8] Other citizens in early modern cities were remote from the worlds of Latin learning and from the barber-surgeon's craft; but they increasingly approached these practitioners to identify their health conditions or to provide them with a useful defence against the accusation that their bodies might spread disease.

Where fears of contamination were concerned, three illnesses outstripped all others as perceived threats to the social order, at least in part because the physical disfigurement or decay they caused was so obvious and disturbing. The most lethal, of course, was bubonic plague, which had been endemic in many parts of Europe since 1348, and which swept across the German lands in periodic waves over the next three centuries. Medical practitioners claimed little ability to treat the disease and by the early sixteenth century almost every community isolated the infected in makeshift quarantine stations beyond the city walls. A similar state of affairs existed for leprosy, also widely regarded as incurable, whose sufferers were often consigned to a semi-monastic existence in houses funded by charitable donations.[9] The incidence of leprosy diminished dramatically after the thirteenth century but physicians were still on the lookout for it three centuries later, armed with lists of the disease's signs that had been compiled by medieval authors. Although in earlier times physicians had shared the responsibility of identifying the affliction (with priests or with lepers themselves), from the late fifteenth century they increasingly controlled the so-called *Schau*, or examination, which determined a suspected leper's fate.[10] Finally, as leprosy waned in influence, the French pox erupted across Europe in 1495, sparking a new wave of anxiety and the founding of houses set aside for the sick. Like leprosy, the French pox was popularly associated with lust and sexual acts, although physicians were generally slower than others to make the connection and did not necessarily consider genital contact the only means of transmission.[11] Moreover, the pox was considered curable with a regimen of mercury salves or with teas and baths concocted with bark from the guaiac tree. Consequently, city facilities for the pox offered short-term medical treatment, not permanent isolation, and differed markedly from the more sacralized paradigm of the leper house. As we will see later, this last distinction is particularly useful for understanding how responses to epidemic disease shifted over the centuries.

These three illnesses, and other conditions that superficially resembled them, inspired a fear of proximity to the sick, in particular because they were considered agents of defilement as well as sickness. Even written references to contagious disease inspired protective measures, the rhetorical formulas 'with honour' (*salva honore*) and 'submitted with respect' (*mit Reverenz zu melden*) that recur throughout the correspondence of the time.[12] This was especially true in the early years of the French pox, when the disease was at its most virulent and had the most frightening effects. In Strasbourg, for example, guest houses were legally barred from accepting the disease's first victims; a chronicler noted that even lepers turned them away, and while the remark is anecdotal, it suggests the level of popular disgust aroused by the disease.[13] But even a century later, in 1615, officials in Munich faced similar problems as they tried to relocate a public ward (*Lazarett*) for the ill, although the facility was not reserved for infectious diseases.[14] Poor relief administrators commonly referred to apparently communicable disorders, including scabies and other severe rashes, as 'not to be tolerated' (*nicht zu dulden*) among the healthy. Hospitals for the elderly or poor excluded anyone with an illness considered contagious, and houses reserved for particular illnesses, such as the French pox, jealously guarded their doors too. Thus, in 1561, the director of Augsburg's orphanage requested that the city remove a sick youth from the building so that he would not infect the other children. Initially, five doctors who examined the boy identified his illness as the 'evil pox', (*bosen Blattern*) and recommended that he be moved to the city facility devoted to its cure. But two other practitioners, including a barber who had treated the boy earlier, disagreed, noting that he had 'an unclean illness, difficult to heal, which will not permit the boy to live very long'.[15] Faced with these conflicting assessments, the city council suggested either that the boy find another charity house or lodge with a poor family who would look after him. Individuals and communities also sought out specific sources of infection and attempted to contain them. For example, in 1585, after a number of deaths thought to be caused by plague, officials in Nuremberg traced the spread of disease to a single schoolmaster and his wife. Their house was quarantined, with all their belongings inside, as interviews with a physician and a pastor confirmed that they both suffered from a 'hot fever' (*hitzig Fiber*) and other signs of plague.[16]

The swirl of controversy that surrounded Anna Löffler, a midwife in Augsburg in the 1550s, is an especially striking illustration of how ordinary people considered the danger of person-to-person contamination.[17] For nearly ten years, Anna had served as a midwife on the city payroll after succeeding her mother, Elizabeth, in 1549. In April 1558, three women accused her of infecting them and their newborns with the 'evil pox' when she had attended their births the previous October. One of them, a cutler's wife named Magdelena Metzger, gave an unusually revealing explanation of her experience in two separate petitions to the city council. As she gave birth, Metzger recounted, Löffler had tried to conceal her fingers when she cut the infant's umbilical cord. In a second petition Metzger further recalled, as others did, that Löffler had not worn protective covers (*Fingerling*) for her fingers as she worked. When other women asked the midwife what was wrong, Löffler responded that she had injured a nail, an answer that Metzger had believed at the time. Thereafter, she and her child fell ill during the six-week period that followed her birth and she suffered pains in her head and body. Although Metzger concealed her distress from the rest of the household, she mentioned her plight to a maid, who suggested that 'perhaps the cellar-girl did not wash properly, and [had] harmed me with damp towels'. Metzger herself discounted this possibility and consulted two other midwives who each gave her remedies that did not work. In the meantime, Metzger heard that "she [Löffler] had the evil pox on her fingers'. She claimed that Löffler 'almost fainted and did not know what to say' when Metzger confronted her, and then gave her another herbal treatment to try. Metzger believed that this remedy had only made her illness worse; it relieved her outward condition, but only by pushing the disease from her skin deeper into her body. On Christmas Day, she consulted a barber-surgeon, and 'he wondered from where such impurity [could] come to me, and [he] openly told me I was full of the evil pox'. Even more tragically, Metzger's child could not recover from the disease and died after living only fourteen weeks.

In two petitions of her own, Löffler responded to these charges on various fronts. In her account, Metzger's infant had an unusually tough umbilical cord, perhaps indicating the child had a defect at birth; and while she acknowledged that 'some black ran out' from two of her fingers before she arrived at Metzger's birth, Löffler noted that many other people, including her own family, had had close contact with her without coming to harm. She had cured her fingers with potions and baths in about a fortnight, during which time she had attended 40 more successful births. Why, then, did Metzger wait ten weeks before approaching her? Löffler claimed that the herb she offered would not push (*treiben*) a disease into or out of the body, as any doctor would confirm, and hence was not responsible for making matters worse. Löffler had even undergone a humiliating physical examination by a group of physicians, who inspected her body from head to toe without finding any reason to doubt her claims. At one point, she noted that many women and children fall sick or die shortly after birth, 'but it is the will of the almighty God that every person submits at his hour, and so it is with sickness too'. No one involved, however, would have mistaken this rhetorical flourish for the issues that mattered: alleged sources of contamination, the curative effects of herbs, and the elapsed time between the transmission of a disease and its visible effect on the body.

The extant records do not indicate what action, if any, Augsburg's council chose to take, although later records show that Löffler retained her position for several years.[18] It is also impossible to know whether Metzger contracted some form of infection, such as puerperal fever, or if she was influenced by other factors. However, her account illustrates a characteristic focus on the physical sources of infection, as well as how ripples of gossip, anecdote and accusation sometimes mirrored the alleged spread of an illness. It was assumed that one might conceal bodily impurity, even if sinful behaviour were not the cause of disease. Sexual mores played no part in Magdalena Metzger's account of the disease, but she believed that Anna Löffler tried to conceal her infected hand, just as Metzger herself refrained from telling her household about her illness. For both women, moreover, an authoritative examination by an educated (male) barber-surgeon or physician was a focal point of their explanation, offering either proof of an illness or vindication of a body free from disease. The official role of such practitioners as judges of bodily purity grew in importance throughout the sixteenth century, especially as the number of academically trained physicians in the Empire increased and as they assumed greater responsibilities within city governments.

The resort to a medical examination to counter an accusation could be especially important for trades people whose professions were already held in low esteem, since an accusation of infection immediately threatened their livelihood and other forms of social contact. In Cologne in 1612, a tanner named Gerhard von Entz complained of his dilemma when he and his wife presented themselves to two barber-surgeons: 'last Sunday among the curtain makers, in the presence and earshot of many upright people, it was openly accused that they might be afflicted with the repulsive French pox disease (submitted with respect).'[19] Thereafter, guild officials had forbidden the couple's associates to come to their house until appropriate measures were taken. As tradesmen who worked with animal skins, and who trafficked with the even lowlier skinners guild, von Entz and his co-workers felt compelled to respond vigorously to any accusations of impurity.[20] In such circumstances, the written testimony of a recognized practitioner, offered under oath, assumed the force of a legal document that an accused person could use to demand readmission into the community, even if communal memory of the slander lingered. City-appointed physicians also intervened with other officials to regulate the traffic of products in and out of a city, or to manage the opening and closing of infected buildings. In one such case, in the city of Augsburg in 1607, a baker pleaded with the physician Raymond Minderer not to close his shop after one of his workers died under suspicious circumstances. Minderer's report to the city council concluded that the apprentice had died of plague, but he also conceded that the man's heavy drinking had contributed to his illness. The council allowed the baker to open the upper floor of his shop but ordered him to keep a glass window barrier in place.[21]

Some city residents pursued guarantees of bodily purity, literally, to the grave, especially in the wake of plague epidemics. Under normal circumstances, both Catholic and Protestant officials refused a public funeral for victims of plague because of fears of poisonous emissions from the corpses. The families of the deceased, however, had the right to request an examination of the bodies of the dead to certify that they were free of infection. During a widespread plague outbreak in 1592–93, citizens in Augsburg enlisted nine physicians for this purpose. Dr Johan Georg Brengger recorded that he had been approached by a certain Master Schwegler 'who asked of me that I give written testimony of his deceased wife, that she did not die of the infection or sickness [i.e. the plague]'.[22] Brengger, who had examined the woman prior to her death, certified 'that I have found her free of the infection'. Another physician named David Wirsung was asked to examine the corpse of a bathmaster named Georg Mayer, a man already of marginal social status because of the lowly position of his guild. Wirsung found 'no suspicious contagious condition'; the cause of death was, in fact, dropsy, and therefore Wirsung concluded that '[Mayer] is admitted to the privilege of a public Christian burial'.[23] In these cases, as in others, the allegation of contamination was not linked directly to concerns about sin or divine punishment. Rather, the stain of infection was an attack on one's personal integrity and honour, and it could only be restored by guarantees that the infection either did not exist or had been redressed.

A similar dynamic existed for leprosy, which sixteenth-century physicians classified within a spectrum of diseases that now included the French pox. While leprosy inspired less concern than it had previously, city-appointed physicians were sometimes asked to identify illnesses that strongly resembled leprosy, or even seemed at risk to transform into a type of the illness. For example, in the small city of Nördlingen in July 1572, a physician named Varius examined a man named Balthassar Kop, who had suffered for twelve years and had severe sores on his arms and hands. Varius concluded that Kop's 'uncleanness' initially derived from frostbite, but if the proper measures were not taken, 'it is to be feared that eventually a leprosy will emerge therefrom'.[24] More frequently, physicians put to rest the suspicions concerning physical conditions that resembled leprosy, while more precisely defining the bodily problem at hand. Hence, in May 1542, another Nördlingen physician named Johan Widman informed the city council of his judgement of Conrad Fursthin, a man accused of harbouring leprosy. 'According to my [rational] faculties', Widman recorded, 'I find few and insufficient signs of leprosy. Therefore I acknowledge and declare him still free of leprosy at this time.'[25] Widman's equivocating 'still' gestured to the fact that Fursthin did have severe sores from the French pox on his hip, which called for some caution with regard to the eventual prognosis.

Likewise, physicians on the medical faculty of the University of Tübingen performed examinations for the Duchy of Württemberg in the southwestern part of the Empire. Between 1550 and 1650, the faculty examined scores of alleged lepers, usually at the request of small town officials who did not have the necessary expertise at hand. Thus, in July 1589, the mayor of the small town of Rosenfeld wrote to the Tübingen medical faculty to request the examination of Anna Scholder, who for several years 'has been kept away from clean people and not allowed to live with us in the community'.[26] The Tübingen faculty concluded that Anna suffered from 'grind' rather than leprosy and recommended purges and a guaiac cure under the supervision of an experienced physician. Through many similar assessments, physicians bolstered their own authority, as arbiters of health and purity as well as therapy, for communities beyond the major urban centres.

When individuals consulted medical practitioners, they usually sought confirmation that their bodies were free from infection. For the desperately poor, however, the identification of a serious disease was an opportunity to persuade a city to extend life-sustaining charity. Juan Luis Vives himself was only one of many observers who cautioned that frauds were a constant pitfall for medical examiners; and as the legal and social standing of examinations became more widely recognized, examiners sometimes faced difficult circumstances as they sought to manage the claims and demands of poor supplicants.[27] An excellent example is the predicament faced by Gereon Sailer of Augsburg who, in 1556, complained to the city council about the 'lazy and unemployed people' (*fauler vnarbaitsam Leut*) who pestered him for help. Among other duties, Sailer controlled access to a shelter on the edge of town (*Siechenhaus*) that had originally been set aside for lepers. Now, however, people of every description clamoured for access: '[the poor] are always pressing their way in', Sailer observed, 'they say "so and so is in the house and they [the current occupants] are also not completely leprous"'.[28] In other words, Sailer found his own expertise under attack, not from a competing authority but from the sheer numbers of poor people who refused to take no for an answer.

The most remarkable example of this problem is the Nurembeg *Leprösenschau* that was conducted annually in the city throughout the sixteenth and early seventeenth centuries, and which actually attracted hundreds of people every year. The city was renowned for the hospitality it extended to lepers for four days every year during Holy Week, in accordance with a charitable bequest provided by a wealthy couple in the early fifteenth century.[29] Those who were examined received free clothing, food and drink for those days, as well as a medical attestation of their condition that they could take home. In the decades after 1550, many people apparently took advantage of this opportunity, although only fragmentary records in other cities survive to confirm it. In Nördlingen, 50 kilometres from Nuremberg, a man named Hans Diethaiss requested that he be placed in a segregated house, 'because I have been recognized at Nuremberg as a leper who should live with my own kind, the lepers, and not with others'.[30] Nuremberg itself funded four houses for people who required isolation, and the prospect of having at least a roof and a nearby chapel undoubtedly encouraged some vagrants to show up during Holy Week.

By the late 1560s, the Nuremberg leper examination had become a mass event in which huge numbers of poor and diseased people assembled at a leper house just inside the city gate, waiting their turn for the promised alms and medical examination. In March 1572, after an especially chaotic assembly, the city council solicited suggestions from the physicians who had participated. One of them estimated the crowd at over 3,000, and while this figure was perhaps exaggerated, the physicians' letters are not the only evidence that the gathering was a large one. An entry in the accounts of the alms office recorded that ten casks of wine were set aside for the event that year.[31] The physicians complained that it was all but impossible to accurately identify the true lepers in such circumstances, especially since there was no time to inspect their urine or to examine a sample of their blood. Another difficulty was that some poor people tried to trick the physicians by wearing bandages or smudging their skin to appear sick. More troublingly, as the physician Volcher Coiter pointed out, the nature of the disease itself made the task of identification extremely difficult. Other diseases, including the French disease (*morbus Gallicus*), had characteristics that were difficult to distinguish from leprosy; and there were different stages and varieties of leprosy itself, some of which might be treatable while others were not.[32] The challenge of identifying the disease prompted at least one Nuremberg physician, Georg Palma, to assemble citations from past and present medical authorities in an effort to codify what was known about leprosy's signs on the body.[33] But in the end, the medical questions were subordinated to more immediate public health concerns. As an exercise of civic charity, the event had become a fiasco, and two years later the city moved the examination to a more remote location and stopped the contributions from the alms office. This episode suggests how the place of leprosy in society had changed by the end of the sixteenth century. As its incidence declined and other severe threats emerged, leprosy had become one disease among many, both as a medical conundrum and as a challenge of civic charity.

The vast majority of people in early modern cities lived far afield from the discussions of Galenic nosology that exercised the physicians in Nuremberg, and elsewhere, for many years to come. Like Juan Luis Vives, most citizens concerned themselves, above all, with maintaining social boundaries, preserving communal integrity and preventing the spread of any polluting forces. In this context, it is useful to consider Charles Rosenberg's observations about perceptions of epidemic disease throughout Western history in regard to contagious illness in general. Rosenberg distinguishes two basic modes of interpreting epidemics which have coexisted, and often been combined, in the West since Antiquity: the configuration view, in which diseases are attributed to the convergence of environmental factors, such as diet, weather and astral influences; and the contamination view, in which attention focuses more narrowly on a physical source of infection and its means of transmission. German city dwellers in the sixteenth century, especially those who were unschooled in medical theory, tended to focus on more concrete theories of contamination, even when the explanation of infection did not have a learned or scientific explanation. Relatively speaking, abstract environmental factors such as sidereal influences or miasmas received less attention, and this extended as well to the role of sin as a cause of individual illness. God was the source of all being, and God's will was acknowledged as the ultimate cause of all events.

But this conviction did not demand inordinate emphasis on divine punishment as an explanation for contagion, nor did it deter ordinary people from a search for the tangible causes of their suffering. Instead, city residents focused on the problems presented by their own bodies and those with whom they had come into contact. To this extent, the evidence from these cities corroborates the literary analysis of Louis Qualtiere and William Slights, who suggest that English satire emphasized human agency in the spread of contagious illness from the later sixteenth century forward.[34]

This thesis is also supported by the changes in the role of leprosy in European society and the contrast between perceptions of leprosy in the later Middle Ages and the French pox after its emergence in 1495. One should not overstate the continuities in earlier medieval approaches to leprosy; as François-Olivier Touati has recently cautioned, widespread belief in the person-to-person transmission of leprosy actually emerged only in the mid-thirteenth century and gained momentum after the Black Death of 1348.[35] However, at least among ordinary people in the later Middle Ages, the disease clearly carried a spiritual stigma, derived from readings of biblical texts, authorized by church decree, and institutionalized by social and sacramental isolation at the communal level. Several centuries later, the French pox was perceived differently as it emerged, even though the disease was similarly associated with sexual excess, and physically resembled leprosy so much that some practitioners admitted difficulty in telling the diseases apart. After 1495, many witnesses testified to their disgust and horror, but the French pox never inspired the same kind of spiritual anxieties or sacramental measures that had accompanied leprosy. Indeed, as the case of both Anna Löffler and Gerhard von Entz suggest, the costs of a pox accusation for social and business status ranked as highly as their spiritual concerns.

By the turn of the seventeenth century, the stigma caused by leprosy and the French pox was not, foremost at least, associated with a disordering of the sick person's relationship to God. Rather, it was perceived as more closely akin to the social 'dishonour pollution' already borne by executioners, skinners and other trades that were widely considered unclean.[36] In 1638, the author Luwig von Hoernigk addressed the issue of honour directly in his *Politia Medica*, a handbook of civic medical practice that codified many German medical regulations from the previous decades. Reminding his readers of a physician's obligation, Hoernigk wrote: 'in the inspection, or examination of those persons who are afflicted with leprosy or the French disease, or held in suspicion of them ... they [the physicians] and the Barbers, or who otherwise may be assessing [the matter], should attend with appropriate diligence and caution, so that no one should innocently suffer an injury to his honour ...'[37] In contrast, for the poorest of city residents, the prospect of charitable aid was a lifeline that was worth the social sacrifice that a 'negative' assessment of their bodies might bring. The efforts of the poor to secure assistance complicated the physical examinations by medical practitioners, as well as the administrative decisions that awarded help to some and denied it to others. For both the poor and the better-off, however, individual responses to disease emphasized the social consequences of infection rather than the discourse of sin and punishment which, more typically, found its place in statements addressed to a general public.

Finally, it is significant that individuals who were accused of harbouring an infection turned to the informed judgement of a medical practitioner to defend themselves from accusations of impurity. In some contexts, this fact may help us to explain why physicians enjoyed relatively high prestige even when their cures were not consistently effective. The medical historian Roger French has attributed this success to the educated physician's ability to cultivate an appreciation of his rational skills and erudition among laypeople.[38] While medical learning was undoubtedly valued in many circles, these civic records suggest that the administrative and legal functions of expert testimony about bodily purity were at least as important in reinforcing the credibility of medical practitioners. For most urban residents of the Empire, barber-surgeons and physicians were necessary not because they offered learned explanations of contagious disease, but because they defined and regulated the boundary between pure and impure bodies, provided recourse for people to defend their reputations, and coordinated the response to epidemic threats. In doing so, they answered to an important need in everyday life and assured themselves a permanent place in the urban social order of early modern Germany.

Notes

1 The translation is by A. Tobriner, in 'A Sixteenth-Century Urban Report', *Social Service Monographs* (1971) p. 36.
2 *Ibid.*, 51.
3 A. Foa, 'The New and the Old: The Spread of Syphilis (1494–1530)', in *Sex and Gender in Historical Perspective*, ed. E. Muir and G. Ruggiero (Baltimore: Johns Hopkins University Press, 1990), p. 27.
4 J. Eck, *Der fünft vnd letst Tail Christenlicher Predig von den Zehen Gebotten ...* (Ingolstadt, 1539), XLVIᵛ; Luther's remarks quoted in *Luther's Works*, ed. J. Pelikan and H. Lehmann, vol. 22 (Philadelphia: Fortress Press, 1955), p. 500.
5 Stadtarchiv Augsburg (hereafter StAA), St. Martin's Stiftung, Nr. 47. Suppliken um Aufnahme in das Blatterhaus, 1552.
6 See the discussion of epidemics as collective events in C. Rosenberg, *Explaining Epidemics and Other Studies in the History of Medicine* (Cambridge: Cambridge University Press, 1992), pp. 285–6.
7 The first of these was the city of Augsburg in 1582, followed by Vienna, Ulm, Nördlingen and Nuremberg. A. Fischer, *Geschichte des deutschen Gesundheitswesens*, Band 1(Hildesheim: Georg Olms, 1965), p. 91.
8 In the medical and philosophical milieu of the time, the interpreting of the 'signs' of a disease and consequent indications for therapy were more important than the 'diagnosis' of an ontologically distinct pathogen or disorder. See the discussion in I. Maclean, *Logic, Signs, and Nature in the Renaissance. The Case of Learned Medicine* (Cambridge: Cambridge University Press, 2002), pp. 297ff.

9 Widespread segregation of lepers was authorized by decree of the Third Lateran Council in 1179. R. Palmer, 'The Church, Leprosy, and Plague in Medieval and Early Modern Europe', in *The Church and Healing*, ed. W. J. Sheils (Oxford: Blackwell, 1982), p. 81.

10 Hence in 1478, an apothecary ordinance for the city of Cologne affirmed the jurisdiction of physicians over that of lepers. A. Schmidt, *Kölner Apotheken* (Cologne, 1931), p. 108.

11 See the detailed discussion in D. Amundsen, *Medicine, Society, and Faith in the Ancient and Medieval Worlds* (Baltimore: Johns Hopkins University Press, 1996), pp. 310–72. Earlier physicians were also circumspect about the sexual causes of leprosy. See L. Demaitre, 'The Description and Diagnosis of Leprosy by Fourteenth-century Physicians', *Bulletin of the Historiy of Medicine* 59 (1985) 327–44, esp. 338.

12 D. Sabean, 'Soziale Distanzierungen. Ritualisierte Gestik in deutscher bürokratischer Prosa der frühen Neuzeit', *Historische Anthropologie*, 4 (1996) 216–33.

13 Krieger, *Topographie der Stadt Strassburg* (Strasbourg, 1889), p. 455.

14 Bayerisches Hauptstaatsarchiv, GL Fasz. 2641, 19 January 1615.

15 StAA, Almosenamt. Das Pilgerhaus betreffend, Tom 1. 22 April 1561.

16 Stadtarchiv Nuremberg, B19, 520. 12 and 13 April 1585.

17 The following account is summarized from StAA, Collegium Medicum (Hereafter CM). Hebammen und Obfrawen, 1548–1813. 22 August 1549, documents from April 1558.

18 *Ibid.*, 24 February 1564.

19 Stadtarchiv Köln, Zunft No. 378. Barbiere vnd Chirurgen, 9 August 1612.

20 See the discussion of tanners and honour in K. Stuart, *Defiled Trades and Social Outcasts: Honor and Ritual Pollution in Early Modern Germany* (Cambridge: Cambridge University Press, 1999), p. 46.

21 StAA,CM. Deputation ad Officium Sanitatio von 1601-27 Tom. II. 6 September 1607.

22 StAA, CM. Deputatio ad Officium Sanitatis von 1556–1600, Tom. I. 3 November 1592.

23 *Ibid.*, 9 November 1592.

24 Stadtarchiv Nördlingen (hereafter StANö), R39 F2, fasz. 14, 25 June 1572.

25 StANö, R39 F5 Nr. 37, 26 May 1542.

26 Universitätsarchiv Tübingen, 20/10, no. 19, 1589.

27 Cited in Tobriner, 'Urban Report', p. 39.

28 StAA, CM. Ärzten, Ordnungen & Dekreta von 1460–1804. 2 May 1556.

29 R. Herrlinger, 'Die Nürnberger Leprösenschau im 16. Jahrhundert', *Ärztliche Praxis*, III/13 (March 1951) 16.

30 StANö. R39 F5 Nr. 37. Undated supplication of Hans Diethaiss.

31 Stadtarchiv Nuremberg. D1 St. Almosen 1279. Entry for 1572, 'Peter Hawsdorffers Stiftung'.

32 Nuremberg Stadtbibliothek. Mss. Cent V, 42, folio 153^v.

33 *Ibid.*, folios 75–85.

34 L. F. Qualtiere and W. W. E. Slights, 'Contagion and Blame in Early Modern England: The Case of the French Pox', *Literature and Medicine*, 22.1 (Spring 2003) 1–24.

35 F.-O.Touati, 'Contagion and Leprosy: Myths, Ideas, and Evolution in Medieval Minds and Societies', in *Contagion. Perspectives from Pre-Modern Societies*, ed. L. I. Conrad and D. Wujastyk (Aldershot: Ashgate, 2000), pp. 179–201.

36 K. Stuart distinguishes taboo and dishonour pollution in *Defiled Trades*, p. 10.

37 L. von Hoernigk, *Politia Medica* (Frankfurt, 1638), p. 16.

38 R. French, *Medicine before Science* (Cambridge: Cambridge University Press, 2003), pp. 1–2.

8

Corruptible Bodies and Contaminating Technologies: Jesuit Devotional Print and the 1656 Plague in Naples

Rose Marie San Juan

Few who contracted the bubonic epidemic in Naples in 1656 lived to tell the tale, but some that did attributed their recovery to an unprepossessing printed portrait of Jesuit saint Francis Xavier (Figure 1). For these survivors, the power of this portrait to arrest the flow of contagion was envisaged as bodies pressed and impressed on each other, conjoined in the act of healing much as they were believed to be in the transmission of deadly disease. Giovanni Battista de Angelis, for instance, testified that he, like many of his neighbours, bought a printed portrait of this saint from a street seller, and always carried it inside his shirt except at night when he placed it under his pillow.[1] One day he found a huge ulcer in the area of his heart, and after seeing death approach him, he took the portrait of the saint and placed it on the affected area. Immediately he fell asleep and awoke one hour later to find the print and his shirt full of blood and festering while the ulcer had disappeared. The portrait became, like a relic or an icon, the carrier of the presence of the saint,[2] but this presence was short-lived and not contained within the materiality of the print, which typically is ignored and even discarded after it has healed the body, usually by extracting corrupt bodily fluids. In fact, Giovanni de Angelis quickly turns his attention from the image of the saint to his own body, as he feels the corrupt liquids trapped inside, measures these in relation to visible body parts, and sees them outside of himself, expelled from his interior and subsumed into the printed body. Virtual and actual bodies are thus intertwined as they mutate from impressed image to bodily presence, from inside to outside, from past to present, and from death to life, and back again.[3] The ability of these bodies to intermingle and become one depended on Christian concepts of the transience of the material body,[4]

107

but it also depended on the transience of the printed replica which enabled the image to constantly reproduce itself within an extended and mutating history.

By the time desire for this particular image of Francis Xavier intensified in Naples,[5] even to the detriment of local saints, its powers to heal had become widely known through wondrous accounts of its effects throughout Europe and Asia. Indeed, in the context of missionary travel, the use of this image was shaped by the accumulation of stories circulated as assiduously as the image itself. It was through these multiple and disconnected journeys that the image of Francis Xavier would re-emerge with new vigour in Naples in 1656, producing a self-awareness that focused on one's vulnerable physicality in relation not only to the sacred body but also to other threatened bodies, including those in distant and unknown places. For the Jesuits the crisis of contagious disease in Naples became an opportune moment to reactivate the city's miracle-working images and relics of its missionary healer, St Francis Xavier. But print technology's recent ways of reproducing and circulating one of these images within missionary journeys held unexpected consequences when it was reactivated in Naples during the plague of 1656, especially through its production of new forms of bodily contact.

By all accounts, the encounter with death in Naples was particularly visceral and persistent.[6] About 150,000 people died during the summer of 1656, compared with 9,500 people in Rome, and within two years the estimated death toll had reached around 270,000. Contemporary writings evoke the intrusive presence of death in the everyday life of the city. Testimonies consistently recoil at unrelenting confrontations with the corpses that filled everyday spaces, especially piazzas and main passageways in which carriages were said to pass only by stepping over 'baptized flesh'.[7] In its evocation of death's assault on urban space, this perspective has remarkable counterparts in surviving printed images. What is distinctive about these prints is not only the unflinching display of suffering bodies that undermines any sense of an organized urban space, but also the focus on the endless multiplication of a diminishing body, a contradiction that is increasingly linked to a certainty about the materiality of the body and an uncertainty about its ability to transcend that materiality after death. [8]

Of course, print was intricately entangled in the ways attitudes towards death were internalized and embodied, and in this respect, it is interesting to compare the deployment of print in Naples and Rome during the summer of 1656. In Rome, urban space was put under the authority of the newly formed Congregation of Health, which strictly enforced new strategies of segregation and quarantine;[9] in fact, the Congregation used print in unprecedented ways to impose its restrictions.[10] Moreover, prints for the market represent a well-organized and contained city, and make claims for the ability of the technology of print to expose that which the eye might

S. Franciſcus Xauerius Indiarum Apost.
R.ᵐᵉPatri, ac Dño mẹo Cọlẹndiß.ᵐᵉP.Antonio Magaglianẹs Indiar. Prọcuratori merit.
Soc. Iẹsu. Nicolaus Perrẹr Dẹuoti∫∫imus ∫ẹrui∫ D.D.

Figure 1

not see but which might be hidden and suppressed within urban space. In effect, the threat of contamination became a way to market an image of the modern and progressive city in which the key boundary between life and death was stabilized in order to keep it out of sight.

In Naples, meanwhile, officials retained the well-established notion of the plague as divine retribution. If no policies of segregation were imposed, it was in part due to political expediency.[11] The Spanish Viceroy refused to interrupt the movement of Spanish troops across the peninsula, and on the encouragement of various religious orders, including the Society of Jesus,[12] increasingly turned to public appeals to divine forces, including street processions and ceremonies. Accordingly, a considerable amount of print was directed to votive functions. Miracle-working imagery may have been censored by church authorities, but in Naples it was nurtured not only by various religious orders, but also by an incipient street market of popular print.[13] In sum, official and unofficial print worked to erase rather than draw divisions, and significantly, unlike in Rome, officials did not use strategies of containment to target specific city areas and communities as the source of urban contamination.

With the threat of contamination thrust to the forefront of people's lives, the passage between life and death, always a site of fear and ambivalence, took on new urgency. Within this context, the propensity to use the votive print in Naples might seem to have merely served religious and civic authorities. Yet, at a time when authorities increasingly drew on fears of contamination to contain not only disease but also urban conflict and dissent, the votive print offered an alternative site to conceive of oneself in relation to exteriority and especially to other bodies. While in Rome, print served to reassert the power of ocular perception to arrest contagion and define a clear boundary between life and death, in Naples print retained the devotional image's bodily idea of the image and opened up a space to reconsider the transience of the body and the productive and destructive aspects of contamination.

Contaminating journeys of the printed image across Europe and Asia

In Naples the desire for an image of Francis Xavier to ward off the threat of contamination was not indiscriminate. On the contrary, it was primarily focused on a particular image of the saint known in the city since the early 1630s. A painted portrait of Francis Xavier of unknown origin was, according to the widely circulated testimony of the Jesuit priest Marcello Mastrilli, instrumental in bringing him back from certain death on 3 May 1634. A pamphlet, published almost immediately, includes the print (Figure 1) by Nicolas Perrey through which the miraculous image would become known in Naples.[14] It tells of the horrendous head injury incurred by Mastrilli from a dropped hammer, of days of delirium during which the priest's internal bleeding could not be accessed by doctors, and of his request that a portrait of Francis Xavier that happened to be kept in a back room be placed by his bed. At the very moment of death, Mastrilli heard his name coming from the vicinity of the image:

he saw in front [of the painting] a figure with a staff in hand very similar to the image except that he was whiter in complexion and more serene in the face ... he demanded: Do you still have the relic of the most holy wood of the cross, responding yes, he ordered him to apply it to the injured part; he took the relic and placed it on the wound, but the saint transferring his staff to the left hand, freed his right hand to show him to place it in the back part of the head where from the beginning he had been tormented by pain.[15]

The embodied saint is like the painting but different: the face is whiter and more serene, suggesting internal animation and goodness. Moreover, the actual saint distinguishes himself from the painted image by moving his pilgrim's staff from one hand to the other in order to guide the relic to the back of the head. Apparently, the doctors had incorrectly assumed that the external wound indicated the location of the internal injury. Francis Xavier, however, does not confuse the superficial symptom with the internal condition.

In practices that attribute magic properties to an image, the distinction between image and person represented is eliminated, and the image acts as the subject itself is expected to act.[16] In this instance, however, the painted image and the sacred body are not the same, and the latter must be coaxed out of the former. The painting is treated as incomplete or deficient and, as with copies that cannot fully claim contact with the sacred original, in need of being supplemented with other carriers of bodily presence such as a relic.[17]

Presence,[18] however, is not left only to the image or even the relic, for before the saint cures Mastrilli he gives him the choice of either dying from his condition or imitating the saint's achievements by retracing his missionary journey to Asia almost one hundred years earlier. In effect, the saint brings Mastrilli fully back to life only once he has chosen death in the form of the saint's journey to India and beyond. In keeping with a Christian notion of the body, death in the form of corrupt fluids is conceived as already present within the flesh of the living,[19] while the spillage of these liquids, especially blood, for the sake of others is believed to bring forth new life. A widely held idea, upheld by new medical arguments, was that certain elements expelled by the body in life or in death held medicinal functions and could cure and revive the body.[20] What was rejected as waste remained part of a larger sense of life, entangling bodies together in an intercorporeality that depended on the continuity of life and death. In Julia Kristeva's view, this is precisely the process of abjection by which one must reject parts of oneself to retain life but which raise anxiety as part of the constant process of bodily diminishment.[21] Thus Mastrilli's journey represents not simply Francis Xavier's unidirectional contamination of the priest and a doubling of himself. Rather, it is a reflexive exchange, in which the sacrificial spillage of Mastrilli's blood feeds the desire for the presence of the sacred body that had not yet been fully realized. After all, while Francis Xavier had produced many miracles of healing throughout his journeys, especially in relation to plague, he had died from natural causes and could not fulfil the promise of contamination through the spilling of his blood. Most of the saint's acts of healing had been done through the presence of his dead body, but his image becomes powerful only when infused with other acts of bodily spillage.

Mastrilli, then, does not simply repeat Francis Xavier's journey, he actually extends it and thus brings new potential to his image. A new kind of image emerges out of Mastrilli's imitation of Xavier's journey to Goa, Manila and Nagasaki. If Mastrilli becomes a kind of pilgrim so does the sacred body impressed on a piece of paper that would be sent out on the same journey: incomplete, transient and in a constant state of becoming. Instead of Jerusalem, this pilgrimage journey is to Asia where the idea of contamination takes on a particular resonance, conjoining the healing of the body with religious conversion. The focus on healing in a journey with Christian missionary aims may not be unexpected but what is unprecedented is the intense focus on the materiality of the body and its survival through intercorporeality.[22] Equally surprising is the new importance of the visual image, which becomes a crucial mirror to consider the self in relation to the limits of the body and the potential in its connection to other bodies. In effect, the accumulation of desire for the revivified presence of the saint in the image is achieved not only through the multiplication of this image, but also through its relocation within a new set of relationships.

Accounts of Mastrilli's journey to Asia, culled from the 1637 trial testimony to verify his martyrdom at the hands of Japanese warlords, were published across Europe and in Jesuit printing presses in Goa, Manila and Macao.[23] In these accounts, the original painting is said to have remained in Naples. The prominent Spanish Jesuit Eusebio Nieremberg, for instance, locates the power of this image exclusively in Naples, explaining that the Jesuits, wanting the entire city to know of the image's effects, decided to install it in a public space and to take it in street processions.[24] But, significantly, it is Mastrilli's longing for the image once he has left Naples that sets off a new process of replication.[25]

The account by Geronimo Perez published in Manila in 1639 reveals a distinctive process of replication.[26] Mastrilli, we are told, acquired copies of the image in Naples, Genoa and Madrid, and although these were made by the greatest painters, he gave them away because he did not recognize the presence of the saint in them. On the night before Mastrilli sailed from Lisbon to Goa, he told a novice painter how to paint the image 'without further models than his own words'. The next morning Mastrilli was amazed to find the presence of the saint in the image, achieved in eight hours while the king's greatest painters could not do it in three months.

The painting becomes the actual embodiment of the saint, with presence produced through the very materiality of the image: the pigment turns to flesh, conveying facial expression, inner animation and even the 'breath of sanctity'. But this is achieved through desire for the saint rather than the skills of the painter or the power of artistic representation; indeed, it is Mastrilli's verbal account that produces this replica and its authenticity can be confirmed by him alone. Only he can recognize the sacred body's ability to express not only love but also grief. While in Manila, Mastrilli notices that the face of the saint is sad, but when he shows it to his companions who are about to depart for Macao, they cannot see the changes in the image. Mastrilli's exchange with the image is verified when his fellow travelers perish in a storm at sea.[27]

Recognition of presence is akin to doubling of the self. In the 1648 Latin edition of Niemberg's life of Mastrilli, the title-page (Figure 2) shows Mastrilli holding the true image of Francis Xavier.[28] The image within the image is defined as a painting through its frame, and as the internal vision of Mastrilli through the appearance of the saint. In his embodied state, Francis Xavier expresses internal emotions through external signifiers. His face glows, his staff is held by the left rather than right hand and he points down towards the edge of the frame, as if recalling the location of Mastrilli's internal head injury. The gesture literally evokes the ruptured boundary between the two bodies by bringing attention to the boundary of the painting.

With the printed replica, however, the sacred body is not captured within the materiality of the image. In some sense the printed image seems to circumvent the problems of replication raised by painting since the mechanical copy implied an unmediated link from copy to copy and even distance from the human hand. But in comparison to the painted image, the mechanical print seems to be lacking in the very components that were deemed to constitute presence. As a carrier of the memory and presence of the sacred body, the mechanical replica seems reduced (Figure 1), with the body merely suggested by the outlines of a body fleshed out elsewhere. With clear and orderly hatchings that define the hard outer boundaries of the body, the sacred pilgrim body must emerge from what is projected onto the blank space between the lines as much as from the lines themselves: ample robes conceal nonexistent flesh, a staff held by the outlines of a left hand imply past and future movements, and a facial schema promises the impending interiorization of this body.

Yet Nieremberg's 1640 Spanish account, combining the testimony of both the Naples and Manila preliminary trials, gives increasing potential for contamination to the printed image. Nieremberg recounts how the printing of accounts of the image's achievements in many cities, including Naples, Madrid and Goa, resulted in the further duplication of the image itself as well as of its effects.[29] The qualities initially ascribed to the original painting – incompleteness, the detachment of the image from its physical support – are now associated more specially with the printed portrait. These distinctions between painted and printed copies were produced through individual moments of exchange, but the mutating print portrait also conjoins these moments.

Figure 2

Figure 3

The death of Marcello Mastrilli in Nagasaki on 14 October 1637 appears on the elaborate title-page (Figure 3) of Antonio Telles De Silva's popular account, published in Lisbon in 1639.[30] This title-page, circulated also as a single-sheet print during the plague of 1656, interweaves the physical suffering through which Mastrilli imitates and extends Francis Xavier's journey to both the past and the future. Inserted within scenes of torture set in 1637 Japan is the internal space of Mastrilli's room in Naples three years earlier. In keeping with the by now ubiquitous published accounts, Francis Xavier emerges from his image to heal Mastrilli; the saint is not an exact double of his image, just as Mastrilli would not be an exact double of the saint. The moment of bodily presence in Naples has within it the demise of the body in Nagasaki. Angels hover carrying the instruments of Mastrilli's future martyrdom, the ladder, wooden supports, ropes, water jugs and sword deployed in the nearby scenes of torture. Different stages of the water torture are displayed, multiplying the body of Mastrilli but also evoking the medical notion that the accumulation of corrupt internal liquids must be purged if life is to continue. And purged they are, transformed into the blood of Mastrilli, soon to emerge and bring forth new Christians. As Julia Kristeva argues, the diminishment of the sacred body is not a death as much as a 'life-giving discontinuity', and while Christianity stresses the loss of the body, it also stresses the communal bond that is forged in the process.[31]

Miraculous images of saints, as Hans Belting has argued, derived their power from the evocation of presence that depended on two different moments, the past achievements of the sacred body through its sacrificial spillage of blood and its promise of an eternal future.[32] According to De Silva's account, the Jesuit priest informed his Japanese captors that he carried with him two images, one a copy of the miraculous image itself, which he wished to show to the Shogun and thus prove its healing powers, and the other a representation of the miracle in Naples.[33] Mastrilli offered to submit himself to any form of torture if the image failed to show its miraculous effects, an implicit acknowledgement of his contribution through imitation and physical suffering to the power of the image. The display of torture in the foreground of the print, imposed without the image being tested by the Shogun, leads one back to the insert and the doubling of the sacred body through the initial event. In effect, the blood about to be spilled in Nagasaki has within it the past and future suffering of Naples as present, past and future intersect and bodies, both suffering and sacred, duplicate. It is the printed image of Francis Xavier that would become the carrier of Mastrilli's blood, constantly evoking paths of struggle and suffering that feed and reproduce it through their accumulation.

Within the missionary context the printed replica served to extend intercorporeal relations through the body's suffering, but it is crucial that it also served to cultivate scattered moments of individual need. It is through this particular combination that the usual conflation of physical and moral well-being became conjoined in a more self conscious way. The testimony of Maria de la Encarnacion, a 25-year-old Portuguese nun who suffered with many tumours, and whose brother in Goa reluctantly sent her his beloved printed portrait of Francis Xavier, is revealing for how it links her internal expenditure with the emergence of a regenerated soul:

> Suddenly she was enveloped in sweat that was extraordinary because it was copious as well as soft ... leaving from inside of me the cause of my malady, which little by little was emitted in the sweating, so that I could see it with my own eyes.[34]

The malady of the body is perceived as something trapped and moving within internal fluids, which require release if the balance between life and death is to continue. The sweat has the purpose of drawing out impurities onto the surface in order for these to be expelled, but it also enables Maria to recognize her interior self, and confront the nature of her suffering. The links between corrupt bodily fluids and impurities of the soul are self-evident. If the death of the nun three days later did not diminish the success of the cure, it is because the body's internal fluids have become conflated with the sins of the soul, and the procedure of evacuation, the aim of most remedies used by popular healers and professional doctors,[35] with the act of penance.

The print's ability to evacuate the constricted body brings life not simply to a single body, but to all that it contaminates with its regenerating force. For this reason childbirth, the bringing of one life through evacuating another, is a recurrent achievement of the printed image of Francis Xavier in Asia:

> A woman in the Philippines being very constrained by a dangerous delivery, and extremely sad because she had been for three days without being able to deliver, at the end of that time her husband placed on her body with great faith an image that the servant of God had left him, and later in time she expelled the child with great facility.[36]

The reappearance of the saint's likeness may imply a repetitive exchange, but the desire for contact invariably introduces alterity by turning the user's attention to her own bodily experience rather than to the image, its authority or authenticity. It is precisely the incomplete character of the replica that demands the viewer's own interiority to complete it, enfolding it with the specificity of a particular narrative account of internal experience.[37] Within this devotional function, print refocused the copy and its ability to reconstitute the memory of the image not simply by taking it back to the authority of the original but by tracing different journeys that were discontinuous but constantly overlapping in both space and time.

To sum up the ways the print replica reconfigured a new relationship between visual likeness and bodily contact: the printed copy, reduced to the bare outlines of the sacred body, functioned as a replica of the essence of that body, that which was thought to remain after death. The print's inability to convey the illusion of bodily flesh, the parts of the body that were deemed to be transient and corrupt but also crucial to presence, prompts users to animate the image by turning to their own interiority. Thus the printed replica enhanced the participation of the user while extending the idea of the interconnectedness of the communal body so crucial to Christian ideas of the afterlife and Jesuit missionary aims. In forging this collective bond, the impoverishment of the printed copy is crucial, for unlike a relic or a painted image, it necessitates presence to be brought forth through the user's awareness of their own material body.

The return of the image to Naples

In the crisis of bubonic disease that struck Naples in 1656, the idea of the contaminating body, both spent and duplicated through the outpouring of infectious fluids, was informed not only by the devastating daily encounter with death but also by a technology of print with new ways of conceiving of bodily replication and bodily contact. The image of Francis Xavier as pilgrim, offering porous boundaries between life and death, would be further extended through distinctive yet overlapping accounts. Operating within a mixture of devotional and medicinal practices, the print portrait refocused the copy and its ability to reconstitute the memory of the saint by keeping in place the intervening missionary journeys of the image and the ways these had accumulated in, and changed, the image. What is suggestive about the reappearance of this devotional print in Naples is the complex chain of memories that interconnected the fear of death prompted by bodily contagion with the continuity of life implied in Christian notions of bodily contact. In effect, the contradictory aspects of the contaminating body became increasingly evident within the image's new journeys through urban space. This body would continue to produce desire for the presence of the saint, and through intertwined yet discontinuous circuits conjoin Naples to Goa, Macao, Manila and Nagasaki, but it would now begin to raise doubts about the possibility of conceiving of presence and of the continuity of life.

When the image of St Francis Xavier re-emerges in Naples, it is already an image in transition. In the testimony of those who like Giovanni de Angelis were healed by the printed portrait of Francis Xavier during the plague, the print was like the human body: transient, incomplete and unpredictable, but also recognizable in spite of being unknowable. It is, I would suggest, as if the recognition of one's interior self, like the recognition of the presence of the sacred body, had become more fraught. Many of the testimonies of people who claimed to be cured by the image imitate the modes of exchange deployed during the image's missionary travels but frequently experience problems bringing forth the presence of the saint. As the testimony of Giovanni de Angelis suggests, the success of the image is always reasserted but its ability to produce presence and thus to complete the doubling of the body becomes increasingly disrupted.

In the context of the plague, the printed image continued to work by expelling trapped corrupt bodily fluids. Contagion manifested itself externally in the forms of visible tumours, for instance Anna Maria Zamblini, a 35-year-old Roman, who suffered from a tumour 'like the egg of a dove'. Invariably, there are physical manifestations, from fever and delirium to the vomiting of vermin, through which the corrupt fluids are extracted from the body.[38] The printed image prompts this outflow and in many instances even consumes it. These testimonies are less likely to conflate the physical and moral implications of expelling internal corruption, and more likely to focus on the visual inspection of one's internal self. In effect, an expanding sense of interiority is accompanied by the recognition of bodily limits and fragility. It is true that the concern with the materiality and fragility of the body is in part an effect of new expectations in the recording of these experiences. Instead of official testimonies from canonization trials, these take the form of private confession and claim to be conveyed in the precise words of the person in question. In the case of Anna Maria Zamblini, the published pamphlet states that the account was taken 'faithfully from the testimony written by the ill woman's own hand' as well as from the testimonies of her confessor, her husband, the barber, a gentleman and long written testimonies by two doctors, one the woman's own and the other a consultant.[39]

Anna Maria Zamblini, whose Jesuit confessor encouraged her to nurture her devotion to Francis Xavier once she became ill, reveals a pragmatic approach to recognizing the presence of the saint through the printed image. She recounts how the confessor brought her a relic of the saint, which she hung around her neck, together with three different types of miraculous images. Two were affixed to the prayer stool next to her bed and to the wall facing the bed while the third she held constantly to her chest. She used these in various devotions and also when she talked to the saint. Since none of these strategies seemed to bring about the moment of encounter, the confessor brought her a printed book of the life of Marcello Mastrilli, and read to her the celebrated miracle in which his health was restored by the saint. He then encouraged Zamblini to renew her vows to Francis Xavier using the same words and gestures that Mastrilli had used to addressed the saint.[40]

It is the imitation of Matrilli's vision, or rather of Mastrilli's description of his vision, which finally brings forth Zamblini's recognition of the saint and her subsequent release from internal corruption. After being cured, the Jesuit confessor prompted Zamblini to describe the embodied saint and he did so by showing her the three printed images that he had brought to her. Zamblini manages to include aspects of all three images in her response, describing the saint as wearing a beret and with hands folded on his midriff in one image, as having the face of another image, although a bit thinner, and as wearing the habit of pilgrim in the third image; it is only the confessor who credits the miracle to the particular image that cured Mastrilli.[41]

The inability to conform to modes of imitation established by the image in its journeys across Europe and Asia has as its counterpart the inability to recognize the saint through his image or his presence. This is the case with Anna Caserta, whose testimony is unusual in that it is given by her neighbour. [42] It is through this intermediary that her account clearly follows the form established by Mastrilli, including being called, just as death approached, by a pilgrim who looked like the one in a framed image that happened to hang near her bed; the imitation of the earlier incident is complete when we are told that the image in question showed the pilgrim standing over the bed of a sick man. But Caserta herself is said not to have known whom the image represented, and only recognized a glow surrounding the face that encouraged her to pray to him.

The inability to recognize the image of Francis Xavier is compounded by Caserta's failure to see the saint, for when she tries to thank him he had already disappeared and his touch on her injury is that of an invisible hand. [43] It is in fact Giuseppe Matina, Caserta's neighbour, who testified that it was he who when told of the incident and shown the image immediately recognized it to be St Francis Xavier with Marcello Mastrilli in his sick bed.

At the very moment when a more individual experience is sought, there is an attempt to contain it within the established format, and to produce the individual experience out of the imitation of others. Increasingly known orally and available through print, the individual narratives enabled the crucial process of imitation to be activated and extended. Yet the constant contamination of one experience by another, of one description of interiority by another, of one conception of the image by another, seems ultimately to put on hold the possibility of the bodily presence of the saint and to encourage a self conscious considesration of the limits of the physical body in relation to both interiority and intercorporeality.

Notes

1   *Racconto d'alcuni de' molti miracoli operati da S. Francesco Saverio in Napoli nel tempo della pestilenza* (Rome, *c.*1660), 2a–2b (unpaginated); also in *Ragguaglio della Miracolosa Protezione di San Francesco Saverio Apostolo delle Indie verso la Città, e il Regno di Napoli nel contagio del MDCLVI* (Naples, 1660), pp. 54–8.

2   On the use of miraculous images, including prints, see Hans Belting, *Likeness and Presence. A History of the Image before the Era of Art*, trans. E. Jephcott (Chicago: University of Chicago Press, 1994), pp. 59–73, 425–32.

3 I draw on Gilles Deleuze, *Cinema2. The Time-Image*, trans. H. Tomlinson and R. Galeta (London: Athlone Press, 1989), pp. 68–97, for the notion of actual/virtual.

4 On the transitional body, see D. Judovitz, *The Culture of the Body. Genealogies of Modernity* (Ann Arbor: University of Michigan, 2001), pp. 4–5.

5 On Francis Xavier during the 1656 plague, see Nicola Rillo, *Napoli e S. Francesco Saverio* (Naples, 1923).

6 On 1656 plague in Naples, see S. De Renzi, *Napoli nell'anno 1656* (Naples, 1867); E. Sonnino, 'Di qui cominciò qualche terrore considerabile nella città di Roma: Popolazione e sanità del XVII secolo', in *Scienza e Miracoli nell'Arte del '600. Alle origini della Medicina Moderna*, ed. S. Rossi (Milan: Electa, 1998), pp. 60–9.

7 Celano quoted in James Clifton, 'Mattia Preti's Frescoes for the City Gates of Naples', *Art Bulletin*, 76 (1994) 479.

8 On these images, see A. Porzio, 'Immagini della peste del 1656', in *Civiltà del Seicento a Napoli* (Naples: Electa, 1984), 2: 51–7, and my forthcoming *Vertiginous Mirrors: Perilous Journeys with Early Modern Jesuit Images of Self Reflection*.

9 On the establishment of the *Congregazione della Sanità* and its actions, see P. Savio, 'Recerche sulla peste di Roma degli anni 1656-1657', *Archivio della Societa romana di storia patria*, III.XXVI (1972) 113–42. On the Congregation's strategies, see my *Rome: A City out of Print* (Minneapolis: University of Minnesota Press, 2001), pp. 219–53.

10 San Juan, *ibid.*, pp. 219–254.

11 Porzio, 'Immagini della peste del 1656', pp. 51–7. E. Nappi, 'Aspetti della società e dell' economia napolitana durante la peste de 1656' (Naples, 1980), vol. 63, n. 148, 12–23; on preventative measures taken in Naples, see Clifton, 'Mattia Preti's Frescoes for the City Gates of Naples', pp. 479–80; S. Watts, *Epidemics and History. Disease, Power and Imperialism* (New Haven, CT: Yale University Press, 1997), pp. 15–25.

12 On urban reorganization and religious orders, see G. Pane, 'L'Urbanistica del Seicento a Napoli' in *Seicento Napoletano.Arte, Costume e Ambiente* (Milan: Edizioni di Comunità, 1984), pp. 51–84; Nappi, 'Aspetti della società e dell' economia napolitana durante la peste de 1656', pp. 9–12.

13 The promotion of miraculous images in Naples is discussed by Romeo De Maio, *Pittura e Contrariforma a Napoli* (Rome: Editori Laterza, 1983), 179–85, 217–22; on the use of sacred images by the Jesuits, see D. Gentilcore, *Healers and Healing in Early Modern Italy* (Manchester: University of Manchester Press, 1998), pp. 14–15; J. D. Clifton, 'Images of the Plague and other Contemporary Events in 17th-Century Naples' (UMI Dissertation, 1989), pp. 60–1.

14 *Breve Raguaglio del miracolo oprato dall'Apostolo dell'Indie S. Francesco Saverio in Napoli li 3 di Gennaro 1634* (Naples, 1634).

15 *Ibid.*, pp. 13–14, 19: 'viddesi innanzi un personaggio col bordone in mano somiglatissimo all'Imagine, se non che'era nella carnaggione piu bianco e più sereno nel viso...gli dimano: Havete ancora reliquia del santissimo legno della Croce, respondendo di sì, l'ordinò che l'applicasse alle parte offesa: quegli prese il Reliquiario, e'l pole sù la ferita, ma il santo transferendo 'l suo bordone all sinistra mano, disoccupò la sua destra per mostrargli che la ponesse nella parte di dietro del capo dove fin da principio havea sentito travagliarsi dal dolore.'

16 On portraits of saints, see E. Kitzinger, *The Cult of Images in the Age before Iconoclasm* (Cambridge, MA.: Dumbarton Oaks Papers, 1954), pp. 100–9; Belting, *Likeness and Presence*, pp. 9-14, 351, 409–19.

17 M. Taussig, *Mimesis and Alterity. A Particular History of the Senses* (London: Routledge, 1993), p. 57.

18 For a useful discussion of presence, imitation and bodily contact, see *ibid.*, pp. 19–23, 44–58; on the use of religious print in healing, see Gentilcore, *Healers and Healing*, p. 21.

19 M. Foucault, *Madness and Civilization. A History of Insanity in the Age of Reason* (New York: Vintage Books, 1973), pp. 15–24.

20 On early modern perception of body and illness, see G. Pomata, *Contracting a Cure: Patients, Healers, and the Law in Early Modern Bologna*, trans. by author (Baltimore: Johns Hopkins University Press, 1998), pp. 129–39; P. Camporesi, *The Incorruptible Flesh. Body Mutation and Mortification in Religion and Folklore* (Cambridge: Cambridge University Press, 1988), pp. 67–89, 106–30.

21 J. Kristeva, *Powers of Horror. An Essay on Abjection* (New York: Columbia University Press, 1982), pp. 1–31.

22 On the transience of the body and its links to knowledge, see Judovitz, *The Culture of the Body*, pp. 1–12.

23 Antonio Telles De Silva, *Historia de la celestial vocacion, missiones apostolicas y gloriosa muerte del Padre Marcelo Franco Mastrilli* (Lisbon, 1639); Geronimo Perez, *Relacion de lo que asta ahora se a sabido de la vida, y martyrio del milagroso Padre Marcelo Francisco Matrili de la Compania de Iesus* (Manila, 1639); *Relacion del Insigne Martyrio que padecio por la Fe de Christo el Milagroso P.Marcelo Francisco Mastrilli* (Manila, 1638).

24 Eusebio Nieremberg, *Vida del Dichoso y venerable Padre Marcelo Francisco Mastrilli* (Madrid, 1640), 17a–17b.

25 *Ibid.*

26 Perez, *Relacion de lo que asta ahora se a sabido de la vida, y martyrio del milagroso Padre Marcelo Francisco Matrili de la Compania de Iesus*, pp. 24–5.

27 Nieremberg, 1640, 42b.

28 Eusebio Nieremberg, *P. Marcellus Mastrillus* (Dilingae, 1648).

29 Nieremberg, *Vida del Dichoso*, 18.

30 This image was also circulated as a single-sheet print in Naples during 1656.

31 J. Kristeva, 'Holbein's Dead Christ', in *Black Sun. Depression and Melancholia*, trans. L. S. Roudiez (New York: Columbia University Press, 1989), pp. 105–38.

32 Belting, *Likeness and Presence*, pp. 9-14.

33 De Silva, *Historia de la celestial vocacion*, p. 112.

34 Nieremberg, *Vida del Dichoso*, 19a–20a: 'de repente se resolvio en unos sudores extraordinarios; porque igualmente fueron copiosos come suaves…dejandome dentro de mi la causa de mi mal, que poco a poco fue saliendo en los sudores, para que yo lo viesse con mis propios ojos.'

35 Pomata, *Contracting a Cure*, p. 131.

36 Nieremberg, *Vida del Dichoso*, 85b: 'Estando una mujer de las Filipinas muy apretada de un parto peligroso, y penossissimo; porque avia estado tres dias sin poder partir, al cabo de ellos le puso su marido con gran Fè una Imagen que el siervo de Dios le avia dexado, luego al punto echò con gran facilidad la criatura.' Another example is reccounted by Perez, *Relacion de lo que asta ahora se a sabido de la vida, y martyrio del milagroso Padre Marcelo Francisco Matrili de la Compania de Iesus*, p. 75.

37 S. Stewart, *On Longing. Narratives of the Miniature, the Gigantic, the Souvenir, the Collection* (Durham, NC: Duke University Press, 1993), pp. 136–7; I draw on Stewart's argument that the souvenir's impoverishment as a duplicate demands an internal narrative that produces desire.

38   Racconto, 1660, 4a; *Relatione d'un Insigne Miracolo operato in Roma da S. Francesco Saverio della Comp. di Giesu* (Rome: Ignatio de'Lazari, 1677), p. 15.
39   *Relatione*, 1677, p. 28.
40   *Ibid.*, p. 10.
41   *Ibid.*, p. 24.
42   *Racconto*, 1660, 4b.
43   *Ibid.*

9
Quarantine and Caress[1]

Frédéric Charbonneau

The experience of contagion generates all kinds of practices, preventive, curative and reflexive; medical, social, ritual and intellectual. Several of these have a strong theoretical component and definitively aim to produce an effect that we will call health, without attempting to define it further. From this dual point of view, narrative is a unique practice because it takes place after events and does not attempt to affect their course; speculation is usually as foreign to it as prescription; and it seeks less to analyse than to describe, less to understand than to touch, less to create than to resurrect. Such evocations of illness can be fictitious as in novels, or historical as in the French memoirs we will study here, without, however, depriving ourselves of calling on fiction when it is able to shed light on the imaginary dimension of history.

Faced with disease, the memorialists are laymen: victims, they know hardly more than what they have suffered; witnesses, only that which they have observed out of curiosity. Perhaps nowhere better than in their work can we read the equivocal nature of experience, its fundamental impurity, its refusal to be reduced to objective facts. From a duality of perspective peculiar to this form of writing, the memorialists understand after the fact – for time has passed and has changed them in passing – their reconstructed experience, a melding of memory and later elaboration. Memoirs are therefore not documents that make it possible to retrace an improbable truth of history, but monuments erected in honour of the individual, subjective and irreplaceable life of men and women whom the trial of contagion, when it took place, modified in their very flesh and in that body of writing that is their work.

Many indeed, are those who speak of it: contracted in childhood or in old age, diseases serve as *topoi*, indicating in the first case the fragility of a life still forming, in the second the imminence of death and the necessity of knowing oneself before it is too late, even inspiring at times the entry into writing. But the great narratives of contagion are generally set during adulthood, when a more active existence, more numerous contacts, more frequent journeys, in short the intensity of social life favoured the transmission of evil.

It is not, however, as though this matter had been resolved. One of the traits that makes the eighteenth century an interesting time for consideration of the way we conceive of the spread of disease is precisely that the very notion of contagion was hotly contested and debated. The evolution of medical theories – the angry debate over inoculation, for example, that raged in France from 1734 to 1774 – and in a more general way the increased importance of scientific research and the new status of the body as an empirical domain, find an echo in the memoirs of the reign of Louis XV. As far as epidemics are concerned, such as smallpox or the plague, two theories collide during this period: one, in the Hippocratic tradition (*On Air, Water and Places*), vigorously put forward by John Arbuthnot (1667–1735) and, in a more measured manner, by the vitalist milieu of Montpellier, purports that if the same causes produce the same effects, then air, water and food must be the principal vectors of epidemics; the other, that of the Italian doctor and poet Girolamo Fracastoro (*c*.1478–1553) (*De contagione, contagionis morbis et eorum curatione*, 1546),[2] current during the second half of the century, distinguishes three modes of transmission: direct, by way of germ-carrying objects; and by distance, using air as the intermediary.[3] According to one's school of thought, the notion of miasma took on a meaning more or less close to that of a specific germ, which at the time was not observable. It is, moreover, the reluctance to accept an occult cause for the transmission of disease that explains in part the resistance of the Enlightenment to the theory of contagion.

Analysis was further complicated by taking into account internal factors – of the greater or lesser receptiveness or vulnerability to miasmas – of which affective and moral dispositions are a part. For to the medical aetiologies, one must add others, moral and religious, that were current at the time. According to the moral perspective, disease originates in the passions (*pathè*):

L'ambition a produit les fièvres aiguës et frénétiques; l'envie a produit la jaunisse et l'insomnie; c'est de la paresse que viennent les léthargies, les paralysies et les langueurs; la colère a fait les étouffements, les ébullitions de sang, et les inflammations de poitrine; la peur a fait les battements de cœur et les syncopes; l'avarice, la teigne et la gale; la tristesse a fait le scorbut; la cruauté, la pierre; la calomnie et les faux rapports ont répandu la rougeole, la petite vérole et le pourpre, et on doit à la jalousie la gangrène, la peste et la rage.[4]

[Ambition has produced sharp and frenetic fevers; envy has produced jaundice and insomnia; lethargy, paralysis and languor come from laziness; anger has caused breathlessness, boiling blood and lung inflammation; fear has caused heart palpitations and blackouts; greed, ringworm and scabies; sadness has caused scurvy; cruelty, kidney stones; calumny and rumours have spread measles, smallpox and scarlet fever, and we owe gangrene, plague and rabies to jealousy.]

Metaphorical causes? Perhaps. However, the Chevalier de Jaucourt, himself a doctor of the Faculty of Leiden, remarks in the article 'Peste' of the *Encyclopédie* that 'it is essential not to be afraid in times of plague; death spares those who scorn it, & pursues those who fear it; all the inhabitants of Marseilles did not die of the plague, & fear caused more deaths than the contagion'.[5] Finally, the belief that it is God who sends illness had a long life and must have contributed to slowing down the progress of inoculation. Even towards the end of the century, among the devout it was said that it was tempting God to provoke disease that might not have come naturally;[6] and during the plague of Marseilles (1720–22), Bishop Belsunce, attributing the plague to divine wrath at the pride of the merchant city, had called his flock to repent.[7]

It must be said that this last episode in Europe of the pandemic of the Black Plague that had claimed 25 million victims between 1346 and 1353 was seen by contemporaries as a paradoxical consequence, and as the very reversal of prosperity and exchange.[8] Marseilles, an open port since 1669 and the nerve-centre for maritime commerce in France, had not been hit by chance: it was the logic of trade and importation that had marked it for a stroke of fate when *Le Grand Saint Antoine*, a ship laden with merchandise, returning from the Levant at the end of May 1720, slipped through the cracks in the quarantine system and introduced the evil into the city. The plague, writes Jean Buvat in his *Journal*, came ashore with the bundles of fabric from Syria that people neglected to 'perfume with lazaret' to rid it of infection.[9] Modernity, which on the whole had been a vast phenomenon of decompartmentalization – the Copernican revolution, large-scale explorations, the growing circulation of goods, people and ideas, the opening of history to the perspective of progress, and so on – and the taste for luxury, which led to the search for distant spices, porcelain, coffee, dyes, those wonders of the Orient doubled by strange corruptions, thus favoured the spread of diseases, their arrival concurrent with increased traffic around the globe.

As for smallpox, 'it concedes nothing to the plague in the disasters that it causes'.[10] In its endemic state in Europe, its incidence and extreme lethality made it one of the primary causes of mortality, especially among infants, and the object of concern all the more pronounced as it did not spare even kings.[11] People turned to calculations of probability in order to decide if inoculation was desirable, for since Voltaire's publication of his *Lettres philosophiques*, this practice, having come from Constantinople to England, had divided French opinion.[12] As early as 1726, a letter by

Father d'Entrecolles advocated adoption of the Chinese method, which consisted of introducing flakes of dried pox pustules into the nostrils.[13] But it was through the use of an incision, a technique adopted from the Turks, that inoculation spread little by little. Among its partisans, one finds Diderot, Rousseau (*Émile*, 1762), La Condamine (*Lettres sur l'inoculation*, 1760, 1764), Tronchin, Bordeu, Tissot (*L'Inoculation justifiée*, 1754) and d'Alembert, despite certain hesitations; among its adversaries, Hecquet, Prévost (*Pour et Contre*, 1740) and La Mettrie (*Traité de la Petite Vérole*, 1749). By 1768, Ménuret de Chambaud had counted more than 800 texts on the subject.[14] In succession, the children of the Duc d'Orléans (1756), the students at the Royal Military School of La Flèche (1769), Empress Catherine of Russia (1771) and then Louis XVI himself and his two brothers (1774) underwent the procedure, a triumph that consecrated the perfection of the vaccination by Edward Jenner (1796).[15]

The memoirs bear the trace of this moment in time: events and knowledge are refracted by contact with subjectivity, appearing as fragments on the surface of the work, detached from strictly medical doctrine and the flawless series of factual causes. Still, thanks to the ambiguity of retrospection and what has been experienced, these narratives put into play the meaning and the consequences of contagion; for if, at first glance, the trial of the plague or smallpox can seem uniquely painful, distressing and costly, one is obliged to note that at second glance, all is much less cut and dried. Inextricably linked to catastrophe, articulated by the threat of suffering and isolation, a positive experience of disease can be drawn, made of rest, loving care and spiritual elevation. Thus, the boundary between the dysphoric and the euphoric is blurred, as temporal perspectives, the symbolic and the real are interlaced in a complex weave that we must now unravel.

Seen from afar by the chronicler, recorded in the security of a study, epidemics are hardly anything other than scourges and their devastating effects appear as the exact reversal of their causes: contagion, made possible by contacts – the two words come from *contigere*, 'to touch' – first brings about their interruption. It is, of course, a question of containing the disease via quarantine (the preventative function), but also perhaps to suppress symbolically the excesses that provoked it (the expiatory function). The consequence is a slowing, then a suspension of activities on all levels, like a progressive paralysis of the individual or the community.[16] The plague of Provence presents in this respect an exemplary outline thanks to its localized nature – unlike smallpox which was prevalent almost everywhere. As of July 1720, the Parliament of Aix had promulgated an edict imposing the death penalty on anyone passing through the walls of Marseilles;[17] and movement was generally made difficult between regions affected and regions spared. The Marquis d'Argens recalls at the beginning of his *Mémoires* that he had had to wait until the end of the contagion before returning to his home in Aix from Strasbourg where he lived in garrison; eighteen months later, still no one was allowed to enter Spain without a passport.[18] The Duc de Saint-Simon writes similarly that during the winter of 1722 the plague contributed to the impossibility of a military campaign against the Emperor 'by destroying men and finances as well through the disruption of commerce'.[19]

With eye witnesses, however, this slightly remote lamentation gives way to something more immediate and intense. We have, for example, from Pierre Prion a breathless page on the arrival of the contagion in the Cévennes and the stampede that followed:

En 1722, il verra le fléau de la peste apporté à la Canourgue, ville du Gévaudan, et dans celle d'Alais, capitale des Cévennes. À l'approche du fléau la peur le saisira. Il quittera Aubais pour aller à Perpignan y rester avec son maître afin de se mettre à couvert des injures épidémiques. ... Lorsqu'il ira dans cette ville, il couchera à cinq heures du matin; une demi-heure après, l'on posera des soldats pour former la ligne contre la peste, dès ce moment, du pays libre il y verra le prohibé. La prudence et la diligence valent un mort ressuscité.[20]

[In 1722, he will see the scourge of the plague brought to the Canourgue, a town of the Gévaudan, and to Alais, the capital of the Cevennes. At the approach of the scourge, fear seized him. He will leave Aubais for Perpignan, there to stay with his master in order to protect himself from the epidemic. ... When he goes to the city, he will go to bed at five o'clock in the morning; a half-hour later, soldiers were stationed to form a line against the plague; from that moment, from the free country he will see the forbidden one. Prudence and diligence pull a dead man back from the grave.]

The *cordons sanitaires* are transformed under Prion's pen into a veritable state of siege conducted by the army: a kind of war where plague victims are the enemies, contained and fettered, as opposed to those whose movement maintains them in 'free country' – and to whom this liberty alone assures the possibility of movement. In the case of a disease that we cannot outrun, fear, inversely, erects barricades, sometimes in vain, as in the case of the elderly Duchesse d'Aumont, sequestered in her home 'in Passy, against smallpox, of which Paris was full. She did not escape it, and in fact died from it.'[21] Limited mobility thus appears in these different contexts as the essential effect of contagion, precipitated by fright or quarantine; and mobility takes on the very character of health, as its stake, its privilege and its protection.

But this is not the only effect. The contagious person, furthermore, could not be seen. In the case of smallpox, this interdiction often lasts longer than the disease because of remaining symptoms that were enough to incite concern. Dufort de Cheverny tells of his wife who 'deprived herself of going out in public, or went with repugnance' because 'the sight of a person still red from this illness troubled her'.[22] He himself, after his inoculation, 'could not stay one instant in Paris, nor see his son, nor address anyone without frightening him'; three months later, he was still red and had to have 'ask permission of the king to present himself to do his service'.[23] Often, it was much worse still and people were disfigured:

Guéri de ma petite vérole, je fus aussi laid de visage que j'avais été beau. Mes traits étaient grossis et absolument changés; mes cheveux, châtain doré et bouclés, étaient tombés; ils revinrent noirs et droits. La première fois que je me vis dans un miroir, ce fut avec une sorte d'horreur. De ce moment je devins plus honteux, plus sauvage; je n'avais plus rien qui me rassurât ... et ce fut peut-être ma laideur qui m'empêcha de chercher à revoir Julie Barbier ...

[When I was cured of my smallpox, my face was as ugly as it had been handsome. My features were coarsened, and completely changed; my curly, chestnut-gold hair had fallen out; it grew back black and straight. I saw myself for the first time in the mirror with a kind of horror, and, from that moment I became even more of a bear. I had now nothing left to give me confidence... Perhaps it was this ugliness which prevented me from seeking out Julie Barbier ...][24]

Restif is so changed that he hides himself: the disease makes him doubly invisible, through metamorphosis and shame. He turns away from the mirror, blinded like Laclos's Madame de Merteuil, who lost an eye to smallpox.[25] The novelist's invention was exemplary; but nature does better: to Jérôme de Pontchartrain, for example, an odious minister hated by Saint-Simon, and whose moral perversion, 'that nothing had been able to soften nor correct the slightest bit, insinuated itself everywhere'.

Sa taille était ordinaire, son visage long, mafflé, fort lippu, dégoûtant, gâté de petite vérole, qui lui avait crevé un œil. Celui de verre dont il l'avait remplacé était toujours pleurant, et lui donnait un physionomie fausse, rude, refrognée [sic], qui faisait peur d'abord, mais pas tant encore qu'il en devait faire. ... [I]l aimait le mal pour le mal et prenait un singulier plaisir à en faire.[26]

[He was of ordinary size, quite thick-lipped, his face was long and full, disgusting, marked by smallpox, to which he had lost an eye. The glass one with which he had replaced it was always running, and gave him a countenance that was false, harsh, wrinkled, frightening at first, but not as much as it should have been. ... [H]e loved evil for evil's sake and took a particular pleasure in doing it.]

Of him as well, one could have said that the 'disease had turned [him] round and that now her soul is in [his] face'.[27] However, that such transformations reveal or conceal in their ambiguity the depths of the soul is secondary here; what we must retain is that in affecting the appearance, the disease also disrupts sight; it forces the eye to look away as the observer recoils in horror.

This is indeed how one experiences the plague when one has the misfortune to look it in the eye. There is a gripping episode in the *Mémoires* of the Comte de Forbin, squadron chief in the Royal Navy, which takes place on his return from a diplomatic mission to Siam in 1687. On board a vessel belonging to the East India Company bound for Masulipatan, a city famous for trade in the Gulf of Bengal, the memorialist describes with unusually graphic details his arrival at the trading post:

> Nous n'étions plus qu'à huit lieues de Masulipatan, lorsque nous vîmes venir du côté de terre un nuage noir et épais, que nous crûmes tous être un orage. ... Le nuage arriva enfin à bord avec très peu de vent, mais suivi d'une prodigieuse quantité de grosses mouches semblables à celle qu'on voit en France, qui mettent des vers à la viande: elles avaient toutes le cul violet. L'équipage fut si incommodé de ces insectes, qu'il n'y eut personne qui ne fût obligé de se cacher pour quelques moments. La mer en était toute couverte ...

> [We were not above eight leagues from Masulipatan, when we saw a thick black cloud come from the land, which we all thought to be a storm ... At length the cloud came aboard us with very little wind, but attended with a vast number of great flies, with tails of a purple colour, like those we see in France which leave maggots on our meat. The ship's company was so pestered with them, that, for a few moments, every man aboard was obliged to hide from them. The sea was quite covered with them ...][28]

This first cloud is followed by a second: dragonflies, probably – 'these had four wings, and resembled those we see about our rivers, whose tails are striped with yellow and black'[29] – so thick that it prevented them from seeing the shore; the sailors had to approach land with a compass. Such circumstances – instability of air, flies explicitly linked to corruption, further on an 'intolerable stench' – are halfway between cause and symptom according to the terms of the time; and this coastline that a fog of insects hides from view, along with the impossibility of seeing anyone at all in the city frozen by a kind of stupor, are like the signature of pestilence.

Ne trouvant personne dans le port, ceux du vaisseau qui connaissaient la ville nous servirent de guides, et nous menèrent à la douane. Personne ne parut dans le bureau, qui était tout ouvert: nous entrâmes pourtant, et nous en parcourûmes toutes les pièces, sans trouver qui que ce soit. Surpris de cette nouveauté, nous marchâmes du côté où était le comptoir de la Compagnie d'Orient; nous traversâmes plusieurs rues sans voir personne. Cette solitude qui régnait par toute la ville, jointe à une puanteur insupportable, nous fit bientôt comprendre de quoi il était question.

Après avoir beaucoup marché, nous arrivâmes devant la maison de la Compagnie. Les portes en étaient ouvertes: nous y trouvâmes le directeur, mort apparemment depuis peu, car il était encore tout entier. La maison avait été pillée, et tout y paraissait en désordre. ...

Nous continuâmes donc à marcher, et nous nous rendîmes au comptoir des Anglais; nous le trouvâmes fermé: nous eûmes beau frapper, personne ne répondit. De là nous passâmes à celui des Hollandais: de quatre-vingt personnes qui le composaient, il n'en restait plus que quatorze; c'étaient des spectres plutôt que des hommes. ... [A]yant des trésors immenses dans leur maison, il leur était défendu, sous peine de la vie, d'en sortir, sans quoi ils ne seraient pas restés

[There being not a soul in the port, such of our ship's company that knew the town were our guides to the customs house. The office was open, so that we went into every part of it, and found nobody in attendance. Surprised at this unusual turn of events, we walked to that part of the town where the office of the [French] East India Company was kept, and crossed several streets without seeing man, woman or child. This desolate aspect of the whole town being accompanied with an intolerable stench, we quickly guessed what the matter was.

After we had covered a good deal of ground, we came to the company's house. The gates were open, and we found the director dead, though we guessed that he had not been long deceased, because every part of him was intact. The house had been ransacked, and everything appeared in disorder. ...

Therefore we marched on till we came to the English factory, which we found close shut; and though we knocked fit to break the door down, nobody answered. From thence we went to the Dutch factory, where, out of fourscore persons, there remained but fourteen alive, who looked more like skeletons than men ... they had an immense treasure in their house, and were forbidden on pain of death to quit it, or else they had not stayed.][30]

There is now in the deserted city none but those held captive by the lure of gain: business people and pillagers, who guarantee in a way the dialectic of prosperity and contagion delineated here. From this point, nothing more has worth and this general suspension of social life,[31] this near-abolition that the memorialists rival each other to describe is nothing if not an obvious phenomenon, a veil thrown over the depths of being, whose corner we must at present raise.

Those who were caught and struck down by the contagion, who speak not only from the position of witnesses, but also as victims, relate their illness in terms such that the discourse of desolation, solitude and fear mixes with a whole other mode of expression that does not contradict the first, but which reduplicates and is a counterpoint to it, as if a foreign voice infiltrates the monody of the narrator to murmur unreasonable things. The sentiment of having been abandoned or punished by God can give way to the certainty of being a chosen one: put to the test, common souls suffer, that of the memorialist shines, a diamond in the rough. Thus the young Agrippa d'Aubigné, who found himself in the city of Orléans during the plague of 1562, 'was the first to taste the plague that caused the death of thirty thousand people. He saw his surgeon and four others die in his room … His servant, named Eschalart, and who since died a minister in Brittany, never left his side and cared for him the whole time without ever taking sick himself, for he always had a psalm on his lips as a preservative.'[32] For this child prone to visions, able to prophesy and whose strong moral fibre was admired by his fellow Protestants, the meaning of the episode is clear. But besides saying clearly that faith saves,[33] this scene stages the unfailing devotion of the servant, a true *topos* in this kind of narrative.

It is true that this servant is not always a domestic, but is very often someone close, sometimes a doctor; in all cases, something that exceeds all requirements is in play, a redemptive zeal, the expression of pure charity, the conjugal bond, a precious friendship, a sublime loyalty. True martyrs, some die on the job:

[Le Grand Condé], aussi bon courtisan qu'habile général, était parti de Chantilly, quoique malade, à la première nouvelle de la maladie de sa belle-fille la duchesse de Bourbon [fille naturelle du roi]: il l'avait trouvée dans la petite vérole; et, méprisant le mauvais air, il ne l'avait point quitté pendant tout son mal; il avait même, malgré sa faiblesse, empêché le Roi d'entrer dans la chambre de la malade, et lui avait dit sur le pas de la porte des choses si fortes et si touchantes, que le Roi s'était retiré, et était parti pour Versailles. La princesse avait été à la dernière extrémité … Sa jeunesse l'avait sauvée; mais M. le prince, qui à son âge, infirme comme il l'était, n'était plus en état de soutenir une pareille fatigue, y succomba …[34]

[[The Grand Condé], as good a courtier as a skilful general, had left Chantilly, though ill, at the first news of the illness of his daughter-in-law, the Duchesse de Bourbon [the King's illegitmate daughter]: he found her with smallpox; and, unmindful of the putrid air, he did not leave her during her entire illness; he had even, despite his weakness, prevented the King from entering the sickroom, and had told him at the doorway such forceful and touching things, that the King had retired, and had left for Versailles. The Princess had been in the final stages ... Her youth had saved her; but Monsieur le Prince, who at his age, debilitated as he was, was no longer able to withstand such a weariness, and succumbed to it.]

Such evocations are particularly moving when the author himself is the object of this attentive care. Contagious illnesses seem to determine to an extreme degree a mode of internal contact, to the extent that they cause social life to cease; and the manifestations of love are all the more precious as the danger should have distanced everyone. By a remarkable subversion of the etymon (*pathos*), suffering changes into a holy passion, such as in this passage from the *Mémoires* of Restif de la Bretonne, where his mother, a veritable Christ figure, saves him, renders him his sight and 'reads his heart':

Une fièvre continue me saisit; l'éruption commença, et je fus trois jours dans le délire le plus effrayant. ... Mon père, tout fort qu'il était, suffisait à peine pour me contenir. On me crut perdu. ... Enfin les souffrances diminuèrent et je dormis; mais l'éruption avait été si abondante qu'on ne parvint à décoller mes paupières, en les humectant, qu'au bout de dix-sept jours. ... Ma mère me soigna infatigablement. Elle fut témoin de mes transports, au premier rayon de lumière que j'entrevis, après qu'elle m'eut longtemps bassiné les paupières avec du jus de lentilles. Messire Antoine Foudriat vint me voir; mais il ne me confessa pas, comme elle l'en priait; ce fut elle qu'il en chargea, 'afin, dit-il, qu'ayant lu dans son cœur, vous sachiez ce qu'il faudra faire pour le préserver'.

[The fever continued; the eruption began, and for three days I fell into the most terrible delirium. ... My father, for all his strength, could hardly hold me down. They gave me up for lost. ... At last my sufferings abated, and I slept; but the eruption had been so abundant that it took seventeen days of sponging to part my lids. ... My mother nursed me indefatigably; she witnessed my transports over the first glimpse of light, which came to me after she had been bathing my eyelids with a decoction of lentils for a long time. Messire Antoine Foudriat came to see me, but would not confess me as she asked him to; he charged her to do this, in order, he said, 'that having seen into his heart you may know what is to do for its preservation'.][35]

A passage in *Histoire d'une Grecque moderne* (1740) by the Abbé Prévost can shed light on this affective use of disease: the narrator, ambassador of France to the *Sublime Porte*, recounts that he fell victim to an epidemic of contagious fever and that this incident had 'resulted in him knowing how dear he was to the lovely Théophé',[36] a young slave whom he had freed and of whom he had become enamoured. Moved into a pavilion in his garden for fear of transmitting the contagion to his servants, the ambassador wanted no one but his valet and his doctor near him; but Théophé forced her way in and 'nothing could discourage her care for an instant. She fell ill herself. My entreaties, my pleas, my complaints could not make her consent to pull herself away. A bed was made for her in my antechamber, where all the force of her illness did not impede her in the least from being continually attentive to mine.'[37] At the level of narration, the episode has the sole function of proving the young woman's tenderness and of giving the ambassador reason to believe that he is loved, the absence of any concrete detail highlighting intimacy even more.

We do not wish to conflate completely the narration of lived events with the artifice of the novelist, but it seems to me, none the less, that in their reconstruction of the past, memorialists do not escape this late hermeneutic by which the chaos of history and the arbitrariness of suffering take on meaning and are converted into positive experience. The *Mémoires* of Saint-Simon offer in this regard a few pages of rare interest concerning the small-pox he had contracted in Spain, and 'which filled the whole country'.[38] He found himself rapidly in 'great danger', even 'to the point of death'; even so, far from expressing anything he might have experienced as disagreeable in these circumstances, from his quarantine Saint-Simon delivers a narrative bathed in an exceptional serenity, without the slightest trace of anguish:[39]

> J'eus ... continuellement cinq ou six personnes auprès de moi, outre ceux de mes domestiques qui me servirent, un des plus sages et des meilleurs médecins de l'Europe, qui, de plus, était de très bonne compagnie ... J'eus une grande abondance partout de petite vérole de bon caractère, sans aucun accident dangereux depuis qu'elle eut paru .... Le premier médecin se précautionnait presque tous les jours de nouveaux remèdes, en cas de besoin, et ne m'en fit aucun que de me faire boire, pour toute boisson, de l'eau dans laquelle on jetait selon sa quantité des oranges avec leur peau, coupées en deux, qui frémissaient lentement devant mon feu, quelques rares cuillerées d'un cordial doux et agréable dans le fort de la suppuration, et, dans la suite, un peu de vin de Rota, avec des bouillons où il entrait du bœuf et une perdrix. Rien ne manqua donc aux soins de gens qui n'avaient que moi de malade, et qu'ils avaient ordre de ne pas quitter, et rien ne manqua à mon amusement quand je fus en état d'en prendre, par la bonne compagnie qui était

auprès de moi, et cela dans un temps où les convalescents de cette maladie en éprouvent tout l'ennui et le délaissement.⁴⁰

[I had ... continually five or six persons around me, besides those of my servants who served me, one of the best and wisest physicians of Europe, who was as well very good company. ... I had a great abundance of smallpox of good character everywhere, without any dangerous accident since it had appeared. ... The first doctor took precautions with new remedies almost every day, in case of need, but the only one he gave me was to make me drink only water into which was thrown oranges with their peel, according to the quantity, cut in two, that simmered slowly before my fire, a few rare spoonfuls of a sweet and agreeable cordial at the height of the suppuration, and, afterwards, a bit of Rota wine, with bouillon containing beef and a partridge. Thus nothing was missing from the care given by people who had only me who was ill, and they had orders not to leave, and nothing was missing from my amusements when I was in a state fit to have any, thanks to the good company that surrounded me, and this at a time when those convalescing from this disease feel all of its boredom and abandonment.]

This inhabited solitude, where neither sensual pleasures – sangria – nor the joys of the spirit – conversation with Higgins, first doctor of Philip V, whose personal friend Saint-Simon will remain – are excluded, is the ideal retreat of *otium*, the pause stolen from the whirlwind of life at court and from the servitude of protocol. In the interval between official functions, the ambassador rediscovers the precept of Epicurus who says that one must hide one's life. The suspension of social life becomes relaxation; the ferocity of political relations is replaced by the happiness of a select society – a 'rêverie du repos' (Bachelard), inadmissible perhaps, and that only illness can excuse.

Notes

1 This study was made possible by a grant from the Fonds québécois de recherche sur la société et la culture (FQRSC).
2 See the chapter in this volume by Isabelle Pantin.
3 R. Rey, 'Contagion ou "constitution épidémique"', in *Naissance et développement du vitalisme en France de la deuxième moitié du 18ᵉ siècle à la fin du Premier Empire* (Oxford: Voltaire Foundation, 2000), pp. 270–1. For the Hippocratic tradition, see for example the article 'Épidémie' in the *Encyclopédie*, signed by Aumont; *L'Encyclopédie ou dictionnaire raisonné des sciences, des arts et des métiers*, ed. Diderot and d'Alembert (Paris: Briasson, David, Le Breton, Durand; followed by Neufchâtel: S. Faulche, 1751–65), V, p. 788. As for contagionist theory, Ménuret de Chambaud, a specialist in skin diseases, serves as its spokesman in the same work; Article 'Miasme' in IV, p. 484.

4 F. de La Rochefoucauld, 'De l'origine des maladies', *Réflexions diverses* XII in *Oeuvres complètes*, ed. L. Martin-Chauffier (Paris: Gallimard, 1950), pp. 375–6.

5 Jaucourt, 'Peste', in *Encyclopédie*, ed. Diderot and Alembert, vol. XII, p. 454: 'l'essentiel est de ne point s'effrayer en tems de peste; la mort épargne ceux qui la méprisent, & poursuit ceux qui en ont peur; tous les habitans de Marseille ne périrent point de la peste, & la frayeur en fit périr davantage que la contagion';. Also in 1785, the doctor Jean-Gabriel Gallot recommends tranquility of spirit against an epidemic of infectious lung diseases in the Poitou region, identified at the time as catarrhal fevers (*Recueil d'Observations ou Mémoire sur l'Épidémie*, 1787). See J.-P. Peter, 'Le désordre contenu: attitudes médicales face à l'épidémie au Siècle des Lumières', *Ethnologie française*, XVII, 4 (1987) 359.

6 R. Favre, 'Vers une politique de la santé publique', in *La mort dans la littérature et la pensée françaises au siècle des Lumières* (Lyon: Presses universitaires de Lyon, 1978), p. 260.

7 *Mandement de Mgr l'Évêque de Marseille, sur la désolation qu'à causé la peste à Marseille, et sur l'établissement de la fête du Sacré-Cœur de Jésus*, 22 October 1720. D. Gordon records the celebration of 1 November 1720, during which Belsunce asked that God sacrifice him for the sins of his fellow citizens, and walked in the city, barefoot, a cord around his neck and wearing a cross, his eyes wet with tears. 'The City and the Plague in the Age of Enlightenment', *Yale French Studies*, 92 (1997) 83.

8 A thesis brilliantly upheld by Gordon, *ibid.*, who quotes a passage of *La contagion de la peste expliquée*, an anonymous text published in Marseilles in 1722 (pp. 81–2): 'Since commerce brings into a kingdom abundance and wealth, it ought to be sustained and favoured as much as possible. But sometimes it is also accompanied by the most distressing reversals, for people thereby communicate to each other their most pernicious diseases.' Ménuret de Chambaud, in his *Avis aux mères sur la petite vérole* (Lyon: Périsse, 1770), 'brosse le tableau d'une société de plus en plus communicable, où le commerce, les voyages, les besoins réciproques, le luxe tissent de multiples liens entre les hommes de toutes les nations, selon une dynamique irrésistible'. / '...paints the portrait of a society growing ever more communal, where commerce, voyages, reciprocal needs, luxury weave multiple links among men of all nations, according to an irresistible dynamic'. See Rey, 'Contagion', p. 293. Regarding the facts surrounding the epidemic, see Gordon, 'The City and the Plague', pp. 78ff.

9 J. Buvat, *Journal*, I, pp. 427–8, quoted by Y. Coirault in a note to his edition of the *Mémoires* of Saint-Simon, VII (Paris: Gallimard, 1988), p. 1484 [p. 700, n. 6]. Furetière, *Dictionnaire universel* (1690): 'Parfumer, se dit aussi en temps de peste, en parlant des soins qu'on prend de chasser le mauvais air des corps qu'on croit infectez, en excitant dans les lieux d'espaisses fumées de bois de genievre, de vinaigre, de poudre à canon & autres qui font de violentes impressions dans l'air, qui le chassent et le renouvellent.' / 'To perfume, said also in time of plague, speaks of the care taken to chase the bad air from bodies that are believed to be infected, by exciting in the environs thick smoke of juniper wood, of vinegar, of gunpowder & others that make violent impressions in the air, that chase it and renew it.' On the cleansing function of perfume, see G. Vigarello, *Le propre et le sale. L'hygiène du corps depuis le Moyen Âge* (Paris: Seuil, 1985), pp. 97–102.

10 Jaucourt, 'Vérole, petite', *Encyclopédie*, ed. Diderot and Alembert, vol. XVII, p. 79: 'elle ne cède rien à la peste par les désastres qu'elle cause.'

11 See, for example, the narrative of the death of Louis XV by Voltaire, *Précis du siècle de Louis XV* in *Œuvres historiques*, ed. R. Pomeau (Paris: Gallimard, 1957), p. 1556.

12 See Favre, 'Vers une politique de la santé publique', pp. 259–65.

13 F.-X. d'Entrecolles, s.j., 'Lettre du 11 mai 1726 au R. P. Du Halde', in *Lettres édifiantes et curieuses de Chine par des missionnaires jésuites, 1702–1776*, ed. I. and J.-L. Vissière (Paris: Garnier-Flammarion, 1979), pp. 330–41.

14 Favre, 'Vers une politique de la santé publique', p. 264.

15 The plague and smallpox are not the only contagious diseases accounted for in the works of the time – syphilis and measles, which had decimated Louis XIV's descendants in 1712, have a notable presence – but the frequency, rate of mortality and ease of diagnosis explain why we have limited ourselves to plague and smallpox in this study.

16 Cf. Stephanson, 'The Plague Narratives of Defoe and Camus: Illness as Metaphor', *Modern Language Quarterly*, XLVIII (September 1987) 235–6: 'Defoe is suggesting a kind of physical confinement whose claustrophobic implications are staggering. Defoe describes a people who; at the height of the plague's mastery, are paralyzed and imprisoned in a city whose activities have been negated and confined by the plague's menacing void.... [N]eighbors are confined to their homes, the sick are tied down to beds and chairs, and there is, finally, the ultimate confinement in the Aldgate pit or in a pine box.'

17 Gordon, 'The City and the Plague', p. 80.

18 J.-B. de Boyer d'Argens, *Mémoires de Monsieur le marquis d'Argens*, ed. Y. Coirault (1735; Paris: Desjonquères, 1993), pp. 32 and 52.

19 Saint-Simon, *Mémoires*, VIII, p. 346. Inversely, at the beginning of 1723, he remarks that 'la peste, qui avait si longtemps désolé la Provence, y fut tout à fait éteinte, et tellement que les barrières furent levées, le commerce rétabli, et les actions de grâce publiquement célébrées dans toutes les églises du Royaume, et au bout de peu de mois le commerce entièrement rouvert avec tous les pays étrangers'; *ibid.*, p. 561. 'the plague that had for so long devastated Provence, was completely extinguished, so much so that the barriers were lifted, commerce re-established, and thanksgiving publicly celebrated in all the churches of the Kingdom, and at the end of a few months commerce was entirely reopened with all foreign countries.'

20 P. Prion, *Pierre Prion, scribe. Mémoires d'un écrivain de campagne au XVIIIᵉ siècle*, ed. E. Leroy-Ladurie et O. Ranum (1744; Paris: Gallimard-Julliard, 1985), pp. 57–8.

21 Saint-Simon, *Mémoires*, VIII, p. 613: 'à Passy contre la petite vérole, dont Paris était plein. Elle ne l'évita pas, et en mourut.'

22 J.-N. Dufort de Cheverny, *Mémoires*, ed. J.-P. Guicciardi (1802; Paris: Perrin, 1990), p. 300: 'se privait d'aller dans les lieux publics, ou y allait avec répugnance'; 'la vue d'une personne encore rouge de cette maladie la troublait'.

23 *Ibid.*, pp. 305–6: 'ne pouvait rester un instant à Paris, ni voir [s]on fils, ni arborer personne sans lui causer d'effroi'; 'demander au roi la permission de [s]e présenter pour faire [s]on service'. Cf. Saint-Simon, *Mémoires*, VIII, p. 299, who discusses 'drugs' ('drogues') that people resorted to using in order to 'unredden' him after his smallpox, before he dared to present himself before the King of Spain.

24 N. Restif de la Bretonne, *Monsieur Nicolas*, ed. P. Testud (1794; Paris: Gallimard, 1989), I, p. 99; *Monsieur Nicolas or the Human Heart Unveiled*, trans. R. C. Mathers, ed. H. Ellis (London: John Rodker, 1930), p. 179.

25 P. A. F. Choderlos de Laclos, *Les liaisons dangereuses* in *Œuvres complètes*, ed. M. Allem (Paris: Gallimard, 1951), Letter CLXXV, p. 398; *Dangerous Acquaintances*, trans. R. Aldington (Norfolk, CT: J. Laughlin, 1952), p. 366.

26 Saint-Simon, *Mémoires*, v. IV, pp. 250-1: 'que rien n'avait pu adoucir ni redresser le moins du monde, perçait partout'.

27 Laclos, *Les liaisons dangereuses*, trans. p. 366; p. 398: 'la maladie l'avait retourn[é], et qu'à présent son âme était sur sa figure'.

28 Comte C. Forbin, *Mémoires*, ed. M. Cuénin (1729; Paris: Mercure de France, 1993), p. 177; *The Siamese Memoirs of Count Claude de Forbin, 1685–1688*, ed. M. Smithies [from the 1731 London trans.] (Chiang Mai: Silkworm Books, 1997), pp. 150–1.

29 *Ibid.*, trans. p. 151.

30 *Ibid.*, pp. 178–80; trans. pp. 151–2.

31 Cf. D. Defoe, *A Journal of the Plague Year*, ed. P. R. Backscheider (1722; New York and London: W. W. Norton, 1992), pp. 28–9 and *passim*: 'All the Plays and Interludes, which after the Manner of the *French* Court, had been set up, and began to encrease among us, were forbid to Act; the gaming Tables, publick dancing Rooms, and Music Houses which multiply'd ... were shut up and suppress'd; and the Jack-puddings, Merry-andrews, Puppet-shows, Rope-dancers, and such like doings ... shut up their Shops, finding indeed no Trade.'

32 A. d'Aubigné, *His Life, to His Children*, trans. and ed. J. Nothnagle (Lincoln, NB: University of Nebraska Press, 1989), pp. 8–9; *Sa vie à ses enfants*, ed. G. Schrenck (1629; Paris: STFM, 1986), p. 58: 'le premier se sentit de la contagion, qui fit mourir trente mille personnes. Il veit mourir son chirurgien et quatre autres en sa chambre.... Son serviteur nommé Eschalart, qui depuis est mort ministre en Bretagne, ne l'abandonna jamais, et sans prendre mal le servit jusques à la fin, ayant un pseaume en la bouche pour preservatif.'

33 See, beyond confessional divergences, the Jesuit Étienne Binet, *Remèdes souverains contre la peste et la mort soudaine*, ed. P.-L. Combet (1628; Grenoble: Jérôme Million, 1998), p. 45: 'Do you want the plague to kill the plague and that it should have no power over you, have a crystalline and very pure conscience, your innocence will crush the plague and this evil will have no power over you ?' ('Voulez-vous que la peste tue la peste et qu'elle n'ait aucune prise sur vous, ayez la conscience cristalline et bien pure, votre innocence fera crever la peste et fera que ce mal n'aura nulle prise sur vous ?').

34 F.-T. Choisy, *Mémoires pour servir à l'histoire de Louis XIV. Mémoires de l'abbé de Choisy habillé en femme*, ed. G. Mongrédien (1719; Paris: Mercure de France, 1966), pp. 168–9.

35 Restif de la Bretonne, *Monsieur Nicolas*, pp. 178–9; trans. pp. 98–9. See Dufort de Cheverny, *Mémoires*, p. 304 for a similar story of personal risk by loved ones.

36 Abbé A.-F. Prévost, *Histoire d'une Grecque moderne*, ed. R. Trousson (1740; Geneva: Slatkine, 1997), p. 255: 'achev[é] de [lui] faire connaître combien [il] étai[t] cher à l'aimable Théophé'.

37 *Ibid.*, pp. 255–6: 'et rien ne fut capable de refroidir un moment ses soins. Elle tomba malade elle-même. Mes instances, mes supplications, mes plaintes ne purent la faire consentir à se retirer. On lui dressa un lit dans mon antichambre, d'où toute la force de son mal ne l'empêcha point d'être continuellement attentive au mien.' Cf. Campion, *Mémoires*, ed. M. Fumaroli (1660; Paris: Mercure de France, 1967) pp. 87–8.

38 Saint-Simon, *Mémoires*, VIII, p. 89: 'dont tout le pays était rempli'.

39 Yves Coirault speaks of 'rayonnement intérieur' in the notes to his edition and evokes memories of Rousseau aux Charmettes. *Ibid.*, p. 765 [p. 90, n. 3].

40 *Ibid.*, pp. 89–90.

10
The Preaching Disease: Contagious Ecstasy in Eighteenth-Century Sweden

Daniel Lindmark

> The consequences of this enthusiasm may be less dangerous as long as the infection is restricted to certain individuals. But since it has really proved to be contagious, just like certain diseases, it is believed that entire crowds of people may be easily infected in the meanwhile, damaging the country and agitating the congregation.[1]

The opening quotation was written by Olof Celsius, president of the Stockholm Consistory. The Chancellor of Justice had consulted the Consistory concerning appropriate measures to be taken against religious enthusiasm, and at the meeting on 28 May 1776, Celsius read his draft reply to the Consistory, which approved the text without changes. The consultation was motivated by ecstatic religious movements in northern Sweden, but also the general presence of Moravianism, Swedenborgianism and free-thinking. The reply reflected the concern expressed by the Chancellor of Justice and proposed some measures to prevent further contagion; among others, stricter censorship and improved education of the clergy.

In the course of the eighteenth century, ecstatic revivals were increasingly perceived as pathological occurrences. The major reason for this pathologization will most likely be found in the physical phenomena characterizing ecstasy: convulsions, stiffness, trance, lost perception of touch, etc. Since such phenomena spread from one individual to the next when ecstatics came together, the supposed disease was often perceived as infectious. Consequently, Olof Celsius' opening image of the diffusion of religious enthusiasm was probably far more than a metaphor.

In this chapter, I will present some of the ideas on ecstatic religion as a contagious disease that could be found in the eighteenth century. The focus will primarily be on the contemporary discussion of the Wiklund Awakening in the Finnish-speaking area of northern Sweden in the 1770s, but the diagnoses and cures proposed by the Swedish authorities will be related to the international debate on religious enthusiasm that had been raging among theologians, philosophers and physicians since the middle of the seventeenth century. I will begin by depicting the major lines of the international debate on religious ecstasy as disease and contagion from a more general perspective of the history of medicine and ideas. When turning to the Swedish scene, I will complement the international perspectives with the more theological arguments put forward in the debate of the Wiklund Awakening.

A conflict between theological and medical explanations

Theologians arrived at different judgements of ecstatic religious phenomena. The diverging views were demonstrated in the major controversy about the *begeisterte Mägde* ('possessed maidens') that took place in Protestant Germany, in 1691–93.[2] The *begeisterte Mägde*, in the vicinity of Halle, claimed to be under divine inspiration during their ecstasy and made apocalyptic and chiliastic prophecies. Orthodox and Pietist theologians agreed on the women's emotional, irrational and passive status, but interpreted the phenomenon in different ways. Orthodox theologians held the view that the women created disorder and risked contaminating others with their rebellious spirit. Many Pietists maintained that the women displayed exemplary piety. Just like the English criticism of enthusiasm,[3] the German discussion focused on the relationship between the Spirit on the one hand, and the Word, the church and the tradition on the other. Through which means did the Holy Spirit act, and how was religious authority legitimated? The Orthodox theologians linked the *begeisterte Mägde* to a destructive contagion of considerable proportions. They saw historical links to the enthusiasts of the Reformation, and regarded the *begeisterte Mägde* as revolutionary as Thomas Müntzer and his likes. Misled and deceived victims of Pietism, the women spread the contagious poison to other people.

Even the leading Pietists themselves did not agree on which perspective should be applied to the *begeisterte Mägde*. While Philipp Jacob Spener never defended the women, August Hermann Francke represented a more positive view, even though he sceptically wrote that ecstatic phenomena might finally be considered an infection. To Francke, a faith manifested in good deeds was a more precious gift than sublime revelations. Admitting that revelations could be illusions of the devil, Francke did not condemn the *begeisterte Mägde*, since he had not discovered anything contradicting the divine Word. Publicly expressing his restrictive and ambivalent attitude, Francke was more wholeheartedly positive in private. This view was shared by many Pietist theologians, who maintained that the emotional, spiritual and physical conditions of the women made them pious examples as well as ideal instruments of God's revelations. According to the representatives of Orthodoxy, the emotional, irrational and passive status of the women made them vulnerable victims of Pietistic poison.[4]

Explaining the ecstatic phenomena was an important bone of contention. The physician Friedrich Hoffmann examined three of the visionaries and interpreted the ecstatic behaviour within the framework of his mechanistic physiological system, which dominated medical thinking at the beginning of the eighteenth century. In Hoffmann's system, invisible spirits served the function of links connecting the soul and body. Still dependent on the Galenic physiology with its *spiritus vitalis*, the vital spirit of the heart and the arteries, *spiritus naturalis*, located in the liver and the veins, and *spiritus animalis*, in the brain and the nerves, Hoffmann expressed the opinion that God as the prime mover imprinted his ideas and will in the metaphysical spirit of man, which was then linked to the body through partly physical and invisible spirits. His detailed examination revealed that he also took into account the classical humoral pathology, even though he denied the prevailing connection between ecstasy and melancholy. Ever since classical Antiquity, proponents of humoral pathology had maintained that enthusiasm was linked to a surplus of black bile: it was simply a form of melancholy. In the more mechanistic views of man that emerged in the seventeenth century, melancholy was explained with reference to animal spirits.[5] Hoffmann compared ecstatic convulsions with tetanus, which was seen to be caused by an agitation of the animal spirits emanating from exaggerated fixations or emotions. He excluded all physical and psychological causes, and in one of the cases he was convinced that ecstatic reactions were caused by reading God's powerful Word, and that God himself had interrupted the connection between body and soul. In another case that he examined later, Hoffmann initiated a more fundamental discussion of the origin of the ecstatic phenomena. Referring to biblical evidence, Hoffmann refuted the idea of divine inspiration: foul curses had been used, Christian people had been defamed, and prophecies had failed. Since evil spirits were the only remaining alternative, Hoffmann developed a sophisticated distinction between demonic and fanatic possession. Even though a demonic spirit only sporadically made itself visible in a person, it was nevertheless permanently residing in a possessed person. A fanatic, on the other hand, could be driven to an ecstatic condition by certain external circumstances. The pro-Pietist Hoffmann defined the woman in question as a fanatic, since her ecstatic condition often occurred in church.[6]

While the physician Hoffmann held the view that ecstatic phenomena had supernatural causes and gave them a theological explanation, the theologian Spener employed a medical interpretation. Discussing various alternatives like self-deception, devilish illusions, emotions in the dark deep of the human soul and divine inspiration, Spener totally dismissed the last possibility. However, he never accused the *begeisterte Mägde* of being in demonic possession, which in the Lutheran theological tradition would have been the only alternative to divine inspiration. Instead, Spener stated that the speeches delivered by the women derived from an agitated imagination. Even the theological faculty at the University of Helmstedt discussed natural causes as possible alternatives to supernatural factors. Even though the theologians were inclined to favour a supernatural explanation by blaming evil spirits for causing confusion and heresy, they submitted the question to be decided by medical expertise.[7]

Enthusiasm as disease and contagion

In the international discussion, religious enthusiasm was often considered a disease, at times even a contagious one. In 1733, the French physician Philippe Hecquet published a pamphlet on the Jansenist ecstatics of Paris who suffered from *les maladies de l'épidémie convulsionnaire*.[8] According to Hecquet, the convulsions were by no means miraculous phenomena, but rather 'hysterical vapours, perhaps truly uterine in nature, [but] caused and sustained by passions of the soul'.[9] Hecquet maintained that the disease was caused by agitation of the capacity of imagination, which was easier to heat among women. The wish to become an instrument of God's miracles through convulsions was reason enough to agitate the passions, in combination with the support lent by the followers. Hecquet had no difficulty in explaining the connection between mental emotions and physical motions. If the simple will could send the animal spirits simultaneously to 'three hundred different muscles when a guitar player touches his instrument while crooning and tapping his feet', the ardent desire of becoming a convulsionary could just as easily agitate a huge amount of muscles in an ecstatic young girl's body.[10]

When it came to cures, Hecquet recommended quarantine. Isolated from their supporters – the defence offered by certain priests, the worship of the followers, and the attention of the public audience – the ecstatics would be deprived of their most effective means of agitation. As soon as the young women had been placed under medical care, they would consequently lose the social support for their behaviour.[11] Hecquet held the same view as many of his colleagues among English physicians: religious enthusiasts should not be treated by theologians, but by medical experts, since enthusiasm is a physiological phenomenon.[12] By joining Earl Shaftesbury in his recommendation to publicly ridicule the convulsionaries in order to make them abandon their ecstatic practice (see below), Hecquet underlined his view of the convulsions as a psychological phenomenon. Thus, public ridicule would be a more appropriate and effective cure than repression.[13]

The German philosopher Christoph Martin Wieland also held the view that religious ecstasy was a disease, 'an actual soul fever' striking individuals but also taking the form of a contagious epidemic.[14] There was no reason to moralize over the enthusiasts, since their condition was no more serious than an ordinary fever. Just like Hecquet, Wieland recommended quarantine for the ecstatics, at least partly to prevent further contagion. By pathologizing the enthusiasts, Wieland dissociated himself from the Lutheran tradition, according to which they were nothing but villains. Instead, the enthusiasts were viewed as innocent victims. Ironically enough, this new understanding did not necessarily lead to more respectful treatment. If the enthusiasts suffered from a disease, there was no need to take their thoughts or speeches seriously. It was useless entering into a discussion with an enthusiast; he neither could nor wanted to listen to reason.[15]

But how were the nature and function of contagion explained? Philippe Hecquet was the first to present a theory of moral contagion. In his pamphlet of 1733, Hecquet drew an analogy between contagions, convulsions and 'the so natural contagion between the two sexes'.[16] Corpuscles continuously escaped in the form of vapour through the pores of the skin, creating an atmosphere around every individual. The corpuscles were understood to be small particles that carried all the characteristics of the individual. When two individuals came close to each other, their atmospheres were mixed, and they inhaled each other's corpuscles. When two people fell in love, the woman's corpuscles transmitted into the male body the female 'inclinations to gentle sweetness', thereby arousing the man's heart and imagination 'in a girlish manner'. In a similar manner, the corpuscles from an ecstatic transmitted the inclination for convulsions to the people nearby.

A couple of decades earlier, similar theories had been presented to explain the ecstatic behaviour of the French Prophets in London.[17] George Keith sought the explanation in the emanation of 'subtle little particles of Bodies of different figures and shapes, with various differing Motions that go from Bodies to Bodies'.[18] Shaftesbury also addressed the contagious behaviour in his influential work *Letter Concerning Enthusiasm* from 1709. He held the view that enthusiasm and panic were related phenomena, which made enthusiasm a very contagious disease which attacked the faculty of imagination. The malady was spread by bad air – a classical explanation for epidemics[19] – and heavy breathing and sweating increased the contagious effect.

Another theory was posited in the eighteenth century to explain moral contagion. In 1784, animal magnetism, represented by Franz Anton Mesmer, was evaluated by a French royal commission, which employed a mechanistic explanation to the practice of mesmerism. In contrast to Hecquet, another proponent of a mechanistic view of man, the commission linked its explanation to the supposed function of the nerves. The nerves could be exhausted by transmitting all the sensations; they became irritated and convulsions could occur among the most sensitive persons. When a convulsionary was close to others, whose nerves were equally strained, the vibrations could be transmitted from one person to the other, just like between two guitars.[20]

Regardless of the explanation, most attempts at explaining ecstatic convulsions and moral contagion seem to have focused on the faculty of imagination, which was considered to be agitated or disturbed. By doing so, they adjusted to a dominating tradition in Western Christendom, according to which true prophecy was linked to reason, while false pretensions were placed in the lower faculty of imagination.[21] Throughout history, critics of religious ecstasy have made a distinction between reason and emotion, thereby defining ecstasy as uncontrolled passion in contrast to behaviour directed by reason. This distinction coincided with the common view of the differences between male and female and should reflect the fact that women dominated many of the ecstatic movements.[22]

The Wiklund Awakening

Nils Wiklund was born in 1732 in a small parish in northern Sweden.[23] As a 28-year-old student at Uppsala University, he converted to Moravianism. Ordained in Stockholm in 1765, Wiklund served as a curate in the parish of Övertorneå for ten years. In 1773, a religious revival began in this Finnish-speaking parish, and Wiklund was perceived as its leader. Accused by his colleagues of creating disorder and undermining the authority of the ministry, Wiklund was repeatedly prosecuted from 1774. After two years of trials, he was deprived of his ministry, but in 1779 he was rehabilitated. He died as an assistant vicar in 1785.

When the revival started, two girls, 11 and 14 years of age, began to preach, warning and admonishing their listeners, urging them to repentance and conversion. Soon enough, the preaching spread to more young people. It often took place in private homes, but could also occur during services in church. On such occasions, the ministers would pause to wait for the so-called 'speakers' to finish. They did not want to challenge those who considered the speakers to be under the influence of the Holy Spirit. The preaching was often performed under convulsive ecstasy, sometimes preceded by stages of trance and visions, many of which presented the pain and agony of condemned sinners suffering eternal punishment.

After repeated complaints by a sceptical colleague of Wiklund's, the Härnösand Consistory initiated an investigation. A clerical commission was appointed to examine the 'preaching disease'. This commission gathered in Övertorneå in December 1775 to interrogate the ministers and speakers alike. The visitation by the commission resulted in serious admonitions. But since the preaching took place under convulsive ecstasy, a medical examination was recommended. During the visitation by the commission, three prominent figures were arrested, two of them being sent to Stockholm to be examined by the National Medical Board.

In January 1776, the County Governor ordered a more thorough investigation. The provincial medical officer was sent to Övertorneå, where in February comprehensive examinations took place. Dr Johannes Grysselius delivered three consecutive reports to the National Medical Board. These include 62 case records referring to 12 male and 50 female preachers. No one was older than 36 years; 49 preachers were younger than 25; and 9 individuals were under the age of 15. Case No. 7 described the symptoms of a 21-year-old Saami woman, Ella Jonasdotter, from the village of Alkulla. Ella was born in Övertorneå in 1755, and during her stay in Nedertorneå had started to preach repentance and conversion. Being weak in health, she had led a quiet life since 1774.[24]

Ella was identified as one of the leading preachers of the Wiklund Awakening, and because of her position she was among those arrested and sent to Stockholm for further examination. Due to her weak health and a letter of recommendation from Vicar Isak Grape in Övertorneå, the county authorities decided to keep her in Umeå. (Grape, however, had meant to convey that Ella was a pious and experienced Christian person of comprehensive knowledge.) Among the charges directed against Ella, the visionary experience was hardly an issue. Instead, she was accused of declaring forgiveness of sins to different persons, and having defined certain persons as blessed and others as condemned. According to official doctrine, this spiritual jurisdiction was restricted to the ordained ministry. Wiklund defended the judgements of the speakers by referring to the Lutheran doctrine concerning universal priesthood, but the authorities could not find such jurisdiction legitimate. Obviously, Ella's popularity had made the problem even worse. It was reported that her adherents were so devoted to her that they had built her a cottage. Since Ella was charged with being an instigator of rebellion, her large following might have contributed to this allegation. Most probably, the Pietistic distinction between true and nominal Christians resulted in the reported polarization between 'two hostile armies'.[25] According to Dr Grysselius, the disruption in Övertorneå was caused by the local ministry's application of the distinction between God's children and unconverted people. The ministers were reported not to allow true Christians to socialize with the infidels.[26] Besides the specific interpretations of universal priesthood and religious conversion, the Härnösand Consistory also identified Pietistic influence in the ministers' preaching. Vicar Grape and Curate Wiklund were reported to have taught that man could restore his righteousness through sincere repentance instead of trusting in God's grace and the merit of Christ. Furthermore, the strict preaching of the Law was supposed to have driven the parishioners to melancholy.[27] Here the classic linkage between enthusiasm and melancholy is indicated.

Judd Stitziel's analysis of the Orthodox ministers' attitudes towards the *begeisterte Mägde* in the vicinity of Halle in the 1690s focuses on the ecstatic visionaries as the antithesis of the Orthodox ministry: uneducated, illiterate, female, of the underclass. By referring to the Lutheran doctrine of the universal priesthood, the *begeisterte Mägde* challenged the well-educated masculine hierarchy. Sometimes this confrontation was articulated in a general discourse of social order, where the women were accused of creating anarchy. Sometimes it was discussed in theological terms, where the Pietist interpretation of universal priesthood and divine inspiration was questioned.[28] By claiming direct communication with God, the *begeisterte Mägde* challenged the ministry's authority, which was indirect and founded on their biblical interpretations formalized in sermons and catechisms. It was simply the power over the Word that was at stake, the right to preach and teach religion in public. There is reason to assume that a similar interpretation could be applied to the Wiklund Awakening, another ecstatic and visionary movement dominated by women.[29]

District Physician Grysselius and the Wiklund Awakening

The most conspicuous feature of the authorities' treatment of the Wiklund Awakening is no doubt the strategy of pathologization. On the regional and central levels alike, the representatives of medical science were employed to examine the phenomenon. The external, physical expressions were described in detail, and Dr Johannes Grysselius reported on 62 individual cases to the National Medical Board. In his conclusions Grysselius suggested a diagnosis that focused on *Mania hypochondriaca*, and recommended a variety of medicaments, including a water cure.[30] First and foremost, Grysselius held the view that the treatment should take place in a Swedish-speaking parish. There were several reasons for the recommended transfer. In the parish of Övertorneå, the speakers were supported by the local ministry and even considered instruments of conversion. When interrupted by speakers during service in church, Curate Wiklund had stated that 'if the people did not want to believe him, they would believe them'.[31] Grysselius meant that Wiklund had forgotten the answer the rich man received when asking for a miracle to persuade his brothers: 'They have Moses and the prophets; let them hear them' (Luke 16: 29, King James Version). Furthermore, Grysselius pointed to the fact that the speakers had many supporters in their neighbourhood. Lots of people considered the speakers blessed by God, since God could choose such plain spokesmen. For that reason, the supporters complied with their admonitions, in itself a good and Christian deed, Grysselius admitted. But among the supporters were many who wanted to become speakers themselves, hoping to get closer to the Kingdom of Heaven by the divine selection that was manifested in the calling to preach repentance and conversion. Grysselius also feared that the support from the neighbourhood and especially the ministry would create an attitude of security and arrogance among the speakers.[32]

Consequently, a transfer to a Swedish-speaking parish would not only deprive the speakers of their support, but the language barrier would also raise obstacles to the recruitment of new adherents. The isolation in a foreign parish would therefore create conditions for a successful medical treatment, especially if the speakers were separated from each other. Grysselius recommended the adjacent Nederkalix parish, since there was a spring; furthermore not many inhabitants would be able to understand the Finnish-speaking preachers, 'who consequently would not be able to contaminate the people there'.[33]

Obviously, Dr Grysselius was of the opinion that the ecstatic visionaries of the Wiklund Awakening were suffering from a disease, even though he also applied theological perspectives to the actions of the speakers and their supporters. Before leaving for Övertorneå, Grysselius wrote a letter to the National Medical Board, where he defined his mission. Using the wordings of the order issued by the County Administration the day before, Grysselius set out to 'investigate the nature of the spasmodic disease with speaking'.[34] In his case records, Grysselius consistently referred to the investigated phenomenon as the 'disease', alternatively the 'fainting', the 'paroxysm' and the 'attacks'.[35] Epilepsy was suggested in a couple of cases, but usually the diagnoses were not that specific. When writing about two men who were prosecuted for having interrupted religious services, Grysselius implied that they were 'affected by the same disease as the rest of the people preaching under fainting'.[36] The assistant vicar of Pajala, Anders Wichman, had accused the two men of being possessed by the devil, and the defence had for the same reason argued for the men's innocence. On account of Grysselius' report, the men were sentenced to confinement in an asylum in Stockholm, a verdict that was never realized. However, this example demonstrates how medical explanations competed with theological perspectives in the contemporary discussion of the Wiklund Awakening.[37]

Without referring to his French colleague, Grysselius placed himself quite close to Philippe Hecquet when prescribing the cures. Hecquet and Grysselius alike were aware of the significance of the support lent by the ministers, the adherents and the public, and recommended isolation of the ecstatic visionaries. Instead of providing international references, Grysselius based his prescriptions on observations of his own patients. He noted that those speakers who had been treated strictly had quit preaching. Ever since the 19-year-old Helena Trast in the village of Turtula had been forbidden by her father to see and listen to other speakers, her own preaching had ceased.

Grysselius related the case of 20-year-old Eva Hindrichsdotter Nicki from the village of Ruskala in greater detail. Most probably, he found her case illustrative with regard to the alleged factors behind the ecstatic phenomena. Eva's father told Grysselius at the examination that he had noticed that the attacks occurred more frequently when she met other people in the village. Therefore he had forbidden her to leave the home, and consequently her attacks had completely disappeared. Eva confirmed her father's view by saying that she had been struck by the attacks, *nolens et volens*, whenever she was in company, but when alone she had been free from the paroxysm. Her father added that the attacks were the more violent the more people were present. He also told Grysselius that the speakers used to condemn all the people who did not accept their message. During their paroxysms, the speakers had visions of the condemned people in hell, while their followers were said to be seen in heaven, wearing crowns. Eva's father feared that easily agitated people could be brought to brink of despair by such condemnations. According to the father, Curate Wiklund had been most dissatisfied with the restrictions placed on Eva's freedom of movement, and wanted her to be able to meet 'converted people and the children of God'.[38]

In his account of the Eva Hindrichsdotter Nicki case, Grysselius clearly demonstrated his view of the symptoms displayed by the speaking people. The ecstatic preaching worked only with the cooperation of the immediate neighbourhood. He indicated that the local ministry not only supported the movement, but was well aware of the significance of its social dimension. The example provides eloquent support to Grysselius' thesis that the symptoms would cease if the speakers were deprived of the support of their followers and isolated from each other. Again, it is worth noting that District Physician Grysselius employed theological arguments, although with a psychological twist. Having pointed to the risk of the speakers falling into a state of certainty and arrogance, he now expressed his concern about the unconverted being brought to despair.

However, the theologians were more restrained in their use of medical arguments, except for the combative assistant vicar of Pajala, Anders Wichman, who employed every possible strategy to fight the Wiklund Awakening, including demonization and pathologization. Eager to discredit the movement, Wichman at the same time wanted to depict the ecstatics as deliberate swindlers. Vicar Isak Grape found these lines of argument contradictory: having stated that his parish had been contaminated, and having signified the speaking as a disease, Wichman later questioned those who wanted to treat the speakers as sick people.[39] Wichman could not easily accept that the defence had changed its strategy by claiming that the speakers could not be sentenced for actions they had performed under the influence of their disease. Evidently, Wichman wanted the speakers to be fully responsible for all their actions instead of treating them as innocent victims. According to Wichman, the speakers claimed to have the power to start and stop their preaching, a fact that was inconsistent with Wichman's concept of a disease, which could not be 'caught or thrown away' on a whim. The fact that strict isolation had made the preaching cease pointed in the same direction: 'A disease cannot be expelled by violent blows.' Instead, the speakers themselves and the evil spirit were responsible for the extraordinary phenomena.[40]

The National Medical Board and the Wiklund Awakening

The National Medical Board engaged their experts to evaluate Grysselius' reports and examine the two preachers, Henric Buskas and Anna Greta Samuelsdotter, who had been brought to Stockholm and were being held in custody in the city jail. Having consulted Mr Avellan, pastor of the Finnish congregation in Stockholm, and his brother, Dr Avellan, who had shared the responsibility for the prisoners' spiritual and medical welfare, City Physician Svensson and Assessor Odhelius made the following diagnosis: 'the sermons emanate from a melancholy temperament, which is shaken and frightened by the local ministry's disputes about the order of salvation on the one hand, and declamatory intimidations on the other.'[41] Developing the latter point, the medical experts indirectly blamed the Pietistic preaching of Nils Wiklund. Preoccupied with reflections on eternal condemnation, the parishioners had been encouraged by those inflammatory sermons. The experts suggested the preachers be separated from their 'seducers' and kept busy working. Furthermore, they recommended isolation, sedative medication and mineral water as a cure for their 'nervous disease'. The physicians referred to the fact that 'sensible remonstrances' had proved efficacious in so far as they had kept the symptoms away, but as soon as one had started preaching, it had immediately infected the other.[42]

Having examined Grysselius' first two reports, including 31 case records and the proposed diagnosis, Professor Roland Martin delivered his opinion. Basically, he maintained that the preachers were suffering from an imaginary disease. Referring to Friedrich Hoffmann, Martin maintained that the movements of the human body could be affected by certain ideas and mental activities. He suggested that exaggerated spiritual exercises, probably encouraged by the ministers, had influenced the lower mental faculties, especially the faculty of imagination. Through an alleged 'contagion of the imagination', the symptoms had passed from one person to the other.[43]

Referring to experiences from examinations of magic and melancholy, Martin stated that in regions characterized by 'heavy and cold air', where the food was coarse and hard to digest, and the drink was bad and cold, etc., people were particularly prone to imaginary diseases. In the hypochondriac state, individuals could develop such a fixation on certain ideas that they either became insomniac or fell into a deep trance, followed by fainting, convulsions and unconscious speaking. According to Martin, such symptoms were usually found among Nordic nations, especially Finland, Lapland and the northern parts of Sweden. This climatic explanation was further vindicated with references to the healthy conditions under which the population in Italy, France and the southern parts of Germany were living. Cultivated senses, civic education, regular and useful work, decent and happy company, more vegetables and a moderate use of natural wines created peoples who were not carried away by their imagination. In order to create 'happy, honest, healthy, and industrious subjects', Martin proposed the removal of the ministers whose 'inconvenient zeal' had contributed to the epidemic, unless they could be ordered to make their preaching more suitable for 'the temperament of a feeble, melancholy, or credulous people'.[44]

On the basis of the expert reports, the National Medical Board resolved to recommend to the King the separation of the ministers from the ecstatic speakers and make arrangements for improved education. Finally, the Board suggested that the affected people be treated by the provincial physician. The two imprisoned ecstatics were released and returned to the parish of Övertorneå. Despite the recommendation issued by the National Medical Board, none of their fellow parishioners seem to have been subjected to medical treatment. The separation of the ministers from the speakers was the only recommendation that was adopted. Curate Nils Wiklund was suspended for three years, and when he was rehabilitated in 1779, the ecstatic phenomena had ceased.

Awakening between theology and medicine

When combining medical and theological arguments, the Swedish discussion connected to its international counterparts. Both theologians and physicians felt free to cross the borders of their respective disciplines. Actually, the discussion of ecstatic phenomena became an ideological battlefield where theologians were challenged by a profession with increasing status, the medical experts, whose engagement in the debate can be seen as promoting their profession. From a certain point of view, the changing interpretations of ecstatic behaviour were part of a general process of secularization. Physical and secular aspects gained ground in areas previously dominated by spiritual and religious explanatory models. The gradual withdrawal of the theologians from the field of ecstasy was probably not undertaken totally reluctantly. The medical interpretations could serve the same purpose of distinguishing between right and wrong behaviour. The medical diagnosis of 'agitated imagination' was certainly an equally efficient strategy for discrediting the ecstatics as the theologically founded demonization and heretization. From the perspective of state control, the medical profession offered a new instrument for defining and treating people who expressed their religiosity in ways that society found inappropriate. Even more, the strategy of pathologization seems to have been more definitive and efficient than the theological definitions. Since theologians were inclined to respond to the ecstatics' claims by developing their theological arguments, the definition of ecstasy was a matter of theological interpretation, depending on the strength of the arguments. When pathologized by the physicians, the ecstatics were dismissed as partners in the discussion. Ideas put forward by insane people with 'stirred imaginations' could not be taken seriously.

In Sweden the pathologizing of ecstasy started in the eighteenth century, but ever since then medical expertise has been used to mark the demarcation between sound and pathological religious behaviour. Even though the technique of quarantine was motivated by concern for mentally disturbed fanatics, there is reason to believe that often it was resorted to in order to prevent further contagion, regardless of how contagion was supposed to work. Ecstatic revivals were seen as serious disturbances of societal order calling for actions from the authorities. It was not only a matter of restoring peace, creating harmony and keeping people busy working, but also of preventing further dissemination of the destructive contagion.

Notes

This chapter presents results from the research project *Domestication of Religion: Popular Religiosity in the Northern Space, 1600–1800*, funded by the Bank of Sweden Tercentenary Foundation and directed by Daniel Lindmark.

1 Minutes of the Stockholm Consistory, 28 May 1776 (No. 253). Transcript in the Archives of Bertil and C. J. E. Hasselberg. Private Archives C42, Vol. 9. The Regional State Archives of Härnösand, Sweden.

2 J. Stitziel, 'God, the Devil, Medicine, and the Word: A Controversy over Ecstatic Women in Protestant Middle Germany 1691–1693', *Central European History* 29, 3 (1996) 309–37.

3 M. Heyd, 'The Reaction to Enthusiasm in the Seventeenth Century: Towards an Integrative Approach', *Journal of Modern History*, 53, 2 (1981) 258–80.

4 Stitziel, 'God, the Devil, Medicine, and the Word', pp. 319–23.

5 J. Goldstein, 'Enthusiasm or Imagination? Eighteenth-Century Smear Words in Comparative National Context', in *Enthusiasm and Enlightenment*, ed. L. E. Klein and A. J. La Vopa (San Marino, CA: Huntington Library, 1998), pp. 29–49.

6 Stitziel, 'God, the Devil, Medicine, and the Word', pp. 326–31.

7 *Ibid.*, pp. 332–3.

8 See B. R. Kreiser, *Miracles, Convulsions, and Ecclesiastical Politics in Early Eighteenth-Century Paris* (Princeton, NJ: Princeton University Press, 1978) and C.-L. Maire, *Les convulsionnaires de Saint-Médard: Miracles, convulsions et prophéties à Paris au XVIIIᵉ siècle* (Paris: Gallimard/Julliard, 1985).

9 Ph. Hecquet, *Le Naturalisme des convulsions dans les maladies de l'épidémie convulsionnaire* (Soleure, 1733), pp. 12–13, quoted from Goldstein, 'Enthusiam or Imagination', p. 39.

10 Goldstein, *ibid.*, pp. 39–40.

11 *Ibid.*, pp. 42–3.

12 *Ibid.*, p. 41.

13 *Ibid.*, p. 45.

14 A. J. La Vopa, 'The Philosopher and the *Schwärmer*: On the Career of a German Epithet from Luther to Kant', in *Enthusiasm and Enlightenment*, ed. L. E. Klein and A. J. La Vopa (San Marino, CA: Huntington Library, 1998), p. 87.

15 *Ibid.*, p. 90.

16 J. Goldstein, '"Moral Contagion": A Professional Ideology of Medicine and Psychiatry in Eighteenth- and Nineteenth-Century France', in *Professions and the French State, 1700–1900*, ed. G. L. Geison (Philadelphia: University of Pennsylvania Press, 1984), p. 187.

17 See H. Schwartz, *Knaves, Fools, Madmen, and that Subtle Effluvium: A Study of the Opposition to the French Prophets in England, 1706–1710* (Gainesville: The University Presses of Florida, 1978).

18 M. Heyd, *'Be Sober and Reasonable': The Critique of Enthusiasm in the Seventeenth and Early Eighteenth Centuries* (Leiden, New York and Cologne: E. J. Brill, 1995), p. 201.

19 See P. Baldwin, *Contagion and the State in Europe, 1830–1930* (Cambridge: Cambridge University Press, 1999), pp. 1–10; and M. Pelling, 'The Meaning of Contagion: Reproduction, Medicine and Metaphor', in *Contagion: Historical and Cultural Studies*, ed. A. Bashford and C. Hooker (London and New York: Routledge, 2001), pp. 15–38.

20 Goldstein, '"Moral Contagion"', p. 189.

21 Heyd, *'Be Sober and Reasonable'*, pp. 202–3; cf. A. C. Fix, *Prophecy and Reason: The Dutch Collegiants in the Early Enlightenment* (Princeton, NJ: Princeton University Press, 1991).

22 D. Lindmark, 'Defining Ecstatic Religion: Female Preachers in 18th-Century Sweden', in *Gender and Generation*, ed. M. A. Mackay (Edinburgh, in press).

23 The following description of the Wiklund Awakening is based on C. J. E. Hasselberg, *Norrländskt fromhetslif på sjuttonhundra-talet* (Örnsköldsvik, 1919), pp. 239–475, and *idem.*, *Under Polstjärnan: Tornedalen och dess kyrkliga historia* (Uppsala: Lindblads, 1935).

24 Hasselberg, *Norrländskt fromhetslif på sjuttonhundra-talet*, p. 327; on her first vision, see D. Lindmark, 'Vision, Ecstasy, and Prophecy: Approaches to Popular Religion in Early Modern Sweden', *ARV: Nordic Yearbook of Folklore* 59 (2003) 177–98.

25 Hasselberg, *Norrländskt fromhetslif på sjuttonhundra-talet*, p. 326.

26 Grysselius' report to the National Medical Board, 26 February 1776. Collegium Medicum [CM], E2:32, 47. National Archives [NA], Stockholm, Sweden.

27 Hasselberg, *Norrländskt fromhetslif på sjuttonhundra-talet*, p. 322.

28 Stitziel, 'God, the Devil, Medicine, and the Word', pp. 334–7; see C. Nordbäck, *Samvetets röst: Om kampen mellan luthersk ortodoxi och konservativ pietism i 1720-talets Sverige* (Umeå: Umeå University, 2004 for similar Swedish examples.

29 An interpretation of female majority in early modern ecstatic movements can be found in Lindmark, 'Defining Ecstatic Religion'. The essentialist viewpoint that ecstasy would be linked to an alleged female sentiment is rejected in favour of a perspective of power. In a society where masculine hierarchies had monopolized official religious offices, religious authority could be obtained by women by claiming direct contact with divinity.

30 *Mania Hypochondriaca periodica, erratica data occasione recurrens, Theologica Locutoria convulsiva vel spasmodica, cum et sine sensatione sensibilitate.* Grysselius' report to the National Medical Board, 26 February 1776. CM, E2:32, 40–8. NA.

31 Grysselius' report to the National Medical Board, 26 February 1776. CM, E2:32, 41. NA.

32 Grysselius' report to the National Medical Board, 26 February 1776. CM, E2:32, 42. NA.

33 Grysselius' report to the National Medical Board, 26 February 1776. CM, E2:32, 44–6; cf. p. 89. NA.

34 Grysselius' report to the National Medical Board, 17 January 1776. CM, E2:32, 72b. NA.

35 Grysselius' report to the National Medical Board, 7 February 1776. CM, E2:32, 3–5. NA.

36 Grysselius' report to the National Medical Board, 21 March 1776. CM, E2:32, 85. NA.

37 The Wiklund Awakening was subjected to several strategies of condemnation, among others pathologization, demonization and heretization. By using typical Pietistic attributes in their description of the Wiklund Awakening, religious authorities employed the strategy of heretization. See Lindmark, 'Vision, Ecstasy, and Prophecy'.

38 Grysselius' report to the National Medical Board, 26 February 1776. CM, E2:32, 37–40. NA.

39 Isak Grape's undated reply to Anders Wichman's written allegations of 26 February 1776. Acta Ecclesiastica 113, 13:2. NA.

40 Anders Wichman's written allegations of 26 February 1776. Acta Ecclesiastica 113, 13:2. NA.

41 Minutes of the National Medical Board, 1 February 1776. CM, A1 A:28. NA.

42 *Ibid.*

43 Minutes of the National Medical Board, 18 March 1776. CM, A1 A:28. NA.

44 *Ibid.*

Part III
Projections

11

A Contagion at the Source of Discourse on Sexualities: Syphilis during the French Renaissance

Guy Poirier

Syphilis appeared in the medical, historical and literary discourse of the French Renaissance as suddenly as the disease itself. During the sixteenth century, discourse on syphilis took on various, colourful and carnivalesque guises, before finding discipline and continuing its odyssey with moral commentary. One must await the libertines and the beginning of the seventeenth century before syphilis wears another mask, that of the baroque image and of 'satyre'.

The first discourses on syphilis; syphilis and medical discourse

It was at the Battle of Fornoue in July 1495, hardly two months after the French retreat from Naples, that Italian doctors noticed the first symptoms of the pox among the soldiers of Charles VIII. The sexual origin of the disease was rapidly established, the first pustules appearing, in a man, under the penis.[1]

One of the first literary works in French to make reference to this new disease is a ballad by Jean Molinet published in the autumn of 1496. In the 'Ballade de la Maladie de Naple', the poet summarizes the incessant suffering provoked by the virulent infection: 'Day and night, is destroyed my rest, / You seduced me through joy and jest.'[2] As for the origin of the contagion, at the end of each verse the lyrical subject censures the person who introduced the disease into France and, more specifically, the woman who transmitted the disease to him:

> Mauldit soit la lubrique
> Fille publique a qui ce mal s'applique,
> Et la bouctique qui tel drogue produit,
> Et le conduit qui m'a a ce con duit,
> Dont le deduit me tourne a desplaisance.[3]

157

[Damned be the lustful
Harlot whose cunt with clap is so damned full,
And the shops producing such a liquor,
And the shaft that drove me to stick her,
Whose amorous game turns to displeasure.]

Molinet also highlights the opprobrium which was then associated with the appearance of symptoms in an individual (pimples, ulcers, and so on), and thus confirms the hypotheses of historians regarding the ostracism of the first patients, ejected from certain hospitals and other care institutions, and even forced into leper hospitals: 'I am loathsome, all run from me ...'[4] Indeed, an echo of this first fear of contagion appears in Erasmus' colloquium on inns. William, one of the interlocutors, protests against the danger of gathering too many people in a restricted space. It is, in this particular case, a criticism of German inns:

But nothing seems to be more dangerous than making for so many persons to breathe the same warm air, especially when their bodies are relaxed and they've eaten together and stayed in the same place a good many hours. Quite apart from the belching of garlic, the breaking of wind, the stinking breaths, many persons suffer from hidden diseasees, and every disease is contagious. Undoubtedly, many have the Spanish itch or, as some call it, French pox, though it's common to all countries. In my opinion, there's almost as much danger from these men as from lepers. Just imagine, now, how great the risk of the plague![5]

Leprosy, plague, syphilis, all these diseases seem to share the same fear of human gatherings. If it is probably too early to speak of a modern conception of hygiene, one could nevertheless reflect on what Erasmus means by 'hidden diseases'. We are still far from the modern conception of 'shameful diseases', and here we should speak instead of diseases that one conceals, deliberately, in order to avoid exclusion. A few lines further on, however, Erasmus' traveller does not hesitate to give us his own interpretation of the disappearance of the bathhouses at the beginning of the sixteenth century. Once again, the concept of a 'hidden' disease verges on that of the disease that can be contracted through sexual promiscuity:

Twenty-five years ago nothing was more customary among the Brabanters than public steam baths. Now these are out of fashion everywhere, for the new pox has taught us to let them alone.[6]

Syphilis thus shared with leprosy and the plague certain similarities as regards the fears associated with their propagation. Other parallels could also be drawn. As we will see, God plays an important role in all cases, at once explaining the origin and the propagation of the disease. The mercury treatment draws its origin from analogous reasoning, as syphilis was first seen as a parasitic disease, a type of leprosy. At least that is the theory of Gérard Tilles and Daniel Wallach, recalling that Arab physicians used this metal for its 'parasiticidal properties'.[7] Other hypotheses also lead to the notion that the mercury treatment was first proposed by barbers and surgeons, notably because the cutaneous symptoms initially disappeared in a spectacular way. Doctors could not but suggest similar practices to their patients.[8]

Even if it seems that the modes of transmission of syphilis were well understood, physicians in the sixteenth century had no qualms about associating the disease with chastisement from God. In the *Vigo en Françoys*, for example, the translator, while indicating that sensuality produces numerous pains, specifies that the disease

a este stagieuse principalement en conjonction de femme vilaine et sale ou au contraire d'homme sale et femme exersant le deduit de dame Venus: et a este sa naissance principalement es parties secretes de l'homme et de la femme ...[9]

[was contagious principally during conjunction with a wicked and unclean woman or on the contrary of an unclean man and a woman who exercises the entertainment of Lady Venus: and had its birth principally in the secret parts of man and of woman ...]

This knowledge of the modes of transmission does not prevent Ambroise Paré from establishing later in the century in an unequivocal fashion two methods of contagion, one attributed to God, the other to sexual behaviour. In addition to sexual contact, Paré indicates cases of patients infected having had no sexual encounters with contaminated individuals: repeated kissing of infants, sharing the same bed or bedclothes with a victim of the pox, using the same dishes as a syphilitic, etc.[10] In this last case, it is not abscesses or wounds that cause contamination, but saliva. This phenomenon establishes a connection between syphilis and the modes of contamination of leprosy and rabies.

Paré is, however, neither the first nor the last author to insist on the importance of the will of God in the appearance and propagation of syphilis.[11] In the seventeenth century, Planis Campy and Thierry de Hery are in full agreement, but will insist also on astral influences:

J'ay dit que c'est le plus souvent par contact venerien, & c. et ce d'autant que ceste maladie n'arrive pas toujours par ceste voye, car elle peut arriver par la propre constellation de la sphere de Venus du petit monde: ou par la constellation de certains Astres du Macrocosme, l'influence desquels excite la constellation des parties gentitalles du petit: et par leur faculté Aymantine eslevent et subliment leur Mercure, lequel cause la verolle et ses accidens.[12]

[I have said that it is most often by venereal contact, & c. and what is more that this disease does not always arrive by this means, for it can arrive by the particular constellation of the sphere of Venus of the small world: or by the constellation of certain Stars of the Macrocosm, whose influence excites the constellation of the genital parts of the small one: and by their magnetic faculty rise up and sublimate their Mercury, which causes the pox and its complications.]

Thierry de Hery gives similar astrological explanations up to the seventeenth century, but nevertheless oriented his argument towards contamination of the breath.[13] According to these doctors and surgeons, certain types of individuals are more vulnerable to the disease and have constitutions which favour contamination. Early on, Paré put forward a hypothesis on this subject:

Les jeunes qui sont de texture molasse, rare et delicate, sont plus disposez à recevoir tels virus, que ceux qui sont de contraires temperatures, et non preparez à recevoir tel venin.[14]

[Young people who are flabby, fragile and delicate are more susceptible to such viruses, than those who are of opposite temperatures, and not prepared to receive such venom.]

As for the internal propagation of the pox, it is perhaps Thierry de Hery who gives the explanation most comprehensible to minds influenced by modern medicine:

Aussi le plus souvent ces parties permierement attouchees sont les permieres affectees de ce mal, et alterees par le venin, qui successivement se communique au foye par les veines, et au cœur par les arteres, toutefois c'est plus tard, parce que le cœur et les parties cordiales resistent plus fort audit venin, et au cerveau par les nerfz, auquel le plus souvent apparoissent les premiers signes de ce mal, d'autant que ledit venin a ce coutume de chercher, et plus aisement infecter les parties spermatiques et moins chaudes ...[15]

[Thus most often those parts first touched are the first affected by this ill, and altered by the venom, which is successively communicated to the liver by the veins, and to the heart by the arteries, however it is later, because the heart and the cordial parts resist the said venom more strongly, and to the brain by the nerves, in which most often appear the first signs of this illness, in as much as this venom usually searches out, and more easily infects the spermatic and less hot body parts ...]

We can thus conclude, in light of these medical descriptions of contagion as the Neapolitan disease, that even if God played the primary role in causing the appearance of the disease, the methods of transmission and of infection were known very early on and formed the basis of certain rules of hygiene that are still present today in the imaginary of the Occident. In the same way, at least according to the characters in Erasmus, certain practices of socialization favouring promiscuity (shared rooms at inns, but also bath-houses) had already been influenced at the beginning of the century by the fear of catching this disease.

These hypotheses are perhaps too daring in the eyes of medical histori-ans, however. Johan Goudsblom makes very clear, in his interpretation of Erasmus' colloquium, that even if there seems to be a link between the fear of contagion and a certain reserve, Erasmus was not motivated by a medical conception of the problem, but rather by a 'repugnance toward crowds assembled in an enclosed space', and a fear of 'social pollution', his physi-cal delicacy inspiring social selectivity.[16] Erasmus also broaches the ques-tion of syphilis in another colloquium, 'A Marriage in Name Only'. In this dialogue, Gabriel and Petronius discuss a 'scabby' wedding, 'ulcerous and festering'.[17] This is the story of the union between a young girl and a syphilitic. The latter is described as moribund, and the two protagonists come to recommend marriage with a corpse rather than with a monster (the word employed) of this kind, who is 'worse than leprous'.[18] Links with the plague are also established in this dialogue: the breath of the infected person is 'sheer poison, his speech a plague, his touch death'.[19] When the two discussants imagine suggesting preventive measures to the authorities, they do not hesitate to compliment the Italians who sequester all those who are ill as soon as the first symptoms of the plague appear. The descrip-tion of the groom's personality given by Gabriel is also worthy of attention. Just as the pox transforms the body of a man into that of a monster, it also gives rise to a variety of vices: gambling, alcoholism, whoremongering, thievery, and so on.[20]

Syphilis and its literary expression: myths, legends and laughter

If syphilis aroused concern and disdain among physicians and philoso-phers, it also lent itself to the lyricism of the different eras of the Renaissance. One of the first myths created about syphilis can be found in the work of Fracastoro. While noting the violence of the disease that appeared at the end of the fifteenth century and insisting on the influence of the stars in its appearance and development, the Italian poet tells us the story of the unfortunate shepherd Syphilus, whose name will henceforth be linked to the disease. This story, told in a *mise en abyme* to Spanish sailors, accentuates first of all the impieties of the nations of the Atlas and of the shepherd Syphilus who incurred the anger of Sirius. Syphilus, seeing his troop decimated, turned his back on his God, preferring to sacrifice to other divinities. The wrath of the Sun was then terrible:

> he hurled his hostile rays and shone with a bitter light. By that glance mother Earth and the sea's flat expanse were attacked, by that venom the air was infected and began to glow hot. Straightaway an unknown pollution was born to flood the blasphemous earth. The first man to display disfiguring sores over his body was Syphilus ... he was the first to experience sleepless nights and tortured limbs, and from this first victim the disease derived its name and from him the farmers called the sickness Syphilis. And soon the evil plague had spread through all the cities among the commons, nor had it spared in its savagery even the king himself.[21]

Fracastoro's work, even if it was a defining element in the construction of the legend of syphilis and its provenance and is informative on the symptomatology of the disease, was not the only document telling a mythical version of the appearance of syphilis. The anonymous work entitled *Triumphe de très haulte et puissante dame Verolle*, published in 1539 and 1540, attributed to both Jean Le Maire des Belges and François Rabelais, presents the introduction of the disease into the kingdom of love in the form of a mythic story. The legend develops around an error, Cupid having mistakenly brought back to Venus the bow and arrow of Death, with whom he had been drinking the previous night. Pleasure (Volupté) accidentally sits on the weapon.[22] Fearing further wounds, Venus demands that the weapon be carried outside the castle. A nymph throws it out the window and into the moat, and the contagion spreads, affecting the animal world as much as the human:

> Mais de la fleiche et de sa grant poison
> Il se perdit des poissons à foison;
> Cignes, canars, laissèrent le repaire,
> Et de plongeons mourut plus d'une paire,
> Tant devient l'eaue amaire et pestilente
> Du fort venin de la fleiche dolente.[23]
>
> [But from the strong poison on the arrow
> Numerous fish were lost in the narrow;
> All swans and ducks left their nests for the air,
> Then diving there perished more than one pair;

Bitter and putrid the water became
When the arrow's deadly venom took aim.]

After the negotiations and the transformation by Death of 'young and ardent lovers' into old men, the narrator of the scene affirms that he does not know the end of the story but does know that the moat, which Venus had none the less attempted to cleanse, is forever the carrier of this venom of death:

> Dame Venus, pour y remedier
> Et la poyson curer et nettoyer,
> Y feist gecter grant nombre de flourettes
> Prinses au clos du jardin amourettes,
> Et, pource que plus amère que fiel
> Estoit au boyre, on y mist force miel;
> Si que par traict de temps l'eaue esclaircist
> Devint fort belle et enfin s'adoulcist,
> Qui pour les gens fut une horrible amorce:
> Car sçavoir faut qu'onc n'en perdit sa force
> Du fort venin portant l'eaue emmiellée. [24]

> [Lady Venus, to make it pure
> And the poison both clean cure,
> Had thrown upon it from above
> Flowers from the garden of love.
> And since it was to drink so vile
> They added honey to the trial;
> So over time the water cleared
> Was beautiful and sweet as mead,
> Which for all was horrible bait:
> For those who drunk there sealed their fate
> So strong was the honeyed venom.]

The ambiguity of the last line cannot but lead the reader to be wary of the pernicious sweetness of the love potion found in 'Venereal moats'. The venom which had seemed to dissolve, a bit like the syphilitic symptoms temporarily treated by mercury, resurfaces in a spectacular manner after a time:

> Tant fort plaisoit aux hommes et aux femmes,
> Mesmes aux hommes, dont ilz sont plus infames,
> Ce très doulx boire et ce joyeulx breuvaige,
> Que maintz beaulx jours ne firent aultre ouvrage;
> Mais en la fin, quand le venin feust meur,
> Il leur naissoit de gros boutons, sans fleur,
> Si très hydeux, si laidz et si enormes,

Qu'on ne vit oncques visaiges si difformes,
N'onc ne receut si très mortelle injure
Nature humaine en sa belle figure.[25]

[So well it pleased the men and women,
Even the men, who are such villains,
This sweet drink and joyous beverage,
That many fair days they did not budge;
But at last, when ripe was the poison,
Large buds erupted on their person,
So grotesque, ugly and enormous,
That they have to be called hideous.
No ill can threaten worse our poor race
Than a deadly insult to the face.]

The story of this legend ends with an unusual piece meant to recall, according to the prefacers, the Roman triumphs. One could also compare this piece to a Royal Entry, where the different individuals, groups or allegorical figures speak in 'celebration of' the sometimes painful heritage of Lady Pox (*Dame Verolle*). This is how the herald, at the head of the procession, presents the group marching toward the Well of Love:

Le Herault.
Sortez, saillez des limbes tenebreux,
Des fourneaulx chaulx et sepulchres umbreux,
Où, pour suer, de grid et verd on gresse
Tous verollez; se goutte ne vous presse,
Nudz et vestuz fault delaisser voz creux
De toutes pars. [26]

[The Herald.
Exit, emerge from limbo dark,
Bearing of fire and tombs the mark,
Where, to sweat, with grey and green are greased
All poxed; if a drop you touch, then cease
To hide, whether clothed or naked stark
Emerge from everywhere.]

Lady Pox is obviously at the end of the procession, calling herself 'queen and princess' of the Well of Love. The last lines express the impatience of the lovers and their blindness caused by love:

Du Puy d'Amour je suis royne et princess,
Tesmoing Venus et Cupido aussi.

La plus grand part du monde en grand humblesse
Rend l'honneur deu à mon triomphe icy;
Si je leur faitz endurer maint soucy
Ce n'est à tort; car, pris de telle ou telle,
Viennent au Puy tout puant et noircy
De mal infaict, sans prendre de chandelle. [27]

[Of the Well of Love I am queen and princess,
Venus and Cupid are my witnesses here.
Many parts of the world with great humbleness
Give honour to my triumph from far and near;
If I force them to suffer so many trials
It is not wrong; for they were caught,
And come to the Well all stinking and so vile,
By disease infected; their prayers come to naught.]

The Epilogue to the reader points out the didactic element of this uncommon triumph. This long poem fulfils two functions. First, the reader must experience pleasure in reading this work. If today our morals prohibit us from mocking people suffering from aesthetic deficiencies, in the sixteenth century this was not the case. The misery of the pox victims cannot but arouse the laughter of those who are well:

il est certain que tu n'y peulx prendre que plaisir pour diverses sortes de verollez qui y sont, les ungs boutannantz, les autres refonduz et engressez, les autres pleins de fistules lachrimantes, les autres tous courbés de gouttes nouées, les autres estantz encores aux faulx-bourgs de la verolle, bien chargez de chancres, pourreaulx, filletz, chaudes pisses, bosses chancreuses, carnositez superflues et autres menues drogues que l'on acquiert et amasse au service de dame Paillardise.[28]

[it is certain that you can only take pleasure in the diverse sorts of syphilitics who are there, some spotty, others deformed and fat, others covered with seeping fistulas, others bent over with knotty gout, others still at the outskirts of syphilis, well-laden with cankers, gangrene, dribbles, the clap, chancrous bumps, superfluous carnosities and other minor liquors that one acquires and amasses in the service of Lady Debauchery.]

This allusion to ridicule and pleasure obviously confirms the perspective developed by Rabelais in the different prefaces to his works. The author or authors of the *Triumphe* moreover did not hesitate to make reference to the 'very precious syphilitic' of Maistre Alcofribas Nasier.[29] In Roland Antonioli's interpretation of the transformations during the century in the representation of the syphilitic, he becomes little by little a kind of crazy

philosopher to whom the physician, notably, addresses himself in his pro-
logues: 'He even incarnates a certain vigour, a confidence in human nature,
rather paradoxically for this ancient brotherhood of the damned ...'.[30] Let
us not forget in this respect that the prologue to *Gargantua* was addressed
to both the 'very illustrious Drinkers' and the 'very precious Syphilitics',
and that the *Tiers Livre* took up almost the same declaration: 'Good people,
very illustrious Drinkers, and you very precious Syphilitics.'[31] If Rabelais
had certainly been in contact with syphilitics during his sojourn at the
Hôtel-Dieu of Lyon, this interpellation on the part of the doctor of laughter
could only serve to integrate these unfortunate debauchees in the cos-
mogony of drinkers and good people.

But let us return to the epilogue of the *Triumphe de Dame Verolle* and to
the 'profit' that 'the man of good understanding' was meant to draw from
it. As well as eliciting pleasure, so this triumph serves as a moral lesson for
the reader, warning him of the dangers that stalk the lascivious man:

Le profit est que, si tu es homme de bon entendement et bien reduict à
honnesteté et raison à l'exemple des malheureux qui tombent par leur
luxure dissolue aux accidents dessus dictz, tu eviteras telz dangiers et
inconveniens de la personne, attendu que l'homme ne faict petite injure
à Dieu quant par sa dissolution et villennie il contamine ce corps tant
parfaict qu'il a reçeu du Createur, joinct que celuy est malheureux qui
par sa volupté desordonné se rend maladif et langoureux pour le demeu-
rant de sa vie, et tombe en telle mesprisance du monde qu'il n'y a nul
qui ne le fuye comme ung ladre et personne contagieuse. C'est doncq le
fruit que recuilliras en lisant ce present œuvre, pour la congnoissance
que tu auras des maulx et misères qui viennent aux verollez.[32]

[The advantage is that, if you are a man of good understanding and well
reduced to honesty and reason by the example of the miserable who fall
by their dissolute lust into the complications told above, you will avoid
such dangers and disadvantages of humanity, given that man does not
give small insult to God when by his dissolution and villainy he contaminates this perfect body that he received from the Creator, also that he
is miserable who by his reckless sensuality renders himself ill and lan-
guorous for the rest of his life, and is the object of such scorn from the
world that there are none who do not flee him as a leper and contagious
person. So this is the fruit that you will harvest by reading the present
work, for the knowledge that you will have of the ills and miseries that
syphilitics suffer.]

In her doctoral thesis on representations of syphilis in French literature of
the sixteenth century, Lesa B. Randall analyses the *Triumphe de Dame*

Verolle as well as *Le Pourpoint fermant à boutons* and *Les Sept Marchans de Naples*. Randall notes that the first agent of propagation of venereal disease, at the beginning of the sixteenth century, is believed to be a woman:

> The close association between women and sex made the connection between syphilis and women seem obvious. As a result, women are portrayed as the source of both sexual desire and syphilis in nearly every literary syphilis text from the period. ... At the same time, however, warnings appear along with explicit suggestions directed toward men, of ways in which to steer clear of the new disease.[33]

This hypothesis can at first appear strange, since the symptoms of the disease were more visible in men, but it confirms the theory which attributes the origin of the disease to the other: the 'Neapolitan disease' for the French, *Morbo Gallico* or French pox for the Italians and the English. Women and especially prostitutes thus become the foil, facilitating an easy condemnation of scapegoats who have no voice, but they are also the source of inspiration permitting the repressed expression of the horror that the disease inspires. Conversely, the expression of suffering in a good number of texts comes first of all from the male syphilitic, who appears to be the only one to experience suffering. So it goes in the *Pourpoint fermant à boutons*, when the lyrical subject describes the diverse symptoms and the presumed remedies of the disease, grouping them after the manner of Latin declensions. The verse ending of this Rabelaisian enumeration nevertheless leaves no doubt as to the perspective adopted by the male poet:

> A cinq cens dyables la verolle
> Et l'ord vaisseau où je la prins!
> Je n'ay dent qui ne bransle ou crolle.
> A cinq cens dyables la verolle!
> La goutte si me rompt et rolle
> Et suis d'ulcères tout esprins.
> A cinq cens dyables la verolle,
> Et l'ord vaisseau où je la prins![34]

> [To five hundred devils the pox
> And the vessel where I caught it!
> All my teeth at risk of loss,
> To five hundred devils the pox!
> Gout will cause me to turn and toss,
> And I am by ulcers all bit.
> To five hundred devils the pox
> And the vessel where I caught it!]

Les Sept Marchans de Naples also takes the masculine viewpoint, this time of seven characters who were swindled in a market that gave them all shares in the disease. As in certain thirteenth-century *fabliaux*, several strata of society are thus affected: the adventurer, the cleric, the student, the blind man, the villager, the merchant and even the braggart, this polite and refined man:

> Marchant je suis de gorre (syphilis) au temps qui court;
> J'en ay payé sus tous la folle enchère;
> En ce marché je me suis monstré lourd;
> Pour ung carcan bien garny, sur le gourd,
> On me bailla soubdain de la plus chière.
> ...
> En mes jambes les gouttes sont enflées,
> Qui difforment de tout en tout la gresve;
> Les aulcunes on les nomme enossées,
> Dont nuyt et jour par icelles je resve.[35]

> [Nowadays I am a merchant of gore;
> I payed the lunatic bargain quite dear;
> In this market I made a trade so poor
> That a large sore on my member I fear
> Was the price paid for contact with a whore.
> ...
> In my legs great swelling from painful gout
> Deforms my old shins in myriad ways;
> Some of the ladies, too, suffer a bout
> I dream of their bodies both night and day.]

After the ephemeral triumph of syphilis

It is rather paradoxical that allusions to syphilis, after the early years of its presence in literature, fade little by little during the rest of the century. In fact, they lose their joyous dimension for the most part, in a move towards moral commentary. This is most noticeable in the travel narratives set in the Americas. The subject of the last chapter of the *Histoire d'un voyage en la terre de Bresil* by Jean de Léry is disease; the traveller insists on the danger of the incurable illness called *Pians*.[36] This disease, whose symptoms Léry claims to have found among children, displays the same purulent manifestations as syphilis.[37] Pustules and cankers nevertheless harbour the same meaning for the Protestant author that they would in France; they signal the presence of a lascivious spirit. Having met an infected Norman among the members of one of the Brazilian tribes, Léry confirms this hypothesis.

The disease becomes a punishment from which no Christian who frequents pagan nations can escape:

Et de fait j'ay veu en ce pays-là un Truchement, natif de Rouen, lequel s'estant veautré en toutes sortes de paillardises parmi les femmes et filles sauvages, en avoit si bien receu son salaire, que son corps et son visage estans aussi couverts et deffigurez de ces *Pians* que s'il eus esté vray ladre, les places y estoyent tellement imprimées, qu'impossible luy fut de jamais les effacer: aussi est ceste maladie la plus dangereuse en ceste terre du Bresil.[38]

[And in fact I saw in that country an Interpreter, a native to Rouen, who had wallowed in all sorts of bawdiness among the savage women and girls, and had reaped his earnings so well, that his body and his face were as covered and defigured by these *Pians* as if he had been a true leper, and in places were so imprinted by them, that it was impossible for him ever to erase them: clearly this disease is the most dangerous in the country of Brazil.]

Frank Lestringant points out that Jean de Léry's description of the disease concurs with André Thevet's observations in the *Singularités de la France antarctique*, as well as in his *Cosmographie universelle*. Thevet points out that the presence of the disease in America comes from the too frequent sexual encounters of the natives:

Reste donc qu'elle provienne de quelque malversation comme de trop fréquenter charnellement l'homme avec la femme, attendu que ce peuple est fort luxurieux, charnel et plus que brutal, les femmes spéciale-ment, car elles cherchent et pratiquent tous les moyens à émouvoir les hommes au déduit.

[Therefore it muste needes bee that it proceedeth of some misgoverne-ment, as to much carnall and fleshely frequentation the man with the woman, considering that thys people is very lecherous, carnal, and more than brutishe, especially the women: for they do seeke and practice all the means to move men to lust.][39]

As Léry will do, Thevet notes, but without being able to cite a witness, that Europeans contract this disease very easily, at the merest contact with the women of the country.[40]

As well as giving information on the way the aboriginals treat this illness with guaiacum wood, Thevet takes the opportunity to develop a commen-tary on the name of the disease. We should, of course, remember that it is the future royal cosmographer speaking. Thevet rejects the terms of the Neapolitan disease and the French pox, preferring instead the term the 'Spanish disease'.[41] Thevet does not hesitate to remind those enemies of France in the sixteenth century of their responsibility as regards the propagation of the disease in the kingdom of the Valois. Moreover, this 'pox' is already widespread, according to the remarks of the cosmographer, in a country that has had up until then few contacts with America.

Even if he did not know as much about America as he leads us to believe, Thevet shows himself nevertheless to be a good observer of French mores. While syphilis was establishing its triumph some twenty years earlier, by the second half of the sixteenth century it had become common. At least that is what historians of medicine believe, noting that the disease would first have appeared in a particularly virulent form.[42] It nevertheless remains a sign of innate lewdness. Michel de Montaigne, in 'L'Institution des enfants', broaches the question of syphilis by suggesting its link with the virile exploits of youth. In his commentary on the necessary education for children, does he not specify that it would be better to teach them to control their desires at a young age in order to preserve the health of students and to permit them, when the time comes, to understand the philosophy of temperance? 'They begin to teach us to live when we have almost done living. A hundred students have got the pox before they have come to read Aristotle's lecture on temperance.'[43] One could therefore surmise that as early as the first edition of the *Essais* (1580), to a Michel de Montaigne syphilis could give the impression of being an accident to avoid on the road of life, a bit like venereal diseases before the appearance of AIDS. This familiarity with the disease, and the philosopher's perspective on the sexuality of adolescents does not, however, square with the accusations levelled at approximately the same date by the detractors of King Henri III.

According to Claude Postel, invectives linked to the body at the time of the Reformation in France were more closely aligned with leprosy, the plague and chancre (which would be different here, according to Postel, from the syphilitic chancre, more reminiscent of that of Job).[44] Few or no accusations linked to syphilis would thus have been used. Is this not another proof that syphilis had suddenly become a commonplace disease, except perhaps within a certain polemical context, when perversity and lasciviousness would be marked? If we look for example at the polemic against Henri III, it must be noted that the king was suspected of having contracted a venereal disease during his stay in Venice, and that the Duc d'Espernon, one of the arch-minions, is qualified by his enemies as being 'poxy'.[45] Pierre Champion speaks particularly of an 'inflammation' contracted from some courtesans that had been inadequately treated and had brought on the sterility of the royal couple, according to the Protestants.[46] René de Lucinge, in the *Miroir des Princes*, reveals the existence of a genital ulcer[47] on Henri III, which seems to grow according to the rhythm of lustful adventures that he imputes to the king.[48]

One must nevertheless await the libertine publications of the seventeenth century to see the images linked with syphilis reappear in a vastly different context. It is not aim in this brief study to establish a repertoire of the allusions in seventeenth century texts, which remains to be done, but we can furnish a few examples of the new context of production of these works. The disease then becomes a source of satire of prostitutes and old women, but also of courtiers.

A few remarks on syphilis and 'satyre'

Let us first quote these lines attributed to Théophile de Viau describing in the *Délices satyriques* and then in the *Quintessence satyrique* an old woman whose body is mutilated by the disease and whose hypocrisy is a moral trait he highlights. Is it a question of false religious devotion or is she a courtesan? The lyrical subject calls her 'slut', but also insists that she was at times falsely pious: 'One day that wicked woman there / In holy water with no care / Distilled tears from her bigot's eye.'[49] As well as the nauseating odour which emanates from the body of this old woman, Viau offers us a vividly imaged description of the ravages of the disease. If he seems to take a certain pleasure in this slightly morbid description, the tone is not the candid laughter of a Rabelais, but rather satirical vengeance directed against a right-thinking old woman:

> Son c. vilain, baveux, suant,
> Et plus que le retrait puant,
> Cizelé de la cicatrice,
> De chaude pisse et des poullins
> Et de mille chancres malins,
> Qui percent jusqu'à la matrice.

> Mille morpions rangez au bords,
> Tous plat pattus et demy-morts,
> Tenoient leur general concile,
> Pour ronger l'onguent verolé,
> Qui leur a quatre fois volé
> Le poil qui leur servoit d'azile. [50]

> [Her c., nasty, runny, sweating,
> More than the deep recess stinking,
> Chiselled by the hideous scar,
> Of hot urine and the clap, marred

By a thousand evil cankers,
That pointed, pierce through to the womb.

A thousand crabs around the edges,
All flat-footed and half-dead,
Held their general assembly,
To gnaw at the ointment poxy,
That had stolen from them four times
The hair that gave refuge sublime.]

The courtiers do not escape the satire of the libertines. Sigogne, in the *Cabinet satyrique* of 1618, publishes a 'Satyre contre un courtisan à barbe rasee'. In this work in verse, he submits to the readers the reasons why courtiers no longer wear a beard. If he makes fun of the virility of the men of the court (who, if they let their beard grow, would have but little to offer in the eyes of the ladies and would not be able to compete with the beards of the hermaphrodites), he also suspects that the beard has suffered the torments of alopecia ('pelade'), the cutaneous malady symptomatic of syphilis:

Et que, naguere, elle eut querelle
Avec le sieur de Verollé,
Qui vous tient pour son enrollé,
Luy donnant une camisade,
La menaçoit de la pelade;
Et, craignant que ce faux garçon
Ne la pelast comme un cochon. [51]

[She had a quarrel long ago
With the Lord of the Pox, whom it may suit
To take you for his loyal recruit.
When giving her a camisade,
Threatened her to be hairless made;
And she feared that this lad so big
Might indeed peel her like a pig.]

These last two examples help elucidate how syphilis can equally become a source of satire and of denunciation on the part of the libertines. It is, in fact, less a matter of pity or mockery of individuals stricken with the illness, and more of bringing to light the presence of this disease which had become 'shameful' in the public eye. If it seems easy thus to attack the reputation of a few moralizing people, what happens if the libertine contracts the disease? It is again Théophile de Viau who answers the question in a sonnet attributed to him that was published at the beginning of the

Parnasse satyrique. He adopts a plaintive tone throughout the piece, and expresses both his horror and the disgust of his close friends faced with the symptoms of the disease. The personal point of view adopted in order to describe the disease and the confession of having contracted it could only please readers unsympathetic to the misfortunes of libertines. On the other hand, the poet makes his sarcastic laughter heard in the last line, where he surprises those right-thinking souls:

> Phylis tout est foutu je meurs de la verolle;
> Elle exerce sur moy sa derniere rigueur:
> Mon Vit baisse la teste et n'a point de vigueur,
> Un ulcere puant a gasté ma parole.
>
> J'ay sué trente jours, j'ay vomy de la colle,
> Jamais de si grands maux n'eurent tant de longueur,
> L'esprit le plus constant fut mort à ma langueur,
> Et mon affliction n'a rien qui la consolle.
>
> Mes amis plus secretz ne m'osent approcher,
> Moy-mesme, en cét estat je ne m'ose toucher,
> Philis [*sic*] le mal me vient de vous avoir foutue.
>
> Mon Dieu je me repans d'avoir si mal vescu:
> Et si vostre couroux à ce coup ne me tuë,
> Je fais le veu desormais de ne foutre qu'en cu.[52]

> [Phylis all is buggered, of the pox I'm dying;
> It exerts upon me its worst now at the last:
> My Member bows its head, its vigour now has passed,
> A stinking ulcer spoils my speech past all trying.
>
> I sweated thirty days, I vomited much paste,
> Never have such huge pains lasted so many days.
> The spirit most constant has died by these ways,
> And my great affliction prays for release in haste.
>
> My most intimate friends dare not to approach me,
> Myself, in this state I don't dare to touch me,
> Philis [*sic*] the ill has come to me from screwing you.
>
> God I truly repent for a life so untrue:
> And if right away I am not killed by your wrath,
> I vow hereafter to screw only in the ass.]

Two questions obviously arise after reading the last two lines. Did the poet believe that anal relations could protect one from the disease or was he making a profession of faith in homosexuality? To answer this question, one must place this piece in the context of 'unnatural' love that one finds in the 'satyrical' writings of the beginning of the seventeenth century. As love between young men was sometimes presented as an alternative to relations with lustful women, there is only a small leap to make in order to see a certain indulgence in the face of homosexual love.[53]

If the principal modes of contagion of syphilis were known early on and gradually integrated, through the works of surgeons, philosophers and physicians of the sixteenth century into what would become a certain hygiene of sexual contact, and also corporeal and intimate contact, the story of this disease in literary discourse remains mysterious. The carnivalesque explosion in the 1530s, when it seemed in good taste to laugh at everything while educating the populace, ended abruptly. Literary syphilis, perhaps following a path similar to the disease's, became more discrete. While medical discourse of the time accumulated case histories and the treatment evolved, it is moral and polemical discourse which took the place of the spectre (already paling) of literary syphilis. It is only with the libertines of the seventeenth century that this discourse regains its strength, this time to denounce the moralizers wanting to stifle once again this sassy lady who had not ended her reign.

Notes

I wish to thank Claire Carlin for organizing the workshop where this volume was conceived; and Rachel Warrington for her translation of this chapter.

1   On this subject, see C. Quétel, *Le Mal de Naples. Histoire de la syphilis* (Paris: Seghers, 1986), pp. 10ff.

2   In *Les faictz et dictz de Jean Molinet* (Paris: Société des Anciens textes français, 1937), II, p. 853: 'De jour, de nuit, de repos suis destruit, / Tu m'as seduit de joie et de plaisance.'

3   *Ibid.*, pp. 853–4.

4   *Ibid.*, p. 854: 'Je suis infect, tout le monde me fuit ...'.

5   Erasmus, *The Colloquies*, trans. C. R. Thompson (Chicago: University of Chicago Press, 1965), p. 150. Regarding these passages, see J. Gouldsblom, 'Les Grandes épidémies et la civilisation des mœurs', *Actes de la recherche en sciences sociales*, 68 (June 1987) 8–9.

6   Erasmus, *ibid.*

7   G. Tilles and D. Wallach, 'Le Traitement de la syphilis par le mercure. Une histoire thérapeutique exemplaire', *Histoire des Sciences Médicales*, XXX, 4 (1996), 501.

8   *Ibid.*, p. 502.

9   *De Vigo en françoys. S'ensuit la pratique et cirurgie de tres excellent docteur en medecine Maistre Jehan de Vigo, nouvellement translatee de latin en françoys* (Lyon: enoist Bounyn, 1525) fol. crrri vo.

10   A. Paré, *Les Oeuvres* (Lyon: Veuve de Claude Rigaud et Claude Obert, 1633), p. 520.

11   On this subject, see R. Antonioli, *Rabelais et la médecine* (Geneva: Droz, 1976), pp. 92–3.

12 D. de Planis Campy, *La Verolle recogneue, combatue et abbatue sans suer, et sans tenir chambre, avec tous ses accidens* (Paris: Nicolas Bourdin, 1623), p. 42.

13 T. de Hery, *La Methode curatoire de la maladie venerienne, vulgairement appellee grosse Verolle, et de la diversité de ses symptomes* (Paris: Jean Dehauy, 1660), p. 13.

14 Paré, *Les Oeuvres*, p. 692. For a similar description, during the seventeenth century, see de Hery, *La Methode curatoire de la maladie venerienne*, p. 17: 'En quoy faut noter que ceux ce texture rare, delicats et mols, seront plus prompts et plus disposez à recevoir ceste affection par tout le corps, et les autres au contraire.' / 'In which one must remark that those of fragicle, delicate and soft texture will be more prompt and more disposed to receive this affliction throughout the body, and the others to the contrary.'

15 de Hery, *La Methode curatoire de la maladie venerienne*, pp. 16–17.

16 J. Gouldsblom, 'Les Grandes épidémies et la civilisation des mœurs', p. 9: 'une répugnance plus grande à l'égard de la foule rassemblée dans un espace restreint'.

17 In E. Rummel, ed., *Erasmus on Women* (Toronto: Univ. of Toronto Press, 1996), p. 146. Rummel cites the translation by C. R. Thompson, *The Colloquies of Erasmus* (Chicago: University of Chicago Press, 1965).

18 *Ibid.*, pp. 147–9.

19 *Ibid.*, p. 147.

20 *Ibid.*, p. 148.

21 *Fracastoro's Syphilis*, trans. and ed. G. Eatough (Liverpool: Francis Cairns, 1984), p. 103.

22 A. de Montaiglon, ed., *Le Triumphe de Dame Verolle*, in *Recueil de poésies françoises*, IV, (Paris: P. Jannet, 1856), p. 231.

23 *Ibid.*, p. 233.

24 *Ibid.*, pp. 240–1.

25 *Ibid.*, p. 241.

26 *Ibid.*, p. 257.

27 *Ibid.*, p. 267.

28 *Ibid.*, p. 268.

29 *Ibid.*, p. 229.

30 R. Antonioli, *Rabelais et la médecine* (Genève: Droz, 1976), p. 94: 'Il incarne même une certaine vigueur, une confiance en la nature humaine assez paradoxale pour cette ancienne confrérie de damnés ...'.

31 François Rabelais, *Les Cinq Livres* (Paris: Le Livre de Poche, 1994), pp. 5 and 541.

32 *Le Triumphe de Dame Verolle*, p. 268.

33 *Representations of Syphilis in Sixteenth-Century French Literature*, thesis, University of Arizona, 1999, p. 28.

34 A. de Montaiglon, ed., *Le Pourpoint fermant à boutons*, in *Recueil de poésies françoises*, IV (Paris: P. Jannet, 1856), p. 281.

35 *Les Sept Marchans de Naples* in A. de Montaiglon, ed., *Recueil de poésies françoises des XVe et XVIe siècles*, II (Paris: P. Jannet, 1855), p. 108.

36 J. de Léry, *Histoire d'un voyage en terre de Brésil* (Paris: Le Livre de Poche, 1994), p. 469.

37 F. Lestringant and S. Lussagnet specify that it is probably another treponema than syphilis, which exhibits the same apparent symptoms but is transmitted by flies or by caterpillars. *Ibid.*, p. 317, n. 2.

38 *Ibid.*, p. 469.

39 A. Thevet, *Les Singularités de la France antarctique* (Paris: La Découverte, 1983), p. 103; *The New Found Worlde, or Antarctike* (1568; rpt Amsterdam: Da Capo Press, 1971), p. 70.

40 *Ibid.*, trans. pp. 70–1; p. 104: 'Quant aux chrétiens qui habitent en l'Amérique, s'ils se frottent aux femmes, ils ne s'en sortiront jamais sans tomber dans cet inconvénient beaucoup plus tôt que ceux du pays.'

41 *Ibid.*, pp. 103–4.

42 See C. Bourrières and O. Devant, 'La Syphilis maligne précoce: une explication de l'épidémie de 1439?' in *L'Origine de la syphilis en Europe* (Paris: Errance editions, 1994), p. 273: 'Les auteurs de l'époque nous ont parlé de lésions cutanées polymorphes, le plus souvent ulcérées, croûteuses, exubérantes, pouvant être le siège d'un écoulement sanieux, ou encore gommeuses et destructrices; elles font suite à un ulcère génital et sont accompagnées de douleurs ostéocopes térébrantes, de céphalées, d'insomnies, dans un contexte de baisse de l'état général. En d'autres termes, il s'agit d'un tableau clinique très proche de celui que nous connaissons aujourd'hui sous le nom de syphilis maligne précoce. En revanche, on est frappé de la fréquence et de la gravité extrêmes qui n'ont plus jamais été décrits par la suite.' / 'The authors of the time spoke of polymorphous cutaneous lesions, most often ulcerous, scabby, exuberant, perhaps being the site of a bloody discharge, or even sticky and deadly; they follow a genital ulcer and are accompanied by osteitic, penetrating pains, by cephalalgias, insomnias, in a context of a deterioration of the general state of health. In other words, it is a clinical picture very close to the one which we know today under the name malignant precocious syphilis. On the other hand, we are stuck by the extreme frequency and seriousness of cases, greater than have been described since.'

43 M. de Montaigne, *Essays of Michel de Montaigne*, trans. C. Cotton (1685), ed. W. C. Hazlitt, Project Gutenberg Release 3600: http://onlinebooks.library.upenn. edu/webbin/gutbook/lookup?num=3600, consulted and downloaded 15 January 2005; *Les Essais de Montaigne*, ed. P. Villey (Paris: Presses Universitaire de France, 1965), I, p. 163 (I, 26): 'On nous apprent à vivre quand la vie est passée. Cent escoliers ont pris la verolle avant que d'estre arrivez à leur leçon d'Aristote, de la tempérence.'

44 C. Postel, *Traité des invectives au temps de la Réforme* (Paris: Les Belles Lettres, 2004), pp. 371ff.

45 *Le Testament de Henry de Valoys, recommande a son amy Jean d'Espernon* (Blois: Jaques Varengles et Denis Binet, 1589), pp. 10–11: 'On dit que pour complaire au vérolé mignon / Ce tyran a baisé du Diable le poitron.' / It is said that to please the poxy minion / This tyrant has embraced the Devil's bottom.'

46 See P. Champion, *Henri III roi de Pologne* (Paris: Grasset, 1951), p. 85, n. 3.

47 See note 43 above.

48 R. de Lucinge, *Le Miroir des Princes, in Annuaire-bulletin de la société de l'Histoire de France* (Paris: Klinckieck, 1995), p. 107.

49 T. de Viau, *Œuvres complètes*, III (Paris: Nizet, 1979), pp. 251–2: 'Un jour ceste vilaine là / Dans un benestier distilla/ Les pleurs de son œil hypocrite.'

50 *Ibid.*

51 Sigogne, *Les Œuvres satyriques du Sieur de Sigogne* (Paris: Bilbiothèque des Curieux, 1920), p. 185.

52 de Viau, *Œuvres complètes*, pp. 267–8.

53 See G. Poirier, *L'Homosexualité dans l'imaginaire de la Renaissance* (Paris: Honoré Champion, 1996), last chapter.

12
Contagious Laughter and the Burlesque: From the Literal to the Metaphorical

Dominique Bertrand

Georges Bataille insisted on the role contagion plays in laughter, an emotional experience in which he discerns 'the specific form of human interattraction'.[1] The philosopher distinguishes between a mediated interattraction and a more immediate one, the first being linked to the presence of a 'trigger' (in this case the comical object), the second to the psychosociological permeability that favours 'contagion' or '*sympathie*':

> Les organismes semblables sont susceptibles, dans de nombreux cas, d'être traversés par des *mouvements d'ensemble* : ils sont en quelque sorte perméables à ces mouvements. Je n'ai d'ailleurs fait ainsi qu'énoncer en d'autres termes le principe bien connu de la contagion, ou si l'on veut encore de la sympathie, mais je l'ai fait je crois avec une précision suffisante. Si l'on admet la perméabilité à des mouvements d'ensemble, à des mouvements continus, le phénomène de la reconnaissance apparaîtra construit à partir du sentiment de perméabilité éprouvé en face d'un autre/socius.

> [Like organisms, in many instances, may well experience group movements. They are somehow permeable to such movements. What is more, I have thus only stated in other terms the well-known principle of contagion, or if you still want to call it that, fellow feeling, *sympathie*, but I believe I have done this with sufficient precision. If one acknowledges permeability in 'group movements', in continuous movements, the phenomenon of recognition will appear to be constructed on the basis of the feeling of permeability experienced when confronted with an other/socius.][2]

This contagious nature, consubstantial with laughter according to Bataille, almost never appears explicitly in the theoretical treatises that examine the 'risible' passion from the Renaissance to the classical age in France. The absence of the notion of contagion is all the more remarkable in that medical discourse exists on the transmission of signs of joy; comic authors, moreover, develop specular representations of wildly multiplied outbursts of laughter. It will be helpful to examine this problematic use of the metaphor of contagion by the burlesque, since the notion of contagion was an obsession among detractors of the genre in the second half of the seventeenth century.

By way of examples from French texts, I propose to demonstrate how, in a context that excludes the imaginary of contagion, the aesthetics of reception incorporate ambiguous references to the interattraction of outbursts of laughter. I will then explore metaphorical discourse on the burlesque as disease.

The contagion of laughter: an unthinkable concept

The absence of contagion in theories of laughter

Sixteenth- and seventeenth-century French treatises on the passions say nothing of the mechanisms governing the contagion of laughter, either in their physiological dimensions or their sociopsychological implications.

In the important treatise he devoted to laughter in 1579,[3] the physician Laurent Joubert endeavours to distinguish laughter's true essence from its counterfeits: pathological rictus, laughter of newborns, laughter provoked by tickling. This perspective leads to the exclusion of automatic and reflexive laughter, growing out of what is literally a tactile contagion (tickling), and a denial of the pathogenic imaginary of contagion.[4] Joubert relativizes the impact of interattraction between laughers in order to emphasize the complex hermeneutic of the ridiculous, conceived as an intellectualized and voluntary process: according to Joubert, the perception of an incongruity must be accompanied by an absence of emotional involvement and be mediated by a verbal interaction that presupposes good linguistic comprehension. Joubert refuses to consider as true laughter one that results from a process of kinaesthetic contagion, divorced from intellectual recognition of the ridiculous:

> Tout ainsi qu'un Français qui est parmi des Allemands, n'entendant aucun mot de leur langage, néanmoins les oit bien et les voit rire: mais s'il rit point avec eux: ou ce sera des lèvres seulement.

> [If a man is among Germans, and does not know their language, he will be able to hear them talking and laughing ... And if perchance he begins to laugh, it will certainly be on credit.][5]

Joubert develops a cognitive model of laughter which he contrasts with purely involuntary emotions, such as fear or anger.

Theorists of the passions in the following century insist more on the involuntary mechanisms of laughter, but they do not take as much account of the phenomena of interattraction. Thus, when Descartes deals with laughter provoked by tickling, he superimposes on the physiological, tactile explanation a subtle psychological mechanism which rests on an unconscious reminiscence of a pleasurable moment.[6] Indeed, Descartes hides the external processes that would reveal contagion or empathy.

Only Cureau de la Chambre, in his *Charactères des passions* of 1643, mentions the role played by the group and attributes equal importance to this hidden dimension and to recognition of the ridiculous: 'It is plausible that the group has a role in its production.'[7] This exceptional attention to a psychosociological dynamic highlights a subtle empathy, characteristic of 'commerce', in other words of human sociability. The notion of contagion remains literally unthinkable to the extent that its dysphoric connotations draw laughter into the realm of pathology.

From a therapeutic interattraction to the fear of excessive permeability

This contradiction between the imaginary of laughter and that of disease in the Renaissance is corroborated by the joyous therapeutics of Rabelais, which works precisely through empathy or interattraction. Maître Alcofribas proposes a principle of 'transfusion of the spirits', the joyful doctor breathing health into his patient, as opposed to the practitioner who maintains a severe and sour demeanour:

De fait, ma pratique de médecine bien proprement est par Hippocrate comparée à combat et farce jouée à trois personnages, le malade, le médecin et la maladie. ... le minois du médecin chagrin ... contriste le malade; et du médecin la face joyeuse, sereine, gracieuse, ouverte, plaisante, réjouit le malade ... par transfusion des esprits sereins ou ténébreux, aérés ou terrestres, joyeux ou mélancoliques du médecin en la personne du malade.

[Indeed, the Practice of Medicine is most properly compared by Hippocrates to a Fight, and also to a Farce played between three Characters, the Patient, the Physician and the Disease; ... the Visage of the Physician, when moping ... depresses the Patient, and the Countenance of the Physician, when joyous, serene, gracious, open and pleasant, elates the Patient ... by the Transfusion of the Spirits – serene or gloomy, aërial or terrestrial, joyous or melancholic – of the Physician into the Person of the Patient.][8]

In this approach, which does not dissociate the mind from the body, joy and laughter encourage the flux of vital energy, 'animal spirits' passing from the doctor to the patient according to a complex model that Rabelais borrows from Hippocrates.[9] This emotional transmission between individuals is inspired by belief in natural magic, which was very much alive among the learned of the Renaissance. Rabelais also inscribes this comic therapy in a theatrical perspective: this permits him to consider indirectly the communication that is established between the comic author and his reader.

Among the facetious authors of the seventeenth century, this curative empathy almost disappears. If the Jesuit Binet calls on the therapeutic virtues of the doctor's good humour, his viewpoint, which is to eradicate ills of a moral and spiritual nature, distances him from Rabelais. A joyous facial expression does not constitute more than a secondary stimulus, the diffusion of animal spirits serving as a diversion so that the 'surgeon' might extract the seed of the infection by way of a much stronger method:

> Je fais comme le bon chirurgien qui cache sa lancette dans du coton musqué, pendant qu'il flatte l'aposthume avec le coton, il vous donne un coup et plonge bien avant la lancette. ... pendant que vos douleurs vous tenaillent, je m'efforce de vous réjouir, et voyant votre cœur, je prends mon temps et vous y fourre les recettes toutes propres à votre mal.[10]

> [I do as would the good surgeon who hides his lance in musk-scented cotton, while he caresses the tumour with the cotton, he makes the cut and plunges the lance well into the body. ...while your pain grips you, I attempt to entertain you, and evaluate your dispositions, I take my time and stuff you with the recipes likely to cure your ills.]

In the para-text of comic stories, therapy by contagious laughter is often reduced to a simple stereotype. Thus Du Souhait is content to invoke self-healing from writer's melancholy through laughter, a metaphoric cure that can be transmitted through reading: 'I spent a few days jotting down these conversations to make you laugh, and to entertain myself, because my inclination was to live without melancholy.'[11]

As confidence in the curative interattraction of laughter diminishes in the seventeenth century, censure of uncontrolled outbursts becomes more and more prevalent in theological discussion and in theories of civility, echoing a Christian anthropology hostile to potential disorder and whose medium is laughter.

Suspicion concerning laughter that 'possesses' individuals is at the convergence of a double, but also Platonic, tradition. We should remember that Plato condemns the inextinguishable laughter of Homer's gods and forbids the keepers of his ideal city from giving themselves over to 'violent laughter', because 'indulgence in violent laughter commonly invites a violent reaction'.[12] Among the Church Fathers, unregulated propagation of laughter takes on demonic connotations. According to John Chrysostom, irresistible outbursts constitute a pathway for the transmission of licentious and blasphemous speech:

Ce qui est encore plus dangereux est le sujet pour lequel éclatent ces ris immodérés. Dès que ces bouffons ridicules ont proféré quelque blas-phème, ou quelque parole déshonnête, aussitôt une multitude de fous se mettent à rire et à montrer de la joie.[13]

[Even more dangerous is the subject of this uncontrolled laughter. As soon as these ridiculous buffoons have proferred some blasphemy or other shameful word, immediately a multitude of fools begins laughing and cavorting.]

Vous, par ce rire hardi, vous imitez les femmes insensées et mondaines, et, comme celles qui paraissent sur les planches des théâtres, vous essayez de faire rire les autres. Voilà le renversement, voilà la destruction de tout bien. Nos affaires sérieuses deviennent des sujets de rire, de plaisanteries et de jeux de mots … Je ne parle pas seulement ici aux séculiers; je sais ceux que j'ai encore en vue; car l'église même s'est remplie de rires insen-sés. Que quelqu'un prononce un mot plaisant, le rire aussitôt paraît sur les lèvres des assistants; et, chose étonnante, plusieurs continuent de rire jusque pendant le temps des prières publiques.[14]

[You, with this hardy laughter, you are imitating unseemly and worldly women and, like actresses on stage, you try to make others laugh. Behold the reversal, the destruction of all that is good. Our serious affairs become subjects of laughter, jokes and puns … I am not speaking only to the secular; I know those whom I still have in my sights; for even the church has filled with unseemly laughter. If someone pronounces a witticism, laughter immediately springs from the lips of the congregation; and, shock-ingly, many continue to laugh even during the time of public prayers.]

This obsessive fear of faltering defences caused by laughter suggests Freudian views about the convergence of the individual unconscious when jokes are told.[15] Associated with the liberation of aggressive and sexual impulses, the dark aspect of irrational laughter is assimilated in theological discussion to a demonic plot, which leads to its general condemnation.

Heightened religious sensibilities during the Counter-Reformation are one cause of mistrust of the latent contagion of laughter on the part of men of the Renaissance and the classical age; another is the evolution of mentalities. The development of subjectivity gives rise to growing uneasiness with collective interattractive phenomena. Montaigne considers threatening the permeability between the self and its environment: 'conta-gion is very dangerous in the crowd'.[16]

Emotional interattraction is discredited in the seventeenth century for another reason: a process of socio-aesthetic 'distinction' assimilates the spread of collective laughter to a popular 'habitus'. The 'civilizing process', to use Norbert Elias's expression,[17] condemns visceral laughter rising simply from the gut, favouring instead more refined forms of jest.[18] In his collection *Des bons mots et des bons contes* (1692), Callières extols a cultivated laughter, the exclusive privilege of an elite capable of allowing the intellectual pleasure of analysis to triumph over the elemental virulence of contagion:

> Le bon goût ... ne se rencontre qu'en un petit nombre de gens capables de juger par eux-mêmes du prix de chaque chose, ceux-là ne suivent pas le torrent des sots rieurs qui bien souvent ne rient que parce qu'ils voient rire d'autres sots comme eux, dignes eux-mêmes de risée d'applaudir à de méchantes choses. [19]

> [Good taste ... is only found in a small number of people capable of judging for themselves the true value of each thing; they do not follow the crowd of laughing fools who very often only laugh because they see other fools like them laughing, worthy themselves of mockery for applauding stupidity.]

Callières, intending to erect barriers against spontaneous movements, deems the purely mechanical transmission of laughter ridiculous. The division of body and mind is attached to a logic of social distinction, whose stakes are to prevent all mixing or dissolution of hierarchies. From this point on, we perceive the unbridled exchange of carnivalesque laughter that exorcised massive fears in the Middle Ages as a socially destabilizing factor.[20]

'Complicity': a refined alternative to basic interattraction

Determined to promote human exchange based on charm, grace and 'amiability', theorists of civility none the less do not discount the benefits of joyous interattraction, and they vigorously defend the pleasures of refined 'complicity' that sophisticated jesting permits. In contrast to passive and uncontrollable contagion, they recommend moderate emotional permeability, reconcilable with the active and conscious participation of each individual. Contributing to the dissemination of a refined good humour compatible with reason, subtle jesting is considered an indispensable ingredient of civil harmony. Antoine de Courtin emphasizes, in his *Traité de la civilité*, that 'an agreeable person, who brings joy and laughter wherever s/he goes, undeniably has the charm to please'.[21] This art of laughter foils the clown's incessant, misplaced farce. Recommended is respect for the essential rule of 'circumstances', which guarantees behaviour in conformity with the specific configuration of time, place and persons.[22]

The spatial problematic is key in the context of this necessary regulation of the positive interattractions of refined jesting among 'peers'. Limiting outbursts of laughter within a restrained and closed circle preserves the social body from the risks of mixing and equalizing. This social and symbolic cloister put in place by the civilizing process intensifies the sociological logic inherent in 'the complicity of laughter'.[23]

Interattraction of outbursts and the aesthetics of comic reception

If normative discourse stigmatizes the uncontrollable propagation of laughter, the practice of comic authors proves to be more ambiguous. To produce laughter, do they not put into practice strategies of interattraction that lead to the unthinking adhesion of their public? The rhetorician Lamy explains a rule of semiotic continuity inherent in the eloquence of the passions: 'The natural signs of the passions make an impression on those who see them, and unless they resist, they end up being carried away.'[24]

Comedy, between the redundancy of laughter and distancing

The collective reception of performed theatre obviously favours an irrational collective dynamic. In fact, Dorante, Molière's mouthpiece in *La Critique de l'école des femmes*, insists on the impetus of a laughter that abolishes social distinctions, and he ridicules the foolish pretensions of those who refuse, out of snobbery, to enjoy themselves along with the popular audience.[25] In the polemic surrounding *L'école des femmes*, Molière's adversaries denounced the tyranny of public laughter: in his *Portrait du peintre*, Boursault, like Joubert, distinguishes true laughter from those forms of kinaesthetic imitation devoid of meaning: 'Any man could be called brutal / If he laughs at the thing, and not the signal.'[26]

Molière placed subtle signposts in his comedies, using 'laughing' characters to point out the ridiculous to spectators: this redundancy, quite noticeable in *Le Misanthrope*,[27] is none the less never systematic and requires active collaboration from the reader. Molière also manipulates distancing effects: thus Arnolphe excuses himself, in *L'école des femmes*, for not having played his role of 'jester' very well.[28] Molière's contemporaries, Montfleury, for example, also exploit the disproportion between outbursts of laughter on stage and the comic reception of their plays. But it is not part of my argument to insist on these gaps:[29] they confirm that an aesthetic of 'contagion' is not operative here.

Novelistic dissonances

It will be more interesting to linger on the thematization of the propagation of outbursts in the novel. Comic fiction reflects on processes of interattraction, which become all the more ambiguous when they also offer a potential mirror of fiction's reception.

It is in Scarron's *Roman comique* that the narrative actualization of the contagion of laughter is most recurrent:

La Garouffière en rit bien fort et donna si bien le branle à toute la Compagnie qu'elle en éclatta à quatre ou cinq reprises. Les valets reprirent où leurs Maistres avoient quitté et en rirent à leur tour; ce que la jeune Mariée trouva si plaisante que, s'ébouffant de rire en commençant à boire, elle couvrit le visage de sa Belle-mère et celuy de son Mary de la plus grande partie de ce qui estoit dans son verre ... On recommença à rire et la Bouvillon fut la seule qui n'en rit point ... Enfin on acheva de rire, parce que l'on ne peût pas rire tousjours.

[La Garouffière laugh'd heartily, and put all the Company in so good a Humour, that they broke out into laughter four or five several times. The Servants began where their Masters left off, and laught in their turn, which the Bride found so Comical, that breaking out into laughter as she was going to Drink, she spurted the greatest part of the Wine, which was in her Glass, on her Mother-in-law, and her Husband's Face ... They all began to laugh again, except Bouvillon ... At last they made an end of laughing, because 'tis not possible to laugh for ever ...][30]

As early as the third chapter of Part II of the novel, Scarron reveals the inner workings of an irresistible comic empathy, whose mediated cause was witty repartee, but there was also to a more immediate effect on the group:

Cette mauvaise rime surprit tout le monde. Le Comédien qui faisait le personnage d'Aymond s'en ésclatta de rire et ne put plus representer un vieillard en colère. Toute l'assistance n'en rit pas moins et pour moy, qui avois la teste passée dans l'ouverture de la tapisserie pour veoir le monde et pour me faire veeoir, je pensay me laisser choir à force de rire. Le Maistre de la maison, qui estoit de ces melancohliques qui ne rient que rarement et ne rient pas pour peu de chose, trouva tant de quoy rire dans le défaut de mémoire de son Page et dans sa mauvaise maniere de reciter des Vers qu'il pensa crever à force de se contraindre à garder un peu de gravité; mais enfin il fallut rire aussi fort que les autres, et ses gens nous avouërent qu'ils ne luy en avoient jamais veu tant faire. Et, comme il s'estoit acquis d'une grande authorité dans le Païs, il n'y eut personne de la compagnie qui ne rît autant ou plus que luy, ou par complaisance ou de bon courage.

[This false Rime surpriz'd every Body; he that acted Aymon's Part burst out a-laughing, and was not able to represent an Angry old Man. All the Assistants laugh'd as well as he; and I my self, who was then peeping through the Hangings to see and be seen, laugh'd also to that degree that I was ready to drop down. The Master of the House, who was one of those Melancholy Persons who laugh but seldom, and never at a small Matter, found his Page's want of Memory, and his aukward way of reciting Verses so laughable a Subject, that he was like to burst by endeavouring to preserve his Gravity; but at last he was fain to laugh as well as the rest; and his Men told us since, that they never knew him so well pleas'd in all thier Lives. Now as he was a Man of great Authority in that Country, there was not one Person in the whole Audience that did not laugh as much as he, or perhaps more, either out of Complaisance, or a natural Inclination.][31]

The extreme case of a character who never laughs corroborates both what is comic in the situation and the impact of immersion in a group. The narrator also notes the decisive role played by the person with dominant status, and who tacitly authorizes the expression of the most natural emotions: 'amiability', the key to civility, underpins emotional interattraction.

In the commentary that follows, Scarron makes obvious, by way of his characters' dialogue, the difference between the oral and immediate context of the transmission of laughter, and its distanced narrative transposition. The narrator, who has not made her listeners laugh much, wonders if, because of her insistence on past outbursts of laughter, she hasn't sapped the element of surprise necessary to produce comic effect: 'I am very much afraid, added Cave, I have now done like those who tell People, *I'll tell you a Story that will make you die with Laughing*, and who seldom or never are as good as their Word: For I must confess I rais'd your Expectation too high about the silliness of my Page.'[32] L'Estoile emphasizes for his part the irreducible gap between the narration of a comical episode and its lived experience, the empirical presence of a group of 'laughers' being indispensable to emotional permeability: ''Tis true the thing may have seem'd more ridiculous to those that saw it, than it will to those who shall hear it related, the aukwardness of the Page contributing much to make it such; and besides, the Time, the Place, and the natural Inclination we have to laugh for Company's sake, are all Advantages it cannot have now.'[33]

Would the gap between lived laughter and narrated laughter have as its inescapable consequence entropy in its transmission? We can in any case ask ourselves if this metatextual commentary does not proceed by wily strategy on the part of the novelist, desirous of covertly exhibiting his own success. Does the exchange between La Caverne and L'Estoile not tend to revive, inside the text itself, the energy lost from direct oral contagion? Like the sublime, comic eloquence is confronted with the challenge of the distance between presence and representation, which reproduces in part the distance between orality and writing.

The function of these *mises en abyme* of the contagion of laughter in the comic novel thus appears complex: such descriptions can play the role of 'triggers', inciting the reader to laughter. But this comic mimetism is not always verifiable and the semiotics of laughter in comic stories can be most equivocal. In Sorel's *Histoire comique de Francion*, the reader's identification with the laughing characters is counterbalanced by the negative image of the latter, who can be vulgar and stupid as are the peasants in Book I or the scoundrel in Book IV. Sorel multiplies the discrepancies, as in the mocking of Hortensius, the target of a farcical game that exposes him to the readers' laughter. These complex distancing strategies go hand in hand with an acute consciousness of the irreducible part played by the reader in an aesthetic of written reception. At odds with the fusional model of interaction in storytelling, Sorel stigmatizes the rudimentary strategy used by Béroalde de Verveille in his *Moyen de parvenir* that consisted of emphasizing the amusing passages by way of outbursts from the protagonists:

[l'autheur mettait quand il fallait pleurer ou quand il fallait rire de peur que l'on ne s'y trompât, si bien qu'après un bon mot, il y avait toujours quelques interjections qui exprimaient une risée générale, mais c'est être falot que de s'amuser à toutes ces brouilleries.][34]

[the author signalled when it was necessary to cry or to laugh from fear that the reader might be misled, to the extent that after a witty remark, there were always a few interjections expressing general amusement, but it's ridiculous to be amused by all of these senseless remarks.]

Reluctant to use such crutches, which he judges ridiculous and useless, Sorel postulates the necessarily mediated nature of recognition of the comic in the context of individual silent reading.

Sorel furthermore discredits the vulgar interattraction of outbursts of laughter during Francion's dream. The monsters, who imitate gestures of hilarity without understanding, substitute an inarticulate infra-language for authentic communication:

Je leur allai répondre que j'étais bon à les faire rire, et pour leur témoigner, je me pris à rire si fort moi-même qu'ils furent contraints de rire aussi, sans savoir pour quelle occasion.

[I was going to tell them that I was capable of making them laugh, and to prove it, I began to laugh so much myself that they too were forced to laugh, without knowing why.][35]

In Sorel's imaginary, this stupid laughter, associated with an animal brutality, is opposed to the intellectualized joy of libertines, who cultivate an enlightened complicity and perfectly control the signs of their joy, and are even able to suppress them if need be.[36] The regulation of laughter in

Francion confirms more generally the relevance of the models of civility in the novel.[37] In his critical work, and in particular in his considerations of the burlesque, Sorel consistently condemns phenomena of irrational inter-attraction.

The metaphor of 'burlesque contagion': a pathology of comic writing

Contagion as unbounded imitation

Sorel explicitly stigmatizes the contagious illness that is the burlesque, repeating a topic much favoured by adversaries of the genre until the nine-teenth century.[38] In the critical appraisal *De la connaissance des bons livres*, Sorel detects the germ of this pathology in the penchant towards facility which encourages the ignorant to write poetry. This unregulated mimetic process perverts the humanistic and classical model of discipline:

> On ne voit rien de si commun aujourd'hui que cette sorte de style; il semble que toute la France soit malade du burlesque; il n'y a personne qui ne s'estime capable d'en faire. Quantité de gens sans étude, et de toutes les conditions, et même des femmes et des files s'entrécrivent des lettres en vers ... On trouve aussi des hommes qui à peine savent lire, lesquels ont la hardiesse de faire imprimer des livres en vers de cette nature, et c'est bien ce qui en montre la facilité.[39]

> [Nothing is more common today than this sort of style; it seems that all France is ill from the burlesque; there is no one who doesn't consider himself capable of reproducing it. Numerous uneducated persons, of all social classes, and even women and girls write each other letters in verse ... One also finds men who can barely read, and who have the nerve to have books of verse of this nature printed, which clearly demonstrates how facile it is.]

Sorel makes a notable exception for Scarron, the inventor of the burlesque genre. But he blames Scarron's successors, whose writing fever is inspired by an uncontrollable impulse. They abandon all requirements of style and disdain current poetic rules and hierarchies:

> Remarquons aussi que plusieurs qui étaient capables de meilleures choses, se sont trop amusés à ce genre d'écrire, et que sa contagion a quelquefois gagné jusques à la prose, quoique son règne principal soit toujours dans les vers. L'accoutumance qu'on a eue à un méchant style va corrompre insensiblement le style noble et sérieux qu'il faut employer aux grands sujets. Ceci a été la cause seule qu'on a commencé de nég-liger la sévérité des lois de la poésie ...[40]

[Notice also that many who were capable of better things, have amused themselves too much with this sort of writing, and its contagion has at times affected even prose. The habit of defective style is going gradually to corrupt the noble and serious style necessary for important subjects. This has been the sole cause of the beginning of our neglect of the strict laws of poetry ...]

The insidious threat linked to burlesque rewriting is the confusion of aesthetic markers. Compromising all authority, this licentious writing is detrimental to the author's voice as much by its irreverence with regard to the venerated models of Antiquity as by the levelling that it engenders between learned writing and that within reach of the vulgar: 'One may have as much esteem as one likes for Burlesque Verse; but it is certain that since it has appeared in France, the people appreciate less those we called Poets and Authors, taking them all for tellers of tall tales.'[41] The contagion of the burlesque inspires a latent anarchy; it induces indistinct writing, dedicated to an automatic form of production and running the risk of anonymity, in total contradiction with the emerging status of the writer.[42]

A medical work at the trial of a deviant aesthetic

Other detractors of the 'burlesque disease' accused more precisely the furor inherent in farcical contagion. An anonymous opuscule published in 1651 and entitled *Contre Satyre ou Réponse aux Cent quatre vers du sieur Scarron, pour lui montrer qu'ayant inventé les vers burlesques, il se peut dire l'auteur des libelles diffamatoires de cette espèce*[43] denounces, like Sorel, excessive imitation, but it establishes a link between this perversion and the contagious effect of laughter and mockery, ingredients essential to the pathology of mimetism. The unbridled 'licence' 'to mock everything' appears indissociable from a frenzy of imitation, which saps all the authority that guarantees order within the social body:

Car l'un d'eux aussitôt fit un Burlesque Enfer,[44]
En se moquant de lui, comme de Lucifer,
Où scandaleusement à toutes nos Provinces,
Il met et Magistrats, et Ministres, et Princes;

Depuis un autre aussi, par imitation
En de semblables Vers a mis la Passion;[45]
Et voilà les effets que nous avons vu suivre
De l'approbation de ton malheureux Livre

Faut-il donc s'étonner si par un même état,
On s'est après moqué des affaires d'Etat?

Car, en autorisant ainsi la raillerie
L'esprit cherche par tout un sujet dont il rie,
Et n'en pouvant trouver en aucun autre lieu,
Il s'attaque à la fin, ou au Prince, ou à Dieu,
Parce que l'homme étant en son propre, risible,
Est à ce mouvement aussi le plus sensible,
Et contrefaisant tout ce qu'il voit pratiquer,
Il s'attache sur tout à rire ou se moquer.

[For one of them made up a Burlesque Hades,
In mocking Lucifer and all men and ladies,
Where scandalously for all of our Provinces
He puts magistrates, and Ministers, and Princes;

Since then yet another, by crude imitation
In similar verse has transcribed the Passion;
So notice the effects upon which we may look
From the approval of your unfortunate Book.

Is it therefore shocking if by the same token,
They mock affairs of State as if they were broken?

For, making mockery thus very visible,
The mind looks everywhere for something risible,
And unable to find elsewhere such levity,
In the end, it attacks the Prince or Deity,
Because man is in essence for mocking meant,
He's also sensitive to this predicament,
And imitating all he sees in practice,
He's especially attracted to laughter or malice.]

This judgement of the burlesque recalls the persistent theological critique of the relaxing of the laugher's defences. The pleasure of imitation crystallizes aggressive antisocial impulses.

Most noteworthy is the performative dimension of this denunciation of burlesque corruption: the pathological metaphor of an ill-defined 'sickness' that spreads insidiously, overlaid by the terminology of poisoning, suggests a judicial rhetoric that calls for the chastisement of a guilty party designated by name: the diseased Scarron is presented as the source of the burlesque illness.

In his defence and illustration of a refined burlesque, Dassoucy insists on the inanity of the metaphor of burlesque contagion, suggesting the clever mechanisms of exclusion that this abuse of language hides. He emphasizes the inappropriateness of this pathological image for the purpose of denouncing the poor aesthetic quality of much burlesque verse: 'the worst lines, as bad as they are, have not infected nor given the plague to anyone'; according to him, criticism should restored to its rightful place, measured by readers' displeasure, since this poetry has simply 'bored and bothered many'.[46] Dassoucy does not, however, resist the temptation to turn the weapon of the pathological metaphor against his adversary Boileau, challenging the defamatory discourse of the super-satirist 'whose venom must have done much more to infect Paris and the provinces than our innocent burlesque'.[47]

From the literal to the metaphorical, the motif of contagion in the representations of laughter and in the reflection on reception and comic creation betrays dissonances, even contradictions. If we cannot discover a coherent rationality in these fragments in order to construct an archaeology of the imaginary of the contagion of laughter, we can none the less highlight the totally pejorative perspective that assimilates the interattraction of outbursts with a principle of excess undermining free will. As in the case of discourse relative to contagion in novels, the background of these negative images of interattraction is that of a Christian anthropology marked by the obsessive fear of the insidious spread of aggressive and sexual impulses.

Clinical analyses of passions from the Renaissance to the seventeenth century are characterized by a significant denial of the medical motif of the contagion of laughter: the idea of a pathological laughter seems obviously incompatible with an obsessive concern for legitimizing its 'healthy' manifestations, and for excluding its inauthentic forms, reflexive and uncontrollable laughter, which tend to be relegated to the confines of disease and madness.

If philosophers hide the transmission mechanisms of laughter from view, comic authors and the authors of treatises on civility develop in an indirect or allusive manner more incisive considerations about the interattraction of laughers and the irrational propagation of hilarity. But they rarely insist upon these ideas, so clear is it that the legitimization of laughter in the classical age must make only discreet allusions to spontaneous transmission. Rather than the risk of chaotic contagion, they seek refined complicity on the part of the intelligentsia. The model of civility thus opposes the unbridled interattraction of vulgar clowning – perceived as an attack against the social body and its hierarchies – with the virtues of a controlled empathy: refined jesting is the proof of harmonious communion among peers.

It is symptomatic that the metaphor of contagion intervenes conversely in an explicit way to stigmatize the fashion of burlesque travesties, conceived of as a pathology of writing. The metaphor of contagion is inscribed in a performative discourse where we find that the obsessive fear of inappropriate laughter points to the danger of social and aesthetic anarchy. We must wait until the eighteenth century to witness an energetic problematic of the fluid of contagion. Marmotel, in the article 'Parterre' of his *Éléments de littérature*, celebrates the mediated and immediate interattraction of the public with the theatre, whether it is being moved to laughter or to tears:

Qu'on se figure cinq cents miroirs se renvoyant l'un à l'autre la lumière qu'ils réfléchissent, ou cinq cents échos le même son; c'est l'image d'un public ému par le ridicule ou le pathétique. C'est là surtout que l'exemple est contagieux et puissant. On rit d'abord de l'impression que fait l'objet risible, on reçoit de même l'impression directe que fait l'objet attendrissant; mais de plus, on rit de voir rire, on pleure aussi de voir pleurer; et l'effet de ces émotions répétées va bien souvent jusqu'à la convulsion du rire, jusqu'à l'étouffement de la douleur. Or c'est surtout dans le parterre, et dans le parterre debout, que cette espèce d'électricité est soudaine, forte et rapide. [48]

[Imagine five hundred mirrors sending one to the other the light that they reflect, or five hundred echos of the same sound; this is the image of a public moved by the ridiculous or the pathetic. It is there most especially that the example is contagious and powerful. One laughs first at the impression that the risible object makes, one receives in the same way the direct impression that a touching object makes; but what is more, one laughs at laughing, one cries as well at crying; and the effect of these repeated emotions goes very often to the point of convulsive laughter or to suffocating pain. Now, it is especially among the people, those without seats, that this kind of electricity is sudden, strong and rapid.]

The critic rehabilitates an exemplary emotional interattraction which the public controls: challenging the criteria of classic taste, Marmontel plunges himself into the breach opened by Molière to make contagion consonant with 'the healthiest reason and the most naive sensibility' of an audience composed of less rich and less refined citizens, in whom 'the natural is the least polished, but also the least altered', and who show themselves to be the most apt to judge works with 'the most common sense' without allowing themselves to be influenced by fashions, or by prejudices of education and of vanity.[49] This apology is the indicator of a new discourse, which shifts the pejorative imaginary of contagion and develops, on the foundation of a durable discipline of laughter,[50] a manifesto in favour of the 'right to laugh' and of the comic:[51] it constitutes an invitation to re-examine critically the sociological and ideological rifts of the 'civilizing process'.

Notes

1 G. Bataille, 'Attraction and Repulsion I: Tropisms, Sexuality, Laughter and Tears', in *The College of Sociology (1937–39)*, trans. B. Wing (Minneapolis: University of Minnesota Press, 1988), p. 107. For the philosopher, 'contagious weeping and erotic contagion are the only things that, subsequently, will be able to deepen human communication' (p. 109).
2 *Ibid.*
3 *Traité du ris* (1579; rpt. Geneva: Slatkine, 1973); *Treatise on Laughter*, trans. and annotated by G. D. de Rocher (Alabama: University of Alabama Press, 1980).
4 In this same spirit, the physician rejects the notion of 'trembling' that suggests a pathological agitation; he prefers the neutral term 'movement'. See the analyses of D. Ménager, *La Renaissance et le rire* (Paris: PUF, 1995), p. 25 ; and B.-R. Vasselin, in *Montaigne et l'art de sourire à la Renaissance* (Paris : Nizet, 2003), pp. 281–6.
5 *Traité du ris*, p. 295; *Treatise on Laughter*, p. 26.
6 R. Descartes, *Traité des passions* (1649; rpt Paris: Bourgois, 10/18, 1965), p. 161: 'plusieurs ne sauraient s'abstenir de rire étant chatouillés, encore qu'ils n'y prennent point de plaisir, car l'impression de la joie et de la surprise, qui les a fait rire autrefois pour le même sujet, étant réveillée en leur fantaisie, fait que leur poumon est subitement enflé malgré eux par le sang que le cœur lui envoie.' *The Passions of the Soul*, trans. and annotated by S. Voss (Indianapolis and Cambridge: Hackett Publishing Co., 1989), p. 134: 'Thus many cannot abstain from laughing when tickled, even though they derive no pleasure from it. For, in spite of themselves, the impression of Joy and surprise which previously made them laugh for the same reason, being awakened in their fantasy, makes their lungs suddenly swell with the blood that the heart sends there.'
7 *Les Charactères des passions* (Paris: J. d'Allin, 1663), vol. 1, ch. IV, 'Du ris', p. 247: 'Il est vraisemblable que la compagnie sert de quelque chose à sa production.'
8 *Lettre à Monseigneur Odet* in F. Rabelais, *Œuvres complètes*, ed. G. Demerson (Paris: Seuil, 1973), p. 563; *The Five Books and Minor Writings*, trans. W. F. Smith (London: Alexander P. Watt, 1893), 2, pp. 14–15.
9 On this subject, see Ménager, *La Renaissance et le rire*, pp. 70–2.
10 É. Binet, *Consolation et réjouissance pour les malades et personnes affligées* (Paris: 1617), p. 3.
11 'Au lecteur', *Histoires comiques*, ed. D. Celce-Muria (Paris: La Pensée universelle, 1978), p. 31: 'J'ai passé quelques jours à tracer ces entretiens pour vous faire rire, et pour me désennuyer, parce que mon inclination était de vivre sans mélancolie.'
12 Plato, *The Republic*, trans. D. Lee (London: Penguin Books, 1987), 388e, p. 144.
13 Jean Chrysostome, *Commentaire sur saint Matthieu* in *Œuvres complètes*, ed. M. Jeannin, (Paris, 1865), 7, pp. 520-1; trans. C. C.
14 Jean Chrysostome, *Commentaire sur l'Épître de saint-Paul aux Hébreux* in *Œuvres complètes*, vol. 11, p. 537.
15 It should be noted, however, that Freud restricts the interaction of jokes to an interindividual and limited exchange. See 'Jokes and Their Relation to the Unconscious' (1905), in *The Standard Edition of the Complete Psychological Works of Sigmund Freud*, trans. and ed. J. Strachey with A. Freud, A. Strachey and A. Tyson (London: The Hogarth Press and the Institute of Psycho-Analysis, 1960), 8.
16 'Of Solitude', *The Complete Essays of Montaigne*, trans. D. M. Frame (Stanford: Stanford University Press, 1958), p. 175. 'De la solitude', *Essais*, I, XXXVIII,

ed. D. Bjaï, B. Boudou, J. Céard and I. Pantin (Paris: Pochothèque, 2001), p. 367: 'la contagion est très dangereuse en la presse'.

17 N. Elias, *The Civilizing Process*, 2 vols, trans. E. Jephcott (Oxford: Basil Blackwell, 1978).

18 On this subject, see my entries 'Raillerie' and 'Ridicule' in the *Dictionnaire de la Civilité et de Savoir-Vivre*, ed. A. Montandon (Paris: Seuil, 1997).

19 F. de Callières, *Des bons mots et des bons Contes, de leur usage, de la raillerie des Anciens et des railleurs de notre temps* (Paris: Barbin, 1692), p. 17.

20 See my book, *Dire le rire à l'Âge Classique* (Presses de l'Université de Provence, 1995).

21 A. de Courtin, *Nouveau traité de la civilité*, ed. M. C. Grassi (Publications de l'Université de Saint-Étienne), p. 98: 'une personne commode, et qui porte la joie et les ris partout où elle va, a des charmes infaillibles pour plaire'.

22 On this subject, see my entry 'Circonstances' in the *Dictionnaire de la civilité et du savoir-vivre*.

23 J. Duvignaud insisted on this preliminary delineation in the social space that all 'jesting complicity' imposes: 'les banquets carolingiens ... les veillées dans des hôtelleries installées le long des routes de pèlerinage, les réunions organisées dans la demeure des notables des villes ou dans le secret des cours princières, des cloîtres, des monastères, des universités où se regroupent les clercs ... les jeux de paume, les salons' ('Carolingean banquets ... long nights in inns on pilgrimage routes, meetings organized in the dwelling of town notables or in the secrecy of princely courts, cloisters, monasteries, universities where clerics mingle ... tennis courts, salons'), *Rire et après, Essai sur le comique* (Paris: Desclée de Brouwer, 1999), p. 142.

24 B. Lamy, *La rhétorique ou l'art de parler* (Paris, 1699), p. 367: 'On a de la joie avec ceux qui rient. Les signes naturels des passions font impression sur ceux qui les voient, et à moins qu'ils ne fassent de la résistance, ils se laissent aller.'

25 *La critique de l'École des femmes*, scene 5.

26 *Le portrait du peintre ou la contre-critique de l'École des femmes* (Genève: Slatkine, 1969): 'tout homme, à moins d'être brutal, / Doit rire de la chose, et non pas du signal'.

27 On this subject, see R. Parish's study, '*Le Misanthrope*: des raisonneurs aux rieurs', *French Studies*, 45, 1 (January 1991) 17–35.

28 Acte III, scene 4.

29 For more precise analyses, refer to my book, *Dire le rire à l'Âge Classique*, Part 3, ch. IV, pp. 241–51.

30 *Le roman comique*, in *Romanciers du XVIIe siècle*, ed. A. Adam (Paris: Gallimard, Bibl. de la Pléiade, 1958), p. 707; *The Whole Comical Works by Paul Scarron*, trans. T. Brown, ed. J. Grieder (1700; rpt. New York and London: Garland Publishing, 1973), 1, pp. 178–9.

31 *Ibid.*, p. 681; trans. pp. 154–5.

32 Trans. p. 155. *Ibid.*: 'J'ay grand'peur, adjousta alors La Caverne, d'avoir fait icy comme ceux qui disent: Je m'en vay vous faire un conte qui vous fera mourir de rire et qui ne tiennent pas leur parole, car j'avouë que je vous ay fait trop de feste de celui de mon Page.'

33 Trans., *ibid*; pp. 681–2: 'Il est bien vray que la chose peut avoir paru plus plaisante à ceux qui la virent qu'elle ne sera à ceux à qui on en fera le récit, la mauvaise action du Page servant beaucoup à la rendre telle, outre que le temps, le lieu et la pente naturelle que nous avons à nous laisser aller au rire des autres peuvent lui avoir donné des advantages qu'elle n'a pu avoir depuis.'

34 *Remarques sur le Berger Extravagant* (Geneva: Slatkine, 1972), p. 748.

35 *Histoire comique de Francion*, in *Romanciers du XVIIe siècle*, p. 146.

36 For a more precise analysis of the antithetical representations of laughter in Sorel's novel, see my article, 'Les représentations du rire dans *L'Histoire comique de Francion*', in *Charles Sorel, Histoire comique de Francion, Cahiers Textuels* (Université Paris VII, 2000), pp. 51–65.

37 See N. Doiron's article in *Charles Sorel dans tous ses états*, ed. E. Bury and E. van der Schueren (Quebec: Presses de l'Université Laval, 2004).

38 In 1906, Ferdinand Brunetière entitled a study 'La maladie du burlesque', *Revue des deux mondes* (July–August 1906), p. 667.

39 C. Sorel, *De la connaissance des bons livres ou examen de plusieurs auteurs*, ed. H. Béchade (1671; rpt. Geneva: Slatkine, 1981), p. 229.

40 *Ibid.*, p. 231.

41 *Ibid.*: 'On fera telle estime qu'on voudra des Vers Burlesques; mais il est certain que depuis qu'on en a vu en France, tout le peuple a fait moins de cas de ceux qu'on a appelés Poètes et Autheurs, les prenant tous pour des conteurs de sornettes.'

42 See Alain Viala, *Naissance de l'écrivain, Sociologie de la littérature à l'Age classique* (Paris: Minuit, 1985).

43 Claudine Nedelec presents this text in the preamble to her work, *États et Empires du burlesque* (Paris: Champion, 2004), pp. 13–17.

44 Allusion to *L'Enfer burlesque ou le sixième de l'Enéide travestie et dédiée à Mademoiselle de Chevreuse; le tout accommodé à l'Histoire du temps*, which appeared in 1649.

45 Reference to *La Passion de Notre Seigneur, en vers burlesques, dédiée aux Ames dévotes* (Paris: Vve J. Rémy, 1649).

46 M. Dassoucy, *Aventures d'Italie*, ed. Colombey (Paris, 1853): 'tout méchants qu'ils soient, [ils] n'ont point infecté ni donné la peste à personne.'

47 *Ibid.*: 'dont le venin ... doit avoir bien plutôt infecté tout Paris et nos provinces que notre innocent burlesque'.

48 J.-F. Marmontel, 'Parterre' in *Éléments de littérature, Œuvres complètes* (Paris: Belin, 1819), 4, Part 2, p. 830.

49 *Ibid.*: 'la raison plus saine et la sensibilité plus naïve'; 'le naturel est le moins poli, mais aussi le moins altéré'.

50 On the persistent discrediting of uncontrolled laughter, see Lise Andriès, 'État des recherches', introduction to the issue of *Dix-Huitième Siècle* on *Le Rire*, 32 (2000) 14–15.

51 I borrow this felicitous phrasing from Dominique Quéro and refer to her article 'Le rire du public du théâtre', in the special issue of *Dix-Huitième Siècle* on *Le Rire*, *ibid.*, pp. 67–83.

13
The Pathology of Reading: The Novel as an Agent of Contagion[1]

Michel Fournier

In order to justify having devoted a chapter of his *Testament ou conseils fidèles d'un bon père à ses enfants* (1648) to 'such a frivolous matter' as the reading of novels, Philippe Fortin de la Hoguette writes to his son: 'novels are a disease of our time; it has been mine, perhaps it will be yours.'[2] If Fortin de la Hoguette uses the term 'disease' in a relatively neutral way to designate the strength of this passion for reading novels, other early modern authors call on this metaphor to condemn the novel strongly. Novel reading is a contagious illness; it is characterized by the epidemic proportions that Jean-Pierre Camus denounced as early as the first decades of the seventeenth century, writing that 'every day the ill grows and spreads, the world is drunk on fables'.[3] The novel not only gains more and more readers, it eventually contaminates the universe of letters.[4]

This corruption of the written object and, through it, of taste in general, is far from being the only effect of contagion by novels. The corruption provoked by the novel surpasses the domain of taste to reach that of mores. Early modern critics who attack bad reading habits make the novel into the contagious genre *par excellence*, as it propagates vice and illegitimate passions. Thus, in his *Peintures morales* (1642–43), after having made of reading a 'silent conversation, that we have with the dead and the absent', Pierre Le Moyne reminds us that 'one must carefully guard against these contagious dead, who augment the diseases of the living, and who poison by imagination and by sight'.[5] Le Moyne then quotes a series of novels as examples of these contagious books. It is once again this risk of contagion that will be at the heart of the Augustinian denunciation of theatre[6] and of novels, which the Academy uses to criticize *Le Cid* in 1637 by recalling that 'bad examples are contagious, even on the stage'.[7] In the following century, in his *Entretiens sur les romans* (1755), Armand-Pierre Jacquin attacks novel reading, with the objective, following Porée, of making visible 'how many traps novels lay for innocence, and how difficult it is to protect oneself from *their contagious corruption*'.[8]

This denunciation of the contagious effects of novel reading is based on the idea of a moral contagion developed by Seneca in *On Tranquillity of Mind*. Appropriated by the moralists, from Montaigne to La Bruyère by way of Charron, this idea appears in the definition of the term 'contagion' given by Furetière in his *Dictionnaire universel* (1690). Contagion, writes Furetière, 'is said figuratively of things moral, such as vices and heresies, that are acquired by communication with those who are infected by them'.[9] Reflection on the novel as genre is influenced by the medical perspective that informs moral thought,[10] among other areas, by way of the problem of the passions. Moral contagion proves all the more problematic since the use of examples is at the heart of the pedagogy of the time; it finds exemplary expression in the imitation that may take over the novel reader. This phenomenon of imitation, staged by *Don Quixote*, is indeed thematized in terms of 'contagion' in the *Berger extravagant* (1627–28) by Charles Sorel who, in the manner of Cervantes, tells the story of a young man gone mad through reading pastoral novels. A character reprimands in these terms the hero of this text who, not content to simply live his madness, seeks to communicate his delirium to all he encounters: 'Kind shepherd I am sorry to see your mind possessed by an infinity of wrong opinions, which moreover, you desire to *make a contagion*, and communicate them to all who approach you.'[11] The idea of contagion accentuates a problematic aspect of the phenomenon of imitation. In effect, imitation not only puts into play a mechanical reproduction of represented behaviours, it also puts a force into play whose action is much more diffuse.

The denunciation of contagion from novels is inscribed in the extension of an organic imaginary which makes reading the food of the soul. In his essay entitled 'That the reading of the books of the *Amadis* is no less pernicious to young people, than reading Machiavelli is to the old' (1587), François de La Noue calls on this imaginary in denouncing the effect of reading the *Amadis*.[12] 'And from this food, he writes, are engendered bad humours which have made souls ill, who did not think from the start that they would accidentally be afflicted by this indisposition.'[13] The medical doxa is thus evoked through humoral theory, and the corruption that this unsound reading brings is conceived of in terms of the imaginary of uncleanliness with which disease is associated in a Christian perspective. 'We would not know how to clean ourselves so well afterwards,' La Noue adds further on, 'that there might remain no stain upon the whiteness of the affections'.[14] By way of the organic metaphor, the imaginary of disease extends into that of poison.[15] In his essay, La Noue uses the resources of this metaphor by attacking the three poisons of novels: superstition, sensuality and violence. In the *Tombeau des romans* (1626), Claude Langlois (known as Fancan), takes up this theme and marries the idea of uncleanliness to that of poison by making novels 'beautiful fountains, but whose water is corrupted'.[16] Jacquin accentuates the epidemic nature of the phenomenon by affirming that it is 'not wrongly that literature complains of Romance, whose poisoned breath necessarily sullies our mores'.[17] In addition to being imbued with religious connotations, 'poisoned breath' echoes the theory of miasmas travelling on air that constitutes one of the hypotheses used to explain epidemics.[18] The notion of contagion is thus a place where the imaginary of sin, uncleanliness, poison and disease meet.

By waking the fears that these phenomena arouse in the reader, the idea of contagion and its extension into the imaginary of poison are certainly marked with a strong axiological charge that makes possible an even more negative view of the danger represented by reading novels, in the manner of the religious polemic that calls upon this metaphor to denounce heresy.[19] But the metaphor of contagion has not only a polemical function, it also implements an anthropology that inspires a precise understanding of the novelistic experience. In the rest of this chapter, I will show that, far from being a simple metaphor, the paradigm of contagion offers critical discourse an imaginary and conceptual framework for comprehension of the functioning of novelistic discourse, one that can orient critical appropriation of the novel.

The anthropology of imaginary contagion

First, the idea of contagion and its extension into the imaginary of poison accentuates a dimension of the sickness of which novelistic discourse is the carrier. It is a question of its diffuse and insidious nature making it so that the illness, in its beginnings, propagates itself in an imperceptible way and thus the stricken individual can, without even knowing it, be a carrier of disease. This almost unconscious force[20] of novelistic discourse is accentuated by Pierre Nicole, in his *Traité sur la comédie* (1667):

> Que ceux qui ne sentent point que les romans et les comédies excitent dans leur esprit aucune de ces passions que l'on appréhende d'ordinaire ne se croient pas pour cela en sûreté, et qu'ils ne s'imaginent pas que ces lectures et ces spectacles ne leur aient fait aucun mal. La parole de Dieu qui est la semence de la vie, et la parole du diable qui est la semence de la mort, ont cela de commun qu'elles demeurent souvent longtemps cachées dans le cœur, sans produire aucun effet sensible.[21]

> [Those who do not feel that novels and comedies arouse in their mind any of these passions that one normally feels should not believe themselves for all that in safety, and they should not imagine that these readings and these spectacles have done them no harm. The word of God who is the seed of life, and the word of the devil who is the seed of death, have this in common that they often long remain hidden in the heart, without producing any noticeable effect.]

In addition to bringing out the insidious effect of the germs of this contagious discourse, the opposition made by Nicole between the seeds of life and the seeds of death accentuates the parallel between the sacred word and fictional discourse. By implementing all the resources of the flesh that lead to sin, novelistic contagion presents itself in a way as a negative double of divine 'communication'. In his *Dictionnaire universel*, Furetière defines communication as an 'action by which one gives to another, one makes him a participant in the good or evil that one possesses'.[22] If the first example given by Furetière refers to the theological acceptation ('it is by means of the Sacraments that God communicates his graces to us'),[23] the second brings us back to contagion: 'Communication of the plague, of leprosy, transpires easily in hot countries.'[24] After having written that 'communication is also said of the contact, of the fellow feeling that we have with someone', Furetière continues by raising the problem of moral contagion with the example that 'communication with heretics is very dangerous to the feeble-minded. Communication with demons has been detested by all peoples.'[25] Physical contagion thus presents itself as one of the manifestations of this phenomenon of 'communication' that, at the time, is conceptualized in the extension of the religious paradigm. The 'popular' aspect of contagion brings out another characteristic of this phenomenon of 'communication' by which the individual 'outside of himself / hors de lui', losing the boundaries that must define him, melts into the crowd or the sensible universe. The epidemic dimension of the illness of which the contagion is a carrier takes into account the disseminated and uncontrolled mode of propagation that governs novelistic 'communication', and that, among other things, distinguishes it from legitimate transmission.

The denunciation of contagion through novels emerges in a context marked by the Judeo-Christian world-view where, far from being an anomaly, this phenomenon of moral 'contagion' refers rather to a generalized state. Marked by the Fall, the human being is caught in the errors of the flesh and in the corrupt state that sin incites. Far from being reduced to a religious denunciation of the sensory world, this contagion refers to an anthropology that permits us to understand better the way in which the critics of the time could grasp the effects of novelistic discourse, as well as the value of discourses making use of that notion. This anthropology finds an exemplary expression in the work of Pierre Charron who, in *De la sagesse* (1601), places the idea of contagion at the centre of his analysis of the human condition. This work seems all the more representative of the shared culture at the beginning of the seventeenth century for as well as having an impressive commercial success, *De la sagesse* presents a veritable synthesis of ideas the period inherits.[26] In this text, Charron evokes several times the idea of a contagion of mores. The moralist makes moral contagion one of the principal impediments to wisdom. 'One of these impediments is external; these are popular opinions and vices, worldly contagion; the other is internal, these are the passions.'[27] In a more 'profane' view, the 'contagion of the world' is associated with the effects of sensual impressions and the mistakes which result from them. But it can also be imbued with religious overtones. Charron makes these impediments into two 'diseases of the mind' and defines the contagion of the world, joined with 'the force of passions', as a 'universal contagion of popular and mistaken opinions accepted in the world'.[28] This 'contagion' thus refers thus to the power of opinion that results from the activity of the imagination and the passions that accompany it. The force of imagination itself gives birth to another form of contagion when this faculty alters the activity of understanding.[29] From this alteration comes superstition, in whose extension, according to critics of the time, the novelistic develops. This denunciation of imaginary contagion is taken up again, later in the century, by Malebranche who, in the second book of the *De la recherche de la vérité* (1674), attacks the excesses of the imagination by insisting that 'there are very few more general causes of men's errors than this dangerous communication of the imagination'.[30] By appealing to the imagination and the passions of the reader, the novel replaces the paradigm of transmission with that of contagion, thus taking its place alongside popular opinion and old wives' tales.

In a humanist perspective, reading good literature has the function of limiting this contagion by creating a distance between the individual and the sensual world, which echoes the passions that live within him. In defining this programme in 1532, Guillaume Budé makes of philosophy, which he places at the centre of *belles-lettres*, an antidote against this state of generalized corruption.[31] 'It is in effect, writes Budé, like the antidote which the well-regulated and prudent life has available against the contagion of crossroads and marketplaces.'[32] Reading profane literature can also protect the individual against this contagion of the world, just as the sacred word could protect the believer against the pitfalls of the flesh and of temptation. In the *Le Tombeau des romans* (1626), Fancan takes up this conception of the curative virtues of good literature and insists on the value of history, this time giving as examples true healings that came about through reading.

C'est à l'histoire que nous devons employer notre loisir; histoire prophétesse et prêtresse de la déesse vérité, à la divinité de laquelle on attribue même la guérison des maladies, on récite entre autres d'Alphonse et de Ferdinand rois d'Espagne et de Sicile, qu'ils recouvrèrent leur santé, que l'art des Médecins ne leur pouvait donner, en se faisant lire les histoires de Tite Live et Quinte Curse. Laurent de Médicis, oyant lire un trait de l'histoire de Conrad troisième empereur, reçut sa santé désespérée des médecins.[33]

[It is for history that we must utilise our leisure; history prophetess and priestess of the goddess truth, to whose divinity we attribute even cure of diseases, we tell for example of Alphonse and of Ferdinand kings of Spain and of Sicily, who recovered their health, which the art of Doctors could not give them, by being read the histories of Titus Livius and Quintus Cursus. Lorenzo di Medici upon hearing read a passage of the history of Conrad the third emperor, regained his health of which the doctors had despaired.]

Fancan concludes with a more metaphorical conception: 'But it would be little if from reading histories came naught but the convalescence of sick bodies. Souls more precious than the bodies where they are imprisoned, receive other cures from them.'[34] Although it is less spectacular than physical healing, the healing of souls points to an even more important phenomenon. Similarly, the curative virtues of good literature are abundantly highlighted by Le Moyne in his *Peintures morales*. In praising the virtue of books, Le Moyne writes that he does not see any 'doctors more gentle, and more accustomed to making bold and unhoped-for cures'.[35] Le Moyne makes Christian speech the first of all remedies; the Church Fathers having, he writes, 'learned from the almighty Doctor what they have prescribed for us'. But as 'all Doctors are not universally pleasant, and all the sick have their own equally diverse tastes for remedies', he adds a series of profane physicians.[36]

If theological discourse tends to organize, indeed to institutionalize the phenomenon of 'communication' by joining the faithful in one 'body'[37] and by orienting the thoughts of the sinner towards the other world, good literature tends in a way to contain the effects of this contagion, by consolidating the boundaries of the individual and by fortifying his capacity to resist the forces of the 'world' and the passions. Presented as a model of this ability to escape the contagion that governs the masses, the figure of the 'sage', having been taken in charge by the moral vulgate, ends up suggesting less a pure ideal of reason than a model of the individuated subject who appears as well in other figures of 'wisdom' that the written culture proposes, and who reproduce a similar process of subjectivization.[38] This is the case, for example, of the educational treatises aiming to develop the 'Christian woman' (Vivès-Changy) or the 'honnête fille' (Grenaille).[39] It is thus at the heart of this imaginary which makes reading an antidote against the contagion of the world and a remedy for the passions that the denunciation of novelistic contagion takes shape, and that these treatises also implement. By proposing to the reader a fictitious identity and by putting into play a practice of reading dominated by the imagination and the passions, the novelistic experience goes against this process of subjectivication. It constitutes a true pathology of reading.

Of novelistic contagion

This phenomenon of imaginary contagion, denounced by the critics, is at the very foundation of the novelistic experience that takes effect by means of images that the text engenders in the mind of the reader. Critical discourse on the novel is inspired by this theory, making the 'species' a vehicle for contagion,[40] and by theories of erotic melancholy.[41] This influence condenses into the notion of 'impression',[42] which is at the centre of the problematic of imaginary contagion. Following Nicole, Jacquin, for example, writes that 'the *impression* that [books] make even in spite of us, on our mind and our heart, always carries a seed of correctness or falsehood, virtue or vice, though at the moment of reading, we feel but very mildly affected'.[43] The Christian idea of 'uncleanliness' thus finds an extension in this notion, which is also used by doctors who attack physical contagion.[44] The 'impression' again accentuates the problem of communication, because it 'is said, according to Furetière, of the qualities that one thing communicates to another, when it acts upon it'.[45]

Through the notion of impression, imaginary contagion is situated at the limits of a 'metaphorical' conception of the phenomenon because this conception brings into play a veritable physics of the image, which conceives of this image as a 'mark' left in the mind. In a lecture entitled 'Des fables, et si elles apportent plus de mal que de bien' (1641), a lecturer from the Bureau d'Adresse refers to this phenomenon in order to condemn the reading of fiction:

Il n'y a que la seule ignorance des enfants qui est capable de les recevoir avec quelque plaisir, mais d'autant plus périlleusement que cette table rase étant une fois *imprimée* de quelque fausse persuasion, elle ne la quitte guère ... Ce qui a fait tenir pour un notable défaut de l'éducation des enfants de leur apprendre des contes; au lieu desquels ces faibles esprits, où comme dans une cire molle s'*imprime* aisément tout ce que l'on veut, se doivent instruire de l'histoire.[46]

[There is naught but the ignorance of children that is capable of receiving them with some pleasure, but even more perilously in that this blank slate once *imprinted* with some false persuasion, will never leave it ... It has been held as a notable error in the education of children to teach them tales; instead of which these weak minds, where all that one wants imprints itself easily as in soft wax, must be instructed in history.]

What certain authors present as belonging to the mind of the child also characterizes a part of the mind of the adult reader: this 'softness' that reading novels accentuates is a property of imagination in general.

The contagious character of the novelistic image comes from the passionate charge which invests it. Another reality at the border of the body and the mind, the passions constitute, for the critics of the time, one of the principal poisons that novels spread. The 'contagious' character of passionate speech in general is highlighted by the rhetorical tradition which requires this strength. In his 1594 treatise *De l'éloquence françoise*, Guillaume du Vair brings out this dimension by writing that

> Toute la force et l'excellence de l'éloquence consiste de vrai au mouvement des passions ... Car la passion s'étant conçue en notre cœur, se forme incontinent en notre parole, et par notre parole sortant de nous entre en autrui, et y donne semblable *impression* que nous avons nous-mêmes, par une *subtile et vive contagion*.[47]

[All the force and excellence of eloquence consists in truth of the movement of the passions. For passion being conceived in our heart, is formed forthwith in our speech, and by our speech leaving us it enters into another, and gives a similar *impression* that we have ourselves, by a *subtle and lively contagion*.]

Du Vair's conception of eloquence is marked by the medical paradigm. This favoured connection with the passionate dimension of discourse engenders a pedagogy based on imitation and emulation, which is itself thought out in terms of 'contagion'. Recourse to the medical paradigm here takes part in a kind of secularization[48] of 'communication' which takes the sacred word as a model. If Gorgias, in his *Encomium of Helen*, was already using the medical metaphor to describe the power of discourse,[49] it was by joining it with magic that, after the example of du Vair, the thinkers of the seventeenth century abandoned it. By inscribing this effect in the medical vein, the notion of contagion is the site of a rationalization, indeed a 'medicalization' of this imaginary possession that permits the growth of the distance that separates religious speech (of which the demonic is still a part, the negative side) from other phenomena of 'communication'.

The contagious effect of passionate speech finds and exemplary expression in the phenomenon that is at work at the moment when the reader lets himself be carried away by fiction. In his *New Reflections on Poetics* (*Nouvelles réflexions sur l'art poétique*) of 1678, Bernard Lamy thus describes this phenomenon:

> Les affections, dont le lecteur se sent animé, le *transportent hors de lui-même*. Tantôt il sent son cœur plein d'un feu martial, et il s'imagine combattre: tantôt agité de mouvements plus doux, il se mêle dans les intrigues du héros de la pièce: il est soldat et amoureux avec lui: et en un mot, il est dans son imagination ce qu'est ce héros, et ce qu'il voudrait être lui-même; ainsi *il n'y a aucun mouvement de son cœur, qui ne soit rendu agissant*; il estime, il aime, il désire, il craint. Il n'y a point de passion dont il ne ressente les agréables émotions.[50]

[The affections, by which the reader is animated, *transport him outside of himself*. Now he feels his heart full of a martial fire, and he imagines himself in combat: now shaken by gentler movements, he is mixed up in the intrigues of the play's hero: he is soldier and lover along with him: and in a word, he is in his imagination what this hero is, and what he would like to be himself; thus *there is no emotion in his heart, that is not made active;* he admires, he loves, he desires, he fears. There is no passion whose pleasant emotions he does not feel.]

Religious, medical and rhetorical perspectives meld together in the notion of 'transport', which Lamy evokes in order to describe the phenomenon that is at the heart of the novelistic experience. This term which, in addition to movement, designates the result of a strong passion, poetic furor, mystical ecstasy and the effect of violent fevers,[51] accentuates the relationship that the novelistic experience has with other phenomena of imaginary contagion, such as religious enthusiasm.[52] Further on, Lamy concludes:

Il ne faut pas s'étonner si les personnes qui lisent les romans reçoivent l'*impression* de tous les sentiments de ceux que le poète fait agir et parler, puisqu'ils y ont un rapport si naturel. Les paroles des personnes passionnées nous troublent et nous agitent quand elles nous trouvent plein de la passion et de la faiblesse dont elles procèdent.[53]

[One must not be astonished if people who read novels receive the *impression* of all the feelings of those whom the poet makes act and talk, because they have such a natural rapport. The words of passionate people disturb and agitate us when they find us full of the passion and weakness from which they proceed.]

Passion, to which the notion of transport leads, designates the contagious agent that the novelistic image carries, and guarantees that this image makes an impression. Novelistic contagion takes place not only because the novel spreads models that are likely to be imitated, but because the novelistic experience, by calling upon the imagination and the passions, constitutes a plunge into the flesh, a change in the borders that must define the individual.

From the antidote of literature to the medicine of books

As shown by Françoise Hildesheimer, the reaction against contagion in the early modern era was a veritable laboratory within which were developed diverse strategies for managing the social body.[54] The paradigm of contagion not only governs the understanding of the phenomenon, it also guides the 'response' of critics to the problem posed by the novel. The two aspects in the fight against contagion, which come both from medicine and from political economy, inform critical discourse on the novel. From the outset, by its denunciation of the novel, this discourse puts into play the first of the two functions of the physician, which are to 'name the disease, and to provide remedies'.[55] The denunciation of the deceitful nature of the novel can certainly have purgative virtues, but it will not suffice to cancel out the contagion that the genre provokes. From 'pharmacological' perspective, the first response of the medical paradigm leads to a metamorphosis of the poison contained in novels into a remedy for other ills.[56] In the 'Eloge des histoires dévotes' that accompanies his novel *Agathonphile* (1621), Jean-Pierre Camus appropriates this strategy, using fiction as a vehicle for the promotion of moral values that would otherwise leave the reader indifferent:

Ce Pharmacien qui avait trouvé le secret de faire prendre ses drogues amères aux malades comme des confitures ne serait-il pas grandement désirable. O! que celui-là serait bien plus à estimer qui pourrait si bien déguiser par des inventions ou historiques ou oratoires, *les antidotes du péché et les préservatifs de la contagion de l'âme*, qu'il puisse ôter aux remèdes de salut cette rudesse imaginaire que les mondains, malades de trop d'aise et de délicatesse, se figurent en la Pénitence et au chemin de la belle Dilection. [57]

[This Pharmacist who had found the secret to making the sick take his bitter drugs as if they were jams would he not be hugely desirable. Oh! how he would be held in much more esteem who could well disguise by historic or oratory inventions, *the antidotes to sin and the prophylactics of the contagion of the soul*, that he might remove from the remedies for salvation this imaginary harshness that the fashionable, ill from too much ease and delicacy, imagine in Penitence and on the road to beautiful Friendship.]

The goal of good literature is transformed and the introduction of the sacred word into fiction, this time, fosters the protection of the reader from the dangers of worldly contagion. The risk of contagion, which remains present in novelistic impressions, gives rise with Camus to a complex poetic aiming to inoculate the reader from the negative effects of the images presented.[58] This 'pharmacological' strategy is taken up by Fancan and eventually constitutes a commonplace of critical discourse, which Huet brings to bear in his *Traité sur l'origine des romans* (1670) and that we still find in Marmontel, at the end of the eighteenth century. The novelist is thus similar to orators, 'doctors of the human passions, who guild their pills to fool the sick usefully'.[59] This logic governs, in particular, the 'virtuous' representation of love, where the introduction of precepts permits an offer of remedies for this ill.[60] The medical metaphor accentuates the legitimacy of the use of artifice, and leads to a definition of the 'good' novel which fools the reader for his own good.

As we have seen, it is chiefly through the question of the passions that the problem of novelistic contagion is broached. The passions are the starting point for theorizing the metamorphosis of poison into remedy. As Camus remarks: '[the] passions are only bad inasmuch as they have a bad objective, but are made good by one who is good.'[61] Thus functions the contagious force of the novel which, according to these critics, can acquire beneficial virtue by being put to the service of another end. The authors of the second half of the seventeenth century complete this strategy by putting forward a poetics aiming to moderate the passions[62] and, through them, this contagious force. In this moderated form, the good novel could serve as an antidote or 'prophylactic' for various ills, for example, by providing knowledge about love and the other passions that permits the reader to arm himself against their negative effects.

The paradigm of contagion leads to another kind of response that belongs, in this case, less to medicine than to political economy. It is indeed less in the discovery of treatments than in the development of measures to circumscribe the illness that progress in matters of managing early modern epidemics is principally achieved.[63] The first critical response inscribed in this perspective concerns the demand for legislation that would put an end to the propagation of the ill. From La Noue to Jacquin, by way of Camus and Fancan, the critics demand a series of measures aiming to fight the contagion of novels. In his discourse against novels (1736), Porée explicitly calls for legislation against the disease:

La loi interdit à quiconque de mettre en vente des aliments susceptibles d'introduire dans l'organisme les germes nuisibles de maladies. Que n'interdit-elle encore de vendre des ouvrages qui, par une nourriture bien plus nocive encore, font pénétrer dans les cœurs les poisons mortels de l'amour?

La loi prescrit que personne n'importe de l'étranger des marchandises auxquelles soit attaché le moindre soupçon de contagion. Que ne prescrit-elle aussi que ne soient pas introduite d'Espagne, d'Italie, d'Angleterre, de Hollande, de Grèce, de Perse, de l'Inde et du Japon des marchandises érotiques qui, bien plus malsaines encore par leur contagion, contaminent la cour, la ville et les provinces par leur virus infâme?[64]

[The law forbids anyone from putting on sale foodstuffs liable to introduce into the organism the harmful seeds of disease. Why not also forbid the sale of works which, by a food much more harmful still, make the mortal poisons of love penetrate the heart?

The law forbids the importation from abroad of merchandise to which the slightest suspicion of contagion is attached. Why not also prescribe that from Spain, Italy, England, Holland, Greece, Persia, India and Japan there not be introduced erotic merchandise much more unhealthy still by its contagion, which might contaminate the court, the city and the provinces with their infamous virus?]

But the role of the critic is not limited to demanding these measures. By inscribing the question of the censorship of books in the medical paradigm which already orients the comprehension of the effects of reading, authors like Le Moyne and Sorel make of the critic a doctor of letters whose task consists of driving out illness. In *De la connoissance des bons livres* (1671), Sorel insists that before permitting the reading of a pleasurable text, 'one must know if poison is not hidden underneath sweet bait, and if all this does not corrupt mores by bad principles'.[65] If the critic, who 'makes a profession of seeing books of all sorts', guarantees himself against the illness by finding in certain books 'the remedy for the ill caused by the others', it is not the same story for the ordinary reader, for whom 'the simple reading of licentious or badly written books would cause too much damage'.[66] The integration of novels into the universe of letters, which Sorel initiates in the *Bibliothèque françoise* (1664) and *De la connaissance des bons livres*, places the genre in the larger realm of discourse, and determines the place that it can legitimately claim to occupy.[67] In doing so, Sorel integrates novelistic discourse within other texts which can serve as remedies or prophylactics against the ill transmitted. As Sorel notes, it is by this logic that Fortin de la Hoguette permits his sons to read novels, even though he makes of novel reading 'a disease of the age'. The author of *Testament ou conseils d'un bon père* (1648), writes Sorel, 'did not believe that those it was his duty to teach should stop at such fables, without receiving prophylactics and adding to them healthy instruction'.[68]

The emphasis placed on the novel as an editorial object, which results in its integration into this economy of 'good books', thus generates a kind of reification of the contagious agent which, from an image conveyed by the novel, is made into a book which editors print and distribute. By associating the circulation of novels with that of exotic merchandise, and in inscribing imaginary contagion in the circuit of physical contagion, Porée pursues this process of reification. This process inspires a veritable institutionalization of the novel as a genre, which develops in the eighteenth century, for example in Lenglet Dufresnoy's *De l'usage des romans* (1734), which again takes up the initiative begun by Sorel and the other critics of the seventeenth century.[69] In constructing the object of this institutionalization, novelistic 'communication' loses bit by bit the scattered and uncontrolled dimension that characterized it. The definition of a good novel, whose contagious force is a means of transmitting virtue, goes hand in hand with the development of this critical institution capable of distinguishing the virtuous novel from the contagious novel, as well as moderating its effects by according it a fair place in the universe of letters.

Far from a mere metaphor, the notion of contagion thus offers to critical discourse a conceptual framework which orients its comprehension of the phenomenon, as well as its response to the problem raised by the novel. While other forms of enthusiasm are suppressed, novelistic 'ecstasy' is the object of regulation by being understood through the paradigm of contagion. The rationalization that accompanies this integration of the phenomenon into the medical paradigm permits the development of a mechanism destined to circumscribe a force which, by appearing as the demoniacal reversal of religious communication, could only be the object of denunciation. The management of the 'pathological' effects of the novel continues in the eighteenth century through the development of criticism as an institution and the work on vapours and nymphomania[70] which take control of the discourse of contagion from novels, all the while making novel reading into one of the causes of these troubles. This cultural control of the reading experience will finally produce, in the second half of the eighteenth century, a transformation of the value given to this contagious force[71], as witnessed by the celebrated 'Éloge de Richardson' by Diderot (1762).[72]

Notes

1 This research for this chapter was made possible by a postdoctoral research fellowship from the Social Sciences and Humanities Research Council of Canada (SSHRCC).

2 *Testament ou conseils fidèles d'un bon père à ses enfants où sont contenus plusieurs raisonnements chrétiens moraux et politiques* (Paris: René Gignard, 1679), p. 112: 'c'est une maladie du temps que les romans; ça été la mienne, ce sera peut-être la tienne.'

3 'Dilude', *Petronille* (1628), in Max Vernet, *Jean-Pierre Camus: Théorie de la contre-littérature*, (Paris-Sainte-Foy: Le Griffon d'argile, 1994), p. 158: '[t]ous les jours le mal s'accroît et s'étend, le monde s'enivre de fables.'

4 In the following century, this propensity is one of the principal complaints that Father Porée addresses to the genre in his famous speech on the novel (1736). The account of this speech appears as the 'Compte rendu du Discours sur les romans de Charles Porée', in the *Journal de Trévoux ou mémoires pour servir à l'histoire des sciences et des arts*, XXXVI (Geneva: Slatkine Reprints, 1968), p. 369.

5 *Les peintures morales, seconde partie de la doctrine des passions, où il est traité de l'amour naturel, et de l'amour divin, et les plus belles matières de la morale sont expliquées* (Paris: Sébastien Cramoisy, 1643), pp. 728–9: 'une conversation muette, que nous avons avec les morts et les personnes absentes'; 'il se faut garder soigneusement de ces morts contagieux, qui augmentent les maladies des vivants, et qui empoisonnent par l'imagination et par la vue'.

6 See L. Thirouin, *L'Aveuglement salutaire. Le réquisitoire contre le théâtre dans la France classique* (Paris: Honoré Champion, 1997), p. 122.

7 J. Chapelain, *Sentiments de l'Académie française sur la tragi-comédie du Cid*, in A. Gasté, La querelle du Cid. *Pièces et pamphlets publiés d'après les originaux avec une introduction* (Genève: Slatkine Reprints, 1970), p. 360: 'les mauvaises exemples sont contagieux, même sur les théâtres.'

8 Armand-Pierre Jacquin, *Entretiens sur les romans. Ouvrage moral et critique, dans lequel on traite de l'origine des romans et de leurs différentes espèces, tant par rapport à l'esprit, que par rapport au cœur* (Geneva: Slatkine Reprints, 1970), p. viii: 'voir combien les romans tendent de pièges à l'innocence, et combien il est difficile de

se garantir de *leur contagieuse corruption*'; my emphasis in English text translation.

9   *Le Dictionnaire universel d'Antoine Furetière*, 3 vol. (Paris: S.N.L.-Le Robert, 1978): '[s]e dit figurément en choses morales, des vices, des hérésies, qui se gagnent par la communication avec les personnes qui en sont infectées.'

10 See P. Dandrey, '*Moralia & medicinalia*. Cadastre, semences et moissons (1977–1997)', *XVIIe siècle*, 202 (1999) 37–53.

11 *Le Berger extravagant, suivi des Remarques*, intro. H. D. Becharde (Geneva: Slatkine Reprints, 1972), p. 276: 'Gentil berger je me fâche de voir votre esprit possédé d'une infinité de mauvaises opinions, dont qui plus est, vous désirez *faire une contagion*, et les communiquer à tous ceux qui approchent de vous.' My emphasis.

12 *Discours politiques et militaires*, ed. F. E. Sutcliffe (Genève: Droz, 1967), pp. 160–76: 'Que la lecture des livres d'*Amadis* est non moins pernicieuse aux jeunes gens, que celle des livres de Machiavel aux vieux'.

13 *Ibid.*, p. 162: 'Et de cette nourriture se sont engendrées des mauvaises humeurs qui ont rendu des âmes malades, qui ne pensaient par aventure pas du commencement arriver à cette indisposition.'

14 *Ibid.*, p. 172: 'On ne saurait si bien se nettoyer après qu'il ne demeure toujours des taches en la blancheur des affections.'

15 On this imaginary, see S. Normand, 'Poison, maladie et métaphore dans le Dictionnaire universel', *Littératures classiques*, 47 (2003) 173–84.

16 Claude Langlois [Fancan], *Le tombeau des Romans où il est discouru I. Contre les Romans; II. Pour les Romans* (Paris: Claude Morlot, 1626), p. 34: 'de belles fontaines, mais qui ont leur eau corrompue'.

17 Jacquin, *Entretiens sur les romans*, p. 296: 'pas à tort que la littérature se plaint de la Romancie, dont le souffle empoisonné rejaillit nécessairement sur nos mœurs'.

18 See J. Ehrard, 'Opinions médicales en France au XVIIIe siècle. La peste et l'idée de contagion', *Annales: Economie, sociétés, civilisations*, 12 (1957) 46–59.

19 On this question, see Greenspan in this volume.

20 See L. Marin, 'La critique de la représentation théâtrale classique à Port-Royal: Commentaire sur le *Traité de la Comédie* de Nicole', *Continuum*, 2 (1990) 81–105.

21 Pierre Nicole, *Traité sur la comédie*, in *Œuvres philosophiques et morales* (Hildesheim and New York: Georg Olms Verlag, 1970), p. 441.

22 Furetière, *Dictionnaire universel* : '[a]ction par laquelle on donne à un autre, on le fait participant du bien ou du mal qu'on possède.'

23 *Ibid.*: '[c]'est par le moyen des Sacrements que Dieu nous fait la communication de ses grâces'.

24 *Ibid.*: 'La communication de la peste, de la lèpre, se fait aisément dans les pays chauds.' Furetière indeed defines contagion as an 'ill that is acquired by communication' ('un mal qui se gagne par communication').

25 *Ibid.*: 'Communication se dit aussi de la fréquentation, de l'intelligence qu'on a avec quelqu'un'; '[l]a communication avec les hérétiques est fort dangereuse aux esprits faibles. La communication avec les démons a été détestée par tous les peuples.' On this last type of communication, see Closson, this volume.

26 See C. Belin, *L'œuvre de Pierre Charron, 1541–1603: littérature et théologie de Montaigne à Port-Royal* (Paris: Honoré Champion, 1995).

27 *Of Wisdome* (1612; rpt Amsterdam: Da Capo Press, 1971), p. 224. *De la sagesse* (Paris: Chaignieau, 1797), pp. 251–2: 'L'un [de ces empêchements] est externe, ce sont les opinions et les vices populaires, la contagion du monde; l'autre interne, ce sont les passions.'

28 *Ibid.*, p. 225; in the original, p. 120: 'contagion universelle des opinions populaires et erronées reçues au monde'.

29 *Ibid.*, p. 226 (p. 122).

30 *On the Search for Truth*, Book II, Part One: *The Imagination*, trans. T. M. Lennon and P. J. Olscamp (Columbus: Ohio State University Press, 1980), p. 161: 'Strong imaginations, writes Malebranche, are extremely contagious: they dominate weaker ones, gradually giving them their own orientation, and imprinting their own characteristics on them.' For the original version, see *De la recherche de la vérité*, Livre II: *De l'imagation*, ed. A. Minazolli (Paris: Presses Pocket, 1990), p. 151: 'il y a très peu de causes plus générales des erreurs des hommes, que cette contamination dangereuse de l'imagination'; 'Les imaginations fortes sont extrêmement contagieuses: elles dominent sur celles qui sont faibles: elles leur donnent peu à peu les mêmes tours, et leur impriment leurs mêmes caractères.'

31 While describing this contagion, Charron and Malebranche still call less upon the antidote of literate culture than upon that of reason.

32 *L'Étude des lettres (De Studio literarum recte et commode instituendis)*, ed. M. de la Garanderie (Paris: Les Belles Lettres, 1988), p. 88: 'Elle est en effet comme l'antidote dont dispose la vie bien réglée et prudente contre la contagion des carrefours et des marchés.'

33 Fancan, *Le Tombeau des romans*, pp. 42–3.

34 *Ibid.*, p. 43: 'Mais ce serait peu si de la lecture des histoires ne provenait que la convalescence des corps malades. Les âmes plus précieuses que les corps, où elles sont ensevelies, en reçoivent d'autres guérisons.'

35 Le Moyne, *Peintures morales*, p. 728: 'médecins plus doux, et plus accoutumés à faire des cures hardies et inespérées'.

36 *Ibid.*, p. 731: 'appris du souverain Médecin ce qu'ils nous ont ordonné'; 'tous les Médecins ne sont pas universellement agréables; et que les malades ont pour eux des goûts aussi divers que pour les remèdes'.

37 On the question of the mystical body put into place by the Church, see H. Merlin, 'Fables of the "Mystical Body" in Seventeenth-Century France', *Yale French Studies*, 86 (1994) 126–42.

38 On this process, see Michel Foucault, 'Deux essais sur le sujet et le pouvoir', in *Michel Foucault. Un parcours philosophique*, ed. H. Dreyfus and P. Rabinow (Paris: Gallimard, 1984), pp. 297–321.

39 J. L. Vivès, *Livre de l'institution de la femme chrestienne tant en son enfance que mariage et viduité, aussi de l'office du mari. Naguère composé en latin par Jehan Loys Vives et nouvellement traduit en langue françoyse par Pierre de Changy* (1523; Geneva: Slatkine Reprints, 1970); F. de Grenaille, *L'Honnête fille*, ed. A. Vizier (1639–40; Paris: Honoré Champion, 2003).

40 See the chapters by Gagnon and Plantin in this volume.

41 See Beecher and Frelick in this volume.

42 On this notion and its usage in Augustinian criticism, see Laurent Thirouin (1997, 122).

43 Jacqin, *Entretiens sur les romans*, pp. 104–5: 'l'*impression* qu[e les livres] font même malgré nous, sur notre esprit et sur notre cœur, y porte toujours un germe de justesse ou de fausseté, de vertu ou de vice, quoique dans le moment de la lecture que nous en faisons, nous ne nous sentions que très médiocrement affectés.' My emphasis.

44 For example, Jacques Moine begins his *Advis sur ce temps contagieux* (Paris: A. Taupinart, 1628), p. 3, by writing: 'Puisque les effets de la peste sont si prodigieux

et si formidables, il n'y a soin ni diligence qu'on ne doive employer pour en prévenir les *mortelles impressions.'* / 'Because the effects of the plague are so formidable and so tremendous, there is no care nor diligence that we must not employ to prevent the *deadly impressions.'* My emphasis.

45 *Dictionnaire universel*: 'se dit des qualités qu'une chose communique à une autre, quand elle agit sur elle'.

46 *Quatrième centurie des questions traitées aux conférences du Bureau d'Adresse* (Paris: Bureau d'Adresse, 1641), pp. 409–10, my emphasis.

47 *De l'éloquence françoise*, in *Œuvres* (Geneva: Slatkine Reprints, 1970), p. 403, my emphasis.

48 On the phenomenon of secularization, see M. Heyd, *'Be Sober and Reasonable'. The Critique of Enthusiasm in the Seventeenth and Early Eighteenth Century* (Leiden, New York and Cologne: E. J. Brill, 1995).

49 Ed. and trans. D. M. MacDowell (Bristol: Bristol Classical Press, 1982).

50 Ed. T. Gheeraert (Paris: Honoré Champion, 1998), p. 182, my emphasis.

51 Furetière, *Dictionnaire universel*: 'Transport, se dit aussi en Médecine. Quand la fièvre est violente, on appréhende le transport au cerveau qui cause le délire' / 'Transport, is also said in Medicine. When the fever is violent we dread its transport to the brain which causes delirium.'

52 On religious enthusiasm, see Lindmark in this volume.

53 Bernard Lamy, *Nouvelles réflexions sur l'art poétique*, ed. T. Gheeraert (Paris : Champion, 1998), p. 183, my emphasis.

54 F. Hildesheimer, *Fléaux et société: de la Grande Peste au choléra, XIVe–XIXe siècle* (Paris: Hachette, 1993).

55 *L'œuvre de Pierre Charron*, p. 135: 'déclarer la maladie, et donner les remèdes'.

56 On the logic of the pharmakon, see J. Derrida, 'La pharmacie de Platon', in *La dissémination* (Paris: Seuil, 1972), pp. 70–197.

57 *Agatonphile, ou les martyrs siciliens* (Paris: C. Chappelet, 1623), p. 845; my emphasis.

58 In *Palombe ou la femme honorable, histoire catalane* (Paris: Chappelet, 1625), p. 144, after having launched a long moral development, Camus presents this poetics in medical terms. On Camus's poetics, see S. Robic-de Baecque, *Le salut par l'excès. Jean-Pierre Camus (1584–1652), la poétique d'un évêque romancier* (Paris: Honoré Champion, 1999).

59 Fancan, *Le Tombeau des romans*, p. 63: 'médecins des passions humaines, qui dorent leurs pilules pour utilement tromper le malade'.

60 The virtues of these 'prophylactics' will be denied by other authors like Nicole and Jacquin, who, on the contrary, take advantage of this to remind us that contagion is all the more dangerous when presented under the mask of virtue.

61 *Parthénice ou peinture d'une invincible chasteté. Histoire napolitaine* (Paris: Robert Bertault, 1637), p. 860: '[l]es passions ne sont mauvaises qu'autant qu'elles ont un mauvais objet, mais elles sont rendues bonnes par un bon'. The novel dates from 1621.

62 On the poetics of the passions in critical discourse of the seventeenth century on the novel, see M. Fournier, 'La poétique des passions dans le discours critique du XVIIᵉ siècle sur le roman', in *Entre théorie et fiction: penser les passions aux 17ᵉ et 18ᵉ siècles*, ed. L. Desjardins and É. Van der Schueren (Québec: Presses de l'Université Laval, forthcoming).

63 See Hildesheimer, *Fléaux et société*.

64 Quoted by G. May, *Le dilemme du roman au XVIIIe siècle. Étude sur le rapport du roman et de la critique (1715-1761)* (New Haven and Paris: Yale University Press and Presses Universitaires de France, 1963), p. 77.

65 *De la connoissance des bons livres, ou examen de plusieurs auteurs*, a cura di L. Moretti Cenerini (Rome: Bulzoni editore, 1974), p. 44: 'il faut savoir si le poison n'y est point caché sous de douces amorces, et si tout cela ne corrompt point les mœurs par de mauvais principes.'

66 *Ibid.*, p. 47: 'fait profession de voir les livres de toutes sortes'; 'le remède au mal causé par les autres'; 'la lecture seule des livres licencieux ou mal faits causerait trop de dommages'.

67 *La Bibliothèque françoise* (Geneva: Slatkine Reprints, 1970).

68 *De la connoissance des bons livres*, p. 100: 'n'a pas entendu que ceux à qui il devait l'enseignement, s'arrêtassent à de telles fables, sans recevoir des préservatifs et y joindre des instructions différentes'.

69 *De l'usage des romans, où l'on fait voir leur utilité et leurs différents caractères; avec une bibliothèque des romans, accompagnée de remarques critiques sur leur choix et leurs éditions* (Amsterdan: Chez la Veuve Poilras, 1734).

70 See J.-M. Goulemot, *Ces livres qu'on ne lit que d'une main: lecture et lecteurs de livres pornographiques au XVIIIe siècle* (Paris: Minerve, 1994); R. Chartier, 'Richardson, Diderot et la lectrice impatiente', *Modern Language Notes*, 114 (1999) 647–66; V. Costa, 'Quand lire à corps perdu devient le corps du délit – Les condamnations sociales du romanesque au XVIIIe siècle', in *Les imaginaires du corps. Pour une approche interdisciplinaire du corps*, ed. C. Fintz (Paris: L'Harmattan, 2000), pp. 75–107.

71 On this transformation of the value given the affective force of the novel, see J. Starobinski, 'Se mettre à la place'. (La mutation critique, de l'âge classique à Diderot)', *Cahiers Vilfredo Pareto*, 38–39 (1978) 364–78, and Chartier, 'Richardson, Diderot et la lectrice impatiente'.

72 Taking into account the limits of this chapter, I will address this question in another article which will be situated within the diverse 'contagious' effects of the present work.

14
Religious Contagion in Mid-Seventeenth Century England[1]

Nicole Greenspan

Surveying the state of religious health in the nation, one pseudonymous 1656 pamphlet pronounced England's condition dire. Disease was reaching epidemic proportions which, left unchecked, would place the commonwealth in grave and perhaps mortal danger:

> should the contagion of those plagues continue to infect persons of all ranks and sexes a few years longer after the same rate of success (and probably it will much more, or some other Judgement befall us in stead of it, if yet the Magistrate lay it not to heart) it will no doubt endanger both Ministry and Magistracy, the Oracles of God, and the Laws of the Land.[2]

As this text demonstrates, the organic model of the body politic lent itself readily to medical analogies.[3] Like the biological body, the collective body politic was subject to infection and disease which required identification and treatment, tasks which fell to the magistrate or 'state physician' charged with preserving the health of the nation. The monitoring of religious health was considered no less important than the social and political, and contemporaries accordingly adopted various strategies designed to curb the spread of religious disease and treat the infected commonwealth.

Some of these methods found parallels in familiar medical practices, including quarantine, purging and amputation. Others were based on the dissemination of information: before preventative or curative action could be taken, both officials and the public had to be apprised of the nature of the contagion and the means by which it could be transmitted. Focusing on Catholicism, this chapter explores the ways in which religious disease was believed to spread through Protestant society and the various remedies prescribed to treat infection. Though the drive to uncover and uproot religious illness spanned the seventeenth century, it assumed additional significance and encountered new challenges during the political, religious and social upheavals of the 1650s. Notably, the provisions for liberty of conscience established by the Commonwealth and Protectorate governments generated passionate debates over the forms of religious contagion, the necessary remedies, and the role of the state physician.

I

For most seventeenth-century Protestant polemicists, the Roman Catholic or 'popish' Church was the main source of religious contagion and the principal threat to the Protestant commonwealth. That papists were engaged in an ongoing plot to eradicate the Protestant faith and return England to the Roman fold was axiomatic, and English Catholics accordingly were construed as potentially treasonous citizens intent on destroying the social, political and religious order.[4] Countless warnings about the dangers of popery issued from the press and pulpit, exhorting the nation to learn from such lessons as the Marian persecutions and the 1641 Irish rebellion that, should Protestants let down their guard, violence and bloodshed inevitably would follow. Nevertheless, it appeared to many pamphleteers, theologians, political theorists and government officials that popery was not losing but gaining momentum in the 1640s and 1650s. Indeed, popery often was likened to a form of communicable disease transmissible through contact: exposure to popish sermons, books, rituals, places of worship and individuals carried the risk of contamination. Curbing its spread and ridding the nation of infection involved identifying risk factors and informing Protestants of the means by which they could protect themselves.

What made popery a particularly virulent form of contagion, polemicists argued, was that encountering it could be enticing and pleasurable for Protestants. Popish worship was characterized as highly sensual and physically gratifying, incorporating images, sumptuous décor, music, elaborate ceremony and ritual formality. This appeal to the flesh and senses and the spectacle of popish rituals and ceremonies was intoxicating, Protestant writers argued, and readily attracted the 'ungodly masses', who could not see past the veil of sensuality and luxury to the malignant core. As the Presbyterian minister Richard Baxter expounded, popery was 'an easie kind of Religion', one 'agreeable to flesh and blood'.[5] Popery was so seductive it was hazardous merely to come into contact with it, especially for those who were ill-equipped to resist its temptations: papists exploited vulnerability and deliberately targeted 'women and ignorant people, who they know are not able to gainsay their falsest, silliest reasonings'.[6]

At the same time, however, polemicists warned that papists effectively exploited the medium of print to reach a broad cross-section of Protestant society. Reading or oral transmission of popish texts was another conduit of contagion. Addressing this issue, a group of London Presbyterian booksellers composed what are commonly referred to as the 'Beacon Fired' pamphlets calling for the suppression of popish, blasphemous, and heretical books, which included lists of titles and brief content synopses to assist the authorities.[7] Not only did the Beacon Firers estimate tens of thousands of copies of these books to be in circulation, but what particularly galled the authors was that composition in the vernacular or availability in English translation ensured accessibility would not be limited to the educated elite. Moreover, the 1652 pamphlet *A Second Beacon Fired by Scintilla* argued that the insertion of images into the Bible and other texts provided visual instruction for the marginally literate or unlettered. A network of popish booksellers and itinerant chapmen then marketed the books in both urban centres and the provincial countryside.[8] These books carried 'a great infection', the Beacon Firers declared, and as purchasers and readers became contaminated they passed the disease on to others, with the result that it 'like a Gangrene spreads more and more'.[9]

Travel was another frequently identified mode of contracting religious disease. The experience of continental Catholicism, where rituals and worship were conducted openly, could be particularly enthralling for Protestants, and once infected the carrier would endanger Protestant society upon his or her return. Protestants accordingly were advised not to visit Catholic Europe before they had received solid religious instruction, otherwise, one Scottish writer observed, 'to travel to France or elsewhere, as I did, and the most part of young men do, is to expose them, not only to the hazard of being tempted to all abominable vices, but to be ensnared in the abominable and gross errors of Popery'.[10] Yet it was the very exoticism of popery that stimulated Protestant curiosity and fascination. Even within England exposure to Catholicism could excite Protestant observers. As Henrietta Maria's confessor Cyprien de Gamache later recalled, when the queen's chapel held open Catholic services in London during Charles I's reign, 'persons of quality, ministers, people of all conditions, who had never been out of the kingdom, came to see them [Capuchins] as one goes to see Indians, Malays, Savages, and men from the extremities of the earth'.[11]

According to some contemporaries, a firm grasp of Protestant doctrine as well as familiarity with popish strategies to gain converts could act as a form of inoculation. During his exile on the continent in the early 1650s, for example, the Anglican John Evelyn found himself enraptured by the opulence and celebratory atmosphere of Catholic churches and rituals in Paris. Evelyn, however, had read enough treatises, heard enough sermons, and observed enough disputations to feel confident he could not only safely hazard exposure but also could pick out what his training had taught him to recognize as the deceptive and misleading arguments of papists.[12] Yet for others, human effort alone offered insufficient assurance that Protestants would emerge unscathed from the experience. As the anonymous author of a 1653 report to the Council of State outlining the dangers of popery admitted, he had been drawn to attend mass at the Spanish ambassador's residence in London. What allowed him to return unharmed was divine intervention alone: 'God in mercy made it an abomination to me, and my soul was grieved to see this state so grossly abused, and so many silly souls cheated.'[13] Contact with popery was fraught with peril, and for ordinary men and women the best protection was avoidance.

II

Following this advice was neither as straightforward nor as practical as it might have sounded. First, the signs of popery had to be exposed. English Catholics generally were obliged to conceal or camouflage their religious practices either to avoid punishment or to participate in communal life. One such means was occasional conformity or attendance at Protestant services.[14] Catholic devotional practices largely took place within the home and, with the exception of cosmopolitan London, Catholic communities tended to centre on local gentry households.[15] To Protestants it seemed that Catholics had innumerable ways of escaping detection: clergy could disguise themselves as labourers, crucifixes could be carried in pockets, and patterns of food consumption and preparation could be hidden behind closed doors. To the extent that it was possible in the seventeenth century, a time in which the boundaries between public and private were ill-defined and permeable, much Catholic devotional practice took place away from the Protestant public eye.[16]

Concealment in turn generated fears that papists could be mistaken for 'peaceable' and loyal Protestants. In the mid-seventeenth century, however, determining just who could be comprehended within the spectrum of orthodox Protestant belief was a subject of heated debate. By the late 1640s the English Parliament had abolished the Church of England hierarchy and proscribed traditional Anglican services, among the most prominent justifications for which was that under Charles I and the Archbishop of Canterbury, William Laud, the Church of England had become contaminated by popery.[17] But while supporters of the Commonwealth and Protectorate celebrated the Parliamentary victory over the popish king and church, most maintained that only the battle had been won and not the war. The attacks on the established church, together with the policies of religious anti-formalism adopted by the Commonwealth and Protectorate governments, facilitated the splintering of Protantism in England, and reports abounded that popery had moved on to infect or spawn various groups including Presbyterians, Independents or Congregationalists, Baptists and Quakers.

Inclusion within the parameters of Protestant orthodoxy was unstable and could shift in response to social and political developments. Accusations of popery aimed to influence public opinion and often appeared at critical moments in the hopes of rallying support for particular government policies and actions.[18] During the Cromwellian invasion of

Scotland, for example, sensationalized accounts linking Presbyterianism and popery sought to rally opinion behind the English army. Like the Church of England before it, detractors claimed, the Kirk had become tainted with popery and Presbyterians, who had led the crusade against popery first in Scotland and then in England, had become its aiders and abettors. Defenders of the invasion argued that Scottish civil and ecclesiastical leaders vowed to impose their popish Presbyterian system on the English Commonwealth, leaving Cromwell's army no choice but to launch a preemptive strike.[19] In these and many other accusations of popish contamination, pamphleteers and journalists professed to bring popery out from its hiding place in the shadows and to strip away its disguises before the broadest possible audience.

The occasional discovery of actual Catholics masquerading as Protestants appeared to lend credence to such allegations, as was illustrated in the widely publicized case of Thomas Ramsay in 1653. A far-reaching investigation conducted by officials in Newcastle together with his own confession established that Ramsay, a young man of Scottish descent, had been enlisted by the Jesuits to pose as a Jewish convert to Protestantism. To support his claim, Ramsay was circumcised. In due course Ramsay joined the Baptist congregation of Thomas Tillam – a former Catholic himself – in Hexham in Northumberland. When persistent rumours alleging that Ramsay was actually a papist began to circulate, Tillam countered that all sceptics would lay their doubts to rest, 'when they shall behold that manifest token in his flesh, the distinguishing Character of circumcision which I also saw before I baptized him'.[20] Though Tillam subsequently conceded his mistake and published an account of the affair both to clear his name and to document the ways in which he had been deceived, the stain of his own Catholic background encouraged some to speculate he may not have fully renounced his popish ties.[21]

The Ramsay case was taken as evidence that papists were willing and able physically to alter their bodies to disguise themselves, and consequently the affair was treated as an important lesson in popish tactics. It also pointed to an effective and necessary method of identifying and combating popery: the vigilance of and cooperation among communities. The resolution of this matter required the combined efforts in Newcastle and Hexham of government officials, ministers, and local residents who acted upon their suspicions. The investigation also offered the positive message that papists might move undetected through Protestant society temporarily, but with enough diligence their presence could and would be revealed. In August 1653 Parliament ordered the details of this investigation to be published accordingly.[22]

III

The spectre of the hidden papist had, from the time of Elizabeth's reign, given impetus to various strategies for uncovering and penalizing

Catholics. It was the traditional role of the state physician – the civil magistrate or others charged with enforcing laws and punishing offenders, including judicial bodies[23] – to oversee the nation's religious health. Rulers, on the top rung of the hierarchical ladder, ultimately were responsible for directing policies and procedures. For supporters of the Commonwealth and Protectorate governments, the reign of Charles I demonstrated that the highest echelons of state and church government could become infected with popery and thereby imperil the religious purity of the nation at large. Executing the king and tearing down the Church of England, however, were only partial steps towards healing the body politic.

Full recovery, numerous pamphlets argued, required both combating religious disease and healing the ruptures within the Protestant faith. As the pseudonymous author of *A Brief and Perfect Journal* explained,

> A Common-wealth of people, is as mans body, some member may be corrupted, and yet the vitals preserved, and the head not impaired; in such cases the skillful Chichurgeon, that takes care to keep and preserve the Microcosm, dismembreth that part from the rest of the body, that might otherwise destroy the whole fabrick.

Recent civil war, the author continued, 'hath yet left some species of malignancy, the sores and corruptions of the Nation are not healed, because not yet cleansed'.[24] Divisions among the godly weakened Protestant unity, which in turn furthered popish aims to crush the Protestant faith. Yet according to the minister Faithful Teate, the festering sores of intra-Protestant dispute were not incurable. Taking the physician as his model, the Protector could work to heal religious wounds and bring about Protestant unity.[25] Mending these breaches would seal the Protestant body and consequently obstruct its penetration by popery and other religious and political contaminants.

From their inception in the sixteenth century, recusancy laws were the most common measure employed to identify and combat religious pollution: persistent refusal to attend Protestant services provided for the sequestration of two-thirds of the nonconformist's or recusant's estate. In theory, over time the provision of financial penalties, whether threatened or implemented, would pressure recusants into conformity. The impact of such laws on Catholic recusants during the 1650s is a matter of some dispute among historians. Most scholars contend that recusancy fines were harshly enforced during the period, but in exchange Catholics were allowed considerable freedom of worship.[26] On the other hand, Terence Smith has concluded from his study of Staffordshire that the condition of Catholics was more severe than historians have allowed. Smith argues that not only were recusancy fines higher between 1649 and 1653 than they had been in previous decades, but also suggests that these stiff financial penalties largely would have neutralized any relief Catholics might have enjoyed from increased liberty of worship.[27]

The level of enforcement of recusancy fines, however, can be a mislead-ing index of the strength of anti-popery in contemporary society. These and other laws against Catholics were intended for occasional use rather than consistent application. Waves of anti-Catholic prosecutions were trig-gered by particular social and political crises, as in 1641–42 with the out-break of civil war and Irish rebellion. As well, recusancy laws did not apply to Catholics who practised occasional conformity. A combination of addi-tional factors determined the degree of enforcement of the laws, including the wealth of the recusant (exemptions were made for poverty) and his/her ability to conceal moveable goods and property, the recusant's degree of local influence, and the diligence of government officials.[28]

Other practical issues made the detection and punishment of Catholics more difficult during the interregnum. To provide for Protestant liberty of conscience, in 1650 the Commonwealth repealed the Elizabethan law mandating Sunday attendance at the local parish church,[29] a measure subse-quently upheld in the two constitutions of the Cromwellian Protectorate, the 1653 *Instrument of Government* and the 1657 *Humble Petition and Advice*. The guiding principle was that 'true Christianity' could flourish only without coercion. While it was still necessary to attend some form of approved Protestant worship, the detection of Catholics could be much more challenging since the local church was no longer required to monitor its parishioners.[30] The Commonwealth and Protectorate governments turned to other strategies at their disposal. One such means was to monitor loca-tions known or suspected to be Catholic meeting places and apprehend clergy and laity who attended Catholic services. Sometimes private resi-dences were singled out for surveillance, as was Count Egmont's house in Holborn in late 1650.[31] The residences of ambassadors or agents from Catholic states were of particular interest to authorities and, on the pre-sumption that they were havens for the 'coming over of the priests and Jesuits from foureigne parts' were subjected to frequent searches and raids.[32]

On one level, by isolating the infected in order to preserve the health of the body politic the imprisonment or confinement of popish and other agents of religious contagion paralleled the medical practice of quarantine. Confinement, however, was believed to be only partially and temporarily effective at best and counterproductive at worst. First, the imprisoned could become the focus of community support and enhance the sense of Catholic solidarity. Moreover, early modern prisons frequently allowed considerable freedom of movement, and consequently prisons could be sites of Catholic interaction and worship.[33] Instead of separating the diseased from the healthy, then, prisons could facilitate the spread of contagion not only within its walls but beyond as prisoners were released and visitors came and went.

To achieve the desired objective – social and spatial freedom from popery – officials often chose another method: banishment, analogous to the medical practice of purging. This was the aim behind the June 1651 order for the expulsion of priests and Jesuits from England,[34] and the regular release of Catholic clergymen from prison upon security (usually a substantial £200) to leave the commonwealth.[35] Most of these expulsions or purges targeted the Catholic clergy: Protestant officials hoped that, deprived of the spiritual nourishment provided by priests, popery eventually would wither and die. Some writers, however, envisaged purging on a much larger scale. As one text argued, the religiously and politically impure should be expelled from the commonwealth and shipped to overseas colonies.[36]

Though deemed less hazardous than imprisonment, purging also carried risks. Thomas Tillam, for example, cautioned that the parliamentary order expelling priests and Jesuits had 'enraged those Locusts' and renewed the determination of papists 'to undermine, poison, and bring to ruin the Churches of Christ'.[37] The available means to treat religious infection were weighed carefully, and it was recognized that each remedy had potential side-effects which occasionally could be worse than the disease. For this reason execution, the most extreme measure was rarely applied: two executions, both of Catholic clergymen, were performed in the 1650s. Like amputation, execution severed the offending member and in so doing risked potentially lethal trauma to the rest of the body. The foremost concern was that execution would create martyrs, and martyrdom could unite papists domestically and perhaps internationally to rise up and attack the commonwealth. Execution, therefore, was a hazardous proposition, particularly during the social and political tumults of the 1650s. Only in exceptional cases was it judged an appropriate remedy and even then did not produce consensus. Indeed, the Protector himself disapproved of the execution of the priest John Southworth in 1654.[38]

Government officials and the judiciary were not the only ones responsible for managing the health of the body politic: as the Ramsay affair demonstrated, detecting and treating religious infection depended heavily upon community cooperation and initiative. Yet ordinary men and women could hinder as much as assist central and local governments and often took the task of monitoring the religious health of the community upon themselves. Many, for example, did not accept the official designation of Anglicans as crypto-papists and refused to notify the authorities of the continuation of Anglican rites. It is also important to note that for the most part, contemporary Protestants did not view papists as a uniform, monolithic group. Catholics were members – albeit marginalized ones – of society: they were neighbours, local notables, landowners, tenants, consumers, producers, and often relatives and friends.

Popery therefore regularly came into contact with Protestant society, and these encounters required negotiation between anti-popish ideology and individual Catholics. Clergy, and especially Jesuits (the order most closely associated with the pope), were deemed the most dangerous. 'Peaceable' lay Catholics, however, could be considered relatively unthreatening, and some received protection from prosecution through the intervention of Protestant relatives, community residents, and even government officials. As the anonymous author of one 1653 pamphlet argued, the clergy and other foreign agents whose principal allegiance was undoubtedly the pope, who were educated in seminaries abroad and returned to subvert the commonwealth by spreading and sustaining their faith, 'merit to suffer as Traytors, Rebels, Incendiaries of the State'. Conversely, those who were 'meerly devout Papists', that is, not actively engaged in undermining the stability of the Commonwealth, 'ought not to be harmed or molested'. [39] For Protestants who viewed the Roman Church as an oppressive and deceitful institution, it was not too great a leap to regard lay Catholics as its victims or dupes. Indeed, Baxter's deeply held, vociferous anti-Catholicism did not prevent him from extending charity to individual Catholics. [40] Anti-Catholicism in seventeenth-century society was, as Anthony Milton recently has described it, a qualified intolerance. [41]

IV

The inconsistent treatment of, and sometimes equivocal attitudes towards, Catholics cannot be reduced to a simple division between ideology and theory on the one hand and reality and practice on the other. It points to the tension between competing theories of the origin and nature of bodily illness. In the pathogenic model of Paracelsus and Fracastoro, which forms the basis of Jonathan Gil Harris's *Foreign Bodies and the Body Politic*, disease was caused by foreign agents penetrating and attacking the body. [42] Preserving health consequently involved protecting the body against invasion by foreign pathogens. If the body became infected, one method for ridding the body of disease was to combat like with like, such as administering one poison to counteract another. In early modern England, Harris demonstrates, dramatists and political theorists portrayed Catholics as foreign agents assaulting the body politic, and counterattack or harsh punishment constituted an appropriate remedy. Yet not all infections were necessarily harmful: pathogens could be expelled by the body, or some of the matter could be retained to serve as a form of inoculation. [43]

The pathogenic model, which increased in influence during the course of the sixteenth and seventeenth centuries, affords a useful framework for interpreting both the prescription of harsh laws against Catholics (counterattacking dangerous foreign invaders) and their inconsistent and intermittent application (some pathogens were less harmful than others). But while Paracelsian theory rivalled it did not decisively replace the traditional Galenic model, which postulated that illness was endogamous and caused by humoral imbalance. The key to health in Galenic medicine was to maintain or, in the case of disease, to recover equilibrium, and achieving balance required treating like with unlike; treating one element or substance with its equivalent would create further imbalance and therefore worsen health. This model allowed contemporary Protestants to regard Catholics not as foreign pathogens but as an organic affliction. Focusing on internal balance, Protestant officials, communities and individuals sometimes determined that measures intended to uproot popery and other forms of religious disease would be akin to treating poison with poison.

The Paracelsian and Galenic systems coexisted alongside one another in the mid-seventeenth century, and indeed the two models were not clearly opposed. Together, these systems offered contemporaries a range of possible responses to religious contagion. Competing theories played out in public debates over the role of the state physician, the extent to which this authority should be granted, and the remedies that ought to be given approval. These debates were tied to the disputes regarding liberty of conscience and principally addressed which groups should merit or be denied freedom of worship and what penalties should be prescribed for those excluded.[44]

On balance, anti-tolerationists favoured the Paracelsian model. Tolerationists, on the other hand, to varying degrees drew on both the Paracelsian and Galenic systems. Though the diversity of positions and concerns can make generalizing difficult, most agreed that papists and other harmful forms of infection were pathogenic in origin. Radical tolerationists, however, generally held that religious disease was endogamous and treatment should be based on the principle of balance: the only suitable remedy for religious error was the delivery of truth.

Moderate tolerationists, who constituted the vast majority, maintained that godly Protestants who subscribed to the fundamentals of 'true Christianity' but differed over such issues as church structure, forms of discipline and modes of worship, should be granted liberty of conscience. Religious uniformity and coercion of consciences, it was argued, were popish practices, and consequently the state physician should not be empowered to wield popish weapons against the godly. While the definition of 'true Christianity' was the subject of much dispute, virtually all moderate tolerationists agreed to exclude papists, popish Anglicans or prelatists, the licentious and disturbers of the public peace. The Commonwealth and Protectorate governments took these provisions for liberty of conscience as the basis of their religious policies, and those excluded from its provisions were subject to penalty on both civil and religious grounds. As John Owen, Independent minister and preacher to the Council of State, declared, papists were both antichristian and disloyal, harbouring in their 'very bowels a fatal engine against all magistracy amongst us', and consequently 'cannot upon our concessions plead for forbearance'.[45] Exclusions were not static, however, and redefinitions of godliness developed in response to the rise of new sects, such as the Quakers, or the increasing prominence of existing groups like the Socinians who denied the divinity of Christ. Most moderate tolerationists barred both these groups, and in 1657 the second constitution of the Protectorate revised the provisions for liberty of conscience to exclude anti-Trinitarians and deniers of the divinity of Christ from freedom of worship.[46]

To radical tolerationists, the shifting boundaries separating 'true Christians' from their enemies and the disagreements over which groups warranted liberty of conscience indicated that definitions of orthodoxy were arbitrary and subject to determination by the current political power. As markers moved, or through the overzealous or capricious actions of those engaged to monitor the nation's religious health, the godly could be crushed by the fist of the civil magistrate. At all costs such a catastrophe must be avoided, the anonymous author of *A True Case of the State of Liberty of Conscience* asserted: 'it's better that a thousand Erronious [sic] persons be tolerated, than one good man be persecuted.'[47] Pamphleteers reminded readers that the application of civil penalties in the religious arena was itself a popish practice designed to suppress religious dissent. According to the Baptist Edward Barber, it was 'the mark of the Church of Antichrist, to terrify others by imprisonments, banishments, and persecutions'.[48] Administering poison to treat poison or utilizing a popish remedy to cure religious infection was both ineffective and perilous.

The proposed solution was to remove spiritual matters from the jurisdiction of the civil magistrate. Attempts to cut out the diseased part of the body endangered its healthy components, Henry Vane Jr explained, rendering the intended cure more detrimental than the illness. The issue here was not only the skill of the magistrate and the care with which remedies were applied: it was that the state physician simply was not competent to diagnose and treat religious disease.[49] This type of illness should be treated with nothing more than reasoned argument and a strong dose of godly Protestantism. As Edward Barber put it, 'sound doctrine, and a good conversation shall be the only meanes to winne upon the contrary minded'.[50] Errors should be refuted candidly and cogently in the press and pulpit, and it was expected that over time the light of true Christianity would cure religious illness.

Most radical tolerationists took care to point out that depriving the civil magistrate of authority in the religious arena was not the same as advocating freedom of all, including Catholic, worship.[51] As Henry Vane Jr carefully noted, 'I doe not intend a necessary Toleration of Papists, much lesse of Priests and Jesuites'.[52] Papists, prelatists and others who disturbed the peace – which frequently was held to include Presbyterians[53] – still fell under the magistrate's civil jurisdiction. Barber, for example, asserted that religious error should incur no penalty, whether fine, banishment or execution, provided that its professors remained 'loyall to our Country'.[54] Any who crossed this line would be subject to penalty, albeit on civil rather than religious grounds.

Some radical tolerationists further distinguished between Catholic laity and clergy and recommended that each receive different treatment. The laity, Henry Vane Jr cautiously suggested, were 'seduced people' who 'have already suffered much' over the previous century and consequently should be handled with 'tenderness'. Recalling the pathogenic model, Vane argues the clergy were much more clearly guilty of civil offences: they were agents who 'fetch their Commission from that foreign power to promote that persecuting and unsociable Religion'.[55] In addressing the religious health of the nation radical tolerationists tended to prefer the Galenic model, though as Vane's text indicates the Paracelsian could be applied to the civil sphere. Catholics thus could be viewed through two different lenses at the same time, which could cause both images to lose focus. Vane seems to have recognized this problem; while he depicted the clergy as dangerous threats to the social order, he left the discussion of penalty open-ended. [56]

Tolerationist writing inspired a torrent of enraged responses contending that depriving the civil magistrate of authority over the religious health of the nation was tantamount to opening the floodgates for religious contagion. Anti-tolerationists, among the most vocal of whom were Presbyterians, urged that religious disease be combated more aggressively. Many protested the absence of uniform measures to monitor and censure unorthodox views and practices and complained that the prohibitions against tolerating papists and prelatists were effectively meaningless. There were in practice no concrete guidelines to prevent these groups from benefiting from freedom of worship, a group of Essex Presbyterian ministers declared, and therefore religious error could flourish under the guise of orthodoxy.[57] Stemming the tide required strict enforcement of the laws against papists and other religious offenders. By putting out books advocating liberty of conscience, the press was held responsible for dispersing contagion, which in turn led to demands for greater press controls.[58] For opponents, even the limited toleration granted by the Commonwealth and Protectorate governments was not only much too lax, it actually encouraged the growth of malady.

V

Contemporary Protestants generally agreed that the body politic suffered from religious disease. The struggles to contain and prevent the spread of contagion, however, provoked fractious debates regarding the origin and treatment of religious illness. In part, these divergences reflected the application of different medical models: designating Catholics as exogamous or endogamous causes of ailment helped guide the selection of remedies. Both systems offered a variety of potential treatments, the efficacy of which was open to challenge and public dispute. Furthermore, when the definition of health itself was open to question, when Anglicans, Presbyterians, Independents and Quakers each claimed for themselves the designation of true Christians and accused others of popery, blasphemy or disturbing the peace, the line separating purity from pollution was far from distinct.

Not only were remedies for religious infection contested and controversial, the preventative measures outlined in the press and pulpit could raise more questions than answers. On the one hand, Protestant men and women were instructed to avoid all persons, activities, texts and objects that were recognizably popish or could lead to exposure to popery. Travel, business and commerce, and the composition of communities and families, to name but a few factors, presented sometimes insurmountable barriers to complying with such directives. On the other, Protestants were reminded that papists were as often as not in disguise, subtly working their way through Protestant society in order to destroy it. Popery was as much intangible as tangible, and it seemed to mutate, assuming new guises and gaining strength and resilience as it spread.

When confronting suspected or avowed Catholics, Protestant men and women could draw from a number of potential responses depending on the level of danger assessed. The process of detecting and responding to disease – metaphorical or organic – was not a uniform one: identifying the symptoms, stage of progression, virulence, risk of contagion and the possible effects of treatment all played important roles in determining a suitable approach. Rather than a disjunction between theory and practice, contemporaries employed a variety of theories and conventions (medical, legal, political, social, religious, as well as conventions of gender, authorship, pamphleteering and controversial literature) which accommodated divergent, even competing, models and interpretations and which helped shape practical responses to religious contagion.

Notes

1 I would like to thank Claire Carlin for organizing the symposium 'Infection without Germs: Christianity and Contagion in Early Modern Europe', held in Victoria, BC in September 2003. I am grateful to Michael Finlayson, Ian Gentles, Barbara Todd and Tanya Hagen for their comments on earlier versions of this chapter.
2 A Friend of true Reformation, *A Lamentable Representation of the Effects of the present Toleration* (London, 1656), p. 3.
3 For a recent study, see J. G. Harris, *Foreign Bodies and the Body Politic: Discourses of Social Pathology in Early Modern England* (Cambridge: Cambridge University Press, 1998).
4 Among the most useful studies of anti-Catholic polemic are A. Milton, *Catholic and Reformed: The Roman and Protestant Churches in English Protestant Thought, 1600–1640* (Cambridge: Cambridge University Press, 1995); and Peter Lake, 'Anti-Popery: The Structure of a Prejudice', in *The English Civil War*, ed. R. Cust and A. Hughes (London and New York: Arnold, 1997), pp. 181–210.
5 R. Baxter, *A Key for Catholicks* (London, 1659), p. 276.
6 *Ibid.*, p. 312.
7 Luke Fawne et al., *A Beacon set on fire* (London, 1652) and *A second beacon fired* (London, 1654).

8 [Michael Sparke], *A second beacon fired by Scintilla* (London, 1652).

9 *Beacon set on fire*, p. 10.

10 John Barclay, ed., *The Diary of Alexander Jaffray* (Aberdeen: G. and R. King, 1856), p. 45.

11 Cyprien de Gamache, 'Memoirs of the Mission in England of the Capuchin Friars of the Province of Paris, from the Year 1630 to 1669', in *The Court and Times of Charles I*, ed. T. Birch, 2 vols (London: H. Colburn, 1849), 2, p. 309.

12 E. S. De Beer, ed., *The Diary of John Evelyn*, 11 vols (Oxford: Clarendon Press, 1955), 3, pp. 23, 26, 34–5, 45–6.

13 T. Birch, ed., *A Collection of the State Papers of John Thurloe*, 7 vols (London, 1742), 1, p. 403.

14 A. Walsham, *Church Papists: Catholicism, Conformity, and Confessional Polemic in Early Modern England* (1993; rpt Woodbridge: Boydell Press, 1999). J. Bossy argues that occasional conformity dwindled by the turn of the sixteenth century in *The English Catholic Community, 1570–1850* (London: Darton, Longman and Todd, 1975), p. 187. J. C. H. Aveling, however, contends that not only did it remain a common practice, especially among male heads of households throughout the seventeenth century but it also ensured the survival of the Catholic faith in England; *The Handle and the Axe: The Catholic Recusants in England from Reformation to Emancipation* (London: Blond and Briggs, 1976), p. 162.

15 That did not, however, mean that the practice of Catholicism was entirely dependent upon the gentry. See B. G. Blackwood, 'Plebian Catholics in the 1640s and 1650s', *Recusant History*, 18 (1986) 42–58.

16 The significance of women's roles and the household in sustaining Catholicism are examined in F. E. Dolan, 'Gender and the "Lost" Spaces of Catholicism', *Journal of Interdisciplinary History*, 32 (2002), 651–64; and *idem.*, *Whores of Babylon: Catholicism, Gender, and Seventeenth-Century Print Culture* (Ithaca, NY: Cornell University Press, 1999); M. B. Rowlands, 'Recusant Women 1560-1640', in *Women in English Society 1500–1800*, ed. M. Prior (London: Methuen, 1985), 149–80; A. F. Mariotti, 'Alienating Catholics in Early Modern England: Recusant Women, Jesuits, and Ideological Fantasies', in *idem*, ed., *Catholicism and Anti-Catholicism in Early Modern English Texts* (New York: St Martin's Press, 1999), pp. 1–34.

17 For Catholicism in the Caroline court, see C. Hibbard, *Charles I and the Popish Plot* (Chapel Hill: University of North Carolina Press, 1983).

18 For the use of print to influence public opinion, see J. Raymond, *The Invention of the Newspaper: English Newsbooks 1641–1649* (Oxford: Clarendon Press, 1996), esp. pp. 80–126; R. Cust, 'News and Politics in Early Seventeenth-Century England', in *The English Civil War*, esp. pp. 254–5; D. Stevenson, 'A Revolutionary Regime and the Press: the Scottish Covenanters and their Printers, 1638–51', in *Union, Revolution and Religion in Seventeenth-Century Scotland* (Aldershot: Variorum, 1997), pp. 315–37.

19 *A Declaration of the Army of England, Upon their March into Scotland* (London, rpt Edinburgh, 1650); Henry Parker, *Scotlands Holy War* (London, 1651), pp. 12, 29–31.

20 Thomas Tillam, *Banners of Love* (London, 1654), p. 27.

21 Thomas Weld, *A False Jew* (London, 1653), sig. Ev; *The Diary of Ralph Josselin, 1616–1683*, ed. A. MacFarlane (London: Oxford University Press for the British Academy, 1976), p. 368.

22 P[ublic] R[ecord] O[ffice] S[tate] P[apers] 25/70/266, published as Weld, *A False Jew*. See also *The Counterfeit Jew* (Newcastle, 1653) and Tillam's defence in *Banners of Love*.

23   *Mercurius Politicus* 26 June–3 July 1651, p. 897.
24   J. S., *A brief and perfect Journal of the Late Proceedings and Successe of the English Army in the West-Indies* (London, 1655), p. 5.
25   Faithful Teate, *The Character of Crueltie* (London, 1656), sig. A8.
26   J. Coffey, *Persecution and Toleration 1588-1689* (Harlow: Longman, 2000), p. 157; R. Hutton, *The British Republic 1649–1660*, 2nd edn (Basingstoke: Macmillan, 1999), pp. 28–9; Aveling, *The Handle and the Axe*, pp. 175–8.
27   T. Smith, 'The Persecution of Staffordshire Roman Catholic Recusants: 1625–1660', *Journal of Ecclesiastical History*, 30 (1979) 334–7.
28   C. Hibbard, 'Early Stuart Catholicism: Revisions and Re-Visions', *Journal of Modern History*, 52 (1980) 20–1; M. Questier, 'Sir Henry Spiller, Recusancy and the Efficiency of the Jacobean Exchequer', *Historical Research*, 66 (1993) 251–66.
29   'The Act repealing Several Clauses in Statutes Imposing Penalties for not coming to Church', in *The Constitutional Documents of the Puritan Revolution, 1625–1660*, ed. S. R. Gardiner (Oxford: Clarendon Press, 1906), pp. 391–4.
30   To make identification of Catholics easier, in 1655 the Protectorate government developed the Oath of Abjuration, which involved the renunciation of papal supremacy as well as central Catholic doctrines, including transubstantiation and purgatory. One year later, the returns were far from complete: whether due to sluggish enforcement or widespread evasion, by April 1656 London and twenty-three counties had failed to return their certificates. These efforts bore little fruit before the Protector's death in 1658. See PRO SP 25/76a/49-50, SP 18/126/15-25.
31   PRO SP 25/11/8; 25/15/68.
32   See for example PRO SP 25/65/191; 25/16/34, 36; 25/35/30; 18/126/37; A. J. Loomie, S.J., 'London's Spanish Chapel Before and After the Civil War', *Recusant History*, 18 (1987) 414.
33   P. Lake with M. Questier, *The Antichrist's Lewd Hat: Protestants, Papists and Players in Post-Reformation England* (New Haven: Yale University Press, 2002), ch. 6.
34   PRO SP 25/20/7.
35   See for example PRO SP 25/64/430; 25/64/440; 25/10/50; 25/39/10; 25/41/80; 25/76/10.
36   *Brief and Perfect Journal*, pp. 5–6.
37   Tillam, *Banners of Love*, p. 1.
38   Coffey, *Persecution and Toleration*, p. 157.
39   *Counterfeit Jew*, pp. 3, 4.
40   W. M. Lamont, *Richard Baxter and the Millenium: Protestant Imperialism and the English Revolution* (London: Croom Helm, 1979), p. 283.
41   A. Milton, 'A Qualified Intolerance: The Limits and Ambiguities of Early Stuart Anti-Catholicism', in *Catholicism and Anti-Catholicism*, pp. 85–115.
42   Harris, *Foreign Bodies and the Body Politic*. See Pantin and Gagnon in this volume.
43   *Ibid.*, chs 1–3.
44   For studies of liberty of conscience in the interregnum, see J. C. Davis, 'Cromwell's Religion', in *Oliver Cromwell and the English Revolution*, ed. J. Morrill (London: Longham, 1990), pp. 193–6, 202; B. Worden, 'Toleration and the Cromwellian Protectorate', in W. J. Sheils, ed., *Persecution and Toleration: Studies in Church History*, 21 (1984) 199–233; C. Polizzotto, 'Liberty of Conscience and the Whitehall Debates of 1648–9', *Journal of Ecclesiastical History*, 26 (1975) 71–2, 79; J. Coffey, 'Puritanism and Liberty Revisited: The Case for Toleration in the English Revolution', *Historical Journal*, 41 (1998) 961–85.

45 John Owen, *A sermon preached to the Honourable House of Commons ... on January 31* (London, 1649), p. 41.

46 *The Humble Petition and Advice* articles 10–11, in *The Constitutional Documents of the Puritan Revolution*, pp. 454–55.

47 Anon., *A True State of the Case of Liberty of Conscience* (London, 1655), pp. 10–11.

48 Edward Barber, *An Answer to the Essex Watchmen's Watchword* (London, 1649), p. 16.

49 [Henry Vane Jr], *Zeal Examined* (London, 1652), pp. 11, 18–21.

50 Barber, *Answer*, p. 10.

51 A minority, however, did explicitly advocate freedom of worship for Catholics. See Coffey, 'Puritanism and Liberty Revisited', 965–9; and N. Carlin, 'Toleration for Catholics in the Puritan revolution' in *Tolerance and Intolerance in the European Reformation*, ed. O. P. Grell and R. Scribner (Cambridge: Cambridge University Press, 1996), pp. 216–30.

52 *Zeal Examined*, sig. A3v.

53 See, for example, Anon., *The Spirit of Persecution Again broken loose* (London, 1656), pp. 2, 6, sig. C3.

54 Barber, *Answer*, p. 13.

55 *Zeal Examined*, sig. A3v.

56 Vane appears to have modified his views over time. In a pamphlet published four years later, he does not include the caveat against the open toleration of papists nor does he distinguish between Catholic clergy and laity. See *A Healing Question* ([London], 1656), esp. pp. 5–8.

57 D[aniel] R[ogers], *The Essex Watchmens Watchword* (London, 1649), pp. 8–9.

58 *Lamentable Representation*, pp. 9, 16, 19.

15
Contagion by Conceit: Menstruosity and the Rhetoric of Smallpox into the Age of Inoculation

David E. Shuttleton

An incident towards the close of Sarah Fielding's sentimental novel *The Adventures of David Simple* (1753) raises questions central to the concerns of this volume regarding how we interpret historical representations of contagion.[1] The novel's eponymous hero and his wife Camilla have received a request from Mr Ratcliff, their rich but autocratic and treacherous relation, demanding a visit from their son Peter, to whom he stands godfather. David's difficult decision to refuse this request at the risk of undermining his son's prospects is made easier when 'young Peter fell ill of the Smallpox'. Camilla persuades David to write Ratcliff a 'civil' letter explaining 'that the Boy was at present too ill to take a Journey, and they were apprehensive was breeding the Small-pox' (p. 387). Affronted, Ratcliff replies with a tirade against their ingratitude and deception, to which he adds this postscript:

> P.S. ... you have rewarded all my dear wife's good Offices to you, with her Destruction; for, by my being abroad, she unfortunately opened your Letter, and I found her in Fits on my return, with the Fright of seeing the name of the Small-pox in your careless letter: and you know too, she has never had that Distemper. (p. 386)

Despite Ratcliff's 'ill-natured-insinuations' that they are lying about Peter's illness, the narrator tells us that the conscientious Camilla might have reason to be 'concerned' that 'by her Means' such 'fatal Consequences' could have arisen, because 'she had heard so many Stories, well attested, of persons being seized with the Small-pox by the Force of their Imaginations, that she would have had some fears, lest that should have been Mrs *Ratcliff's* case' (p. 386).

228

As modern readers encountering this episode almost thirty years after a combination of virology, vaccination and isolation measures have effected the eradication of smallpox from the global population, how are we to make sense of Camilla's concerns over her own culpability in 'Mrs Ratcliffe's Case'? Although the postscript affirms that it was generally known that you could suffer smallpox only once, her husband's problem over the infectious communication does not rest in a charge that the letter itself is contagious in the materialist sense we would now understand (indeed Ratcliff is insisting that they are lying about Peter's illness); rather it is the very word 'smallpox' and its potential effect on Mrs Ratcliff's mind that threatens to prove fatal. Given the particularly gruesome symptoms of smallpox and recent bio-terrorist anxieties, we might readily identify with the apprehensions earlier generations inevitably felt at the mere mention of such a virulent and destructive disease, but at the same time feel smug in our scientific understanding that in fact it would take more than simply fear to cause an infection by any of the ten or so strains of the albeit highly contagious virus *variola major* identified by the World Health Organization in its concluding report of 1988.[2] But would the original readers of Fielding's novel, itself published thirty years after the first adoption of inoculation for smallpox in England, have considered Camilla's understanding of smallpox contagion a sign of her ignorant adherence to an already outmoded superstition or taken it as a reasonable concern in the light of contemporary disease concepts? How did such claims for the 'Force of their Imaginations' accord with contemporary medical theory? My ensuing discussion seeks to provide answers to these questions.

The belief that smallpox could be prompted by fear needs to be considered in the context of what Genevieve Miller, in her still standard study of *The Adoption of Inoculation for Smallpox* (1957), once called 'the confused and conflicting theories of the cause and mechanisms of smallpox which prevailed when inoculation was begun in Western Europe and the British colonies in America'.[3] Miller talks of 'the lack of orthodoxy in medical theory and the confused multiplicity of causes and underlying mechanisms which were assigned to smallpox' at a time when 'a mixture of ancient and modern thought ... resulted in a very intricate complex of ideas that cannot be reduced to a few simple statements'.[4] She does, however, suggest that post-1720 'the new experience of inoculation' prompted 'a significant shift in emphasis as the eighteenth century progressed' whereby 'the material origin of the disease shifted from a location inside the body to one without' and 'the increasing conviction that smallpox is a specific disease attributable to a specific material cause'.[5] Adopting the terms employed by Charles E. Rosenberg in *Explaining Epidemics*, this implies a move away from a 'configuration' model, pre-dating the knowledge of specific infectious agents, which attributed an epidemic to 'a unique configuration of circumstances, a disturbance in the "normal" health-maintaining and health-constituting arrangement of climate, environment, and communal life', towards what Rosenberg terms a 'contamination' model: this often 'reduced itself down to person-to-person contagion, of the transmission of some morbid material from one individual to another'.[6] But even as Rosenberg was describing 'contamination' as 'logically alternative' to 'configuration', he was having to concede that 'historically the two have often been found in relatively peaceful, if not logical co-existence'.[7] Such 'co-existence' certainly chimes with Miller's sense that '[v]estiges of the old and elements of the new persisted side by side'[8] and my own reading of pre-modern accounts of smallpox which often draw upon several explanatory models, sometimes in the same text.[9] As we shall see, claims for the causative role of the imagination further confound any neat division between 'configuration' and 'contamination' explanations.

Early modern scholars were divided over whether the classical authorities ever encountered smallpox, but the consensus was that it had been carried into Europe from its assumed origins in the over-heated climate of Northern Africa by Islamic invaders or, in what amounted to the same racially motivated slur, by returning Christian crusaders.[10] This often demonizing association with an Islamic 'other' exploited the fact that the first writer to offer a fully codified account of the disease was the tenth-century Arabian physician Rhazes (Abu Bakr Muhammad ibn Zakariya al Razi). He treated it, alongside measles, as a relatively mild depuratory fever; a natural form of purgation caused by a 'ferment' of the blood producing corrupt matter analogous to the must of fermenting wine. The drastic impact of smallpox on Native American populations, where it was previously unknown, reinforced a prevalent view that it was a disease of modern luxury. Orthodox treatments, based on Rhazes's 'Hot Method', were designed to assist this natural process by inducing perspiration through keeping the patient wrapped in blankets in a closed room and by the use of purges, diuretics and phlebotomies aimed at drawing the disease outwards and away from the vital organs.[11] Though long recognized as communicable, in Britain smallpox emerges as an endemic, frequently epidemic killer only from the Restoration onwards when it overtakes plague, leprosy and syphilis as the most common pathogenic cause of premature death. Various causes have been offered for this increased virulence, but greater mobility of populations and associated urban expansion at a time of rapid growth in trade were probably major factors in accelerating the spread of infection.

In his treatise *De Contagione et Contagionis morbis et eorum curatione* (1546) Girolomo Fracastoro of Verona was the first to describe smallpox and measles as distinct disease entities as part of his influential hypothesis that specific diseases could be contracted through exposure to specific ('seeds' or '*semina*'), either through intimate, one-to-one contact or by breathing in particles ('*fomites*') lingering in bedding, clothes or transported goods (as detailed in Isabelle Pantin's contribution to this volume). Airborne 'seeds' of disease might infect the air of sick-chambers (or linger in sealed crypts), generate local pestilential climates (miasmas and so-called 'constitutions of the air'), or be carried over long distances on the wind. But Fracastoro also acknowledged some validity in an established claim that smallpox originated as a so-called 'innate seed' of corruption implanted in the blood of the foetus during gestation as a result of the blocked menses of the mother. This remains dormant until triggered by external 'contagious' conditions or by a humoral imbalance in the individual resulting from a surfeit or general bad regimen.[12] Later theorists, drawing analogies with the action of poisons, employed different vocabularies – corpuscularian, animalculist and Newtonian – to explain the precise mechanisms at work.[13]

Examining case histories from contemporary medical treatises attributing smallpox to fearful imaginings will not only illustrate how physicians sought to accommodate such claims within the conceptual models outlined above while hinting at other, subtle and possibly immaterial mechanisms for contagion at a distance, but also serves to expose the largely imaginative and rhetorical basis of such proffered explanations, as already glimpsed in the Levitican concept of menstrual pollution. In doing so, we should be alert to Vivian Nutton's reminder that 'we are dealing with descriptions of the invisible, with hypothetical reconstructions of how things are or act, based only on the observance of "macrophenomena"' for 'no ancient doctor ever saw the seeds, animalcula, or effluvia that were said to cause ... disease'.[14]

Isbrand van Diemerbroeck's *A Particular Treatise of the Smallpox and Measles* (1689; 1694) concludes with some intriguing 'Case Histories', including several concerning the force of the imagination.[15] Diemerbroeck's 'History VI' tells of 'a certain Apothecary that was a strong man about Thirty Years of Age' who entering 'into a Citizens House, when he found and saw of a suddain his Patient all over covered with the Small Pox upon his face, he trembled a little at the sight of so much deformity and so departed'. 'To drive the Whimsey out of his head' the apothecary starts to 'drink very hard', but 'nevertheless all he could do could not put the Fancy out of his thoughts, which the sight of such an Object had imprinted in his Mind'. Though he 'were otherwise a Man of undaunted Courage', within six days of this sighting 'a fever seized him' and he sank into a delirium 'attended with the red spots that usually fore-run the Small Pox'. This fever lasted a mere 24 hours and the patient 'being restor'd to his Health, went abroad again in three weeks'.[16] Anticipating the objection that the apothecary 'might not be [already] touched with any Infection, or whether he might not contract the Distemper from some other Cause', Diemerbroeck observes that the man had often

visited at other times, several persons that lay Sick of the same distemper, without any prejudice; and therefore the cause seems rather to be that suddain conturbation of his Mind and Sprits, with which he was stricken upon the unexpected Sight of this same Sick Person, and continually ran in his thoughts; from which idea such a disposition arose in his Body which at length produced the Small pox.[17]

Diemerbroeck suggests that the fear actually caused the disease by prompting a disturbance to the balance of body and mind in the apothecary and as a consequence advises against 'timerous' persons going 'near those that are Sick of the pestilence or Small-Pox' when 'the sight of one ill' of the disease 'could move a Man of that courage as this Apothecary was'.[18]

In Thomas Fuller's *Exanthematologia, or an Attempt to Give a Rational Account of Eruptive Fevers, especially of the Measles and Smallpox* (1730), we find a related group of cases in which contagion over a distance is attributed to contact with mediating objects bearing fearful associations. A young woman is sent a gold chain 'which another had worn in the Small-Pox, to keep them ... out of their Throat', but it 'chanced to bring Terror, and caused her Fancy to work her up into a kindly Small-Pox'.[19] Fuller makes no suggestion that this talismanic chain carried infected particles, but rather the narrative curiosity is focused on how an object that had been used to provide protection becomes a trigger. Moreover – in what now reads like a narrative chain of contagious tales – the woman in this last case in turn tells Fuller of like events, including the story of how a gentleman who 'lay sick of the Small-pox' instructed his servant to send a key 'which he had not lately touch'd and lay in a chamber far distant from him to his Mother': she then 'conceited it brought Infection' and soon fell fatally ill with the disease.[20] Again, contrary to any modern assumptions concerning infectious contamination, in this case – which resembles that of the frightening epistle in *David Simple* – Fuller clearly emphasizes that the mediating object had never been in direct contact with the original victim. It does not even trigger the disease at a distance by mediating between two physical touches but merely by prompting a contaminating 'conceit'.

Related histories are more specifically concerned with triggering by shocking sights. Diemerbroeck's 'History V' records the story of how two young sisters 'encounter a young lad ... newly cured of the Small Pox' who 'was got abroad, and coming along in the street, at least thirty paces distant from them, having his face all spotted with red spots, the remainders of the footsteps of the disease; with which sight they were so scared that they thought themselves infected already'.[21] Diemerbroeck 'endeavour'd by many arguments to dispel these idle fears', prescribing a purge and ordering them 'to walk abroad, visit Friends, and by pleasant Discourse and Conversation, and all other ways imaginable to drive those vain conceits out of their Minds'. But 'all that I could do signified nothing, so deeply had this conceit rooted itself in their Imagination' until 'at length, *without occasion of Infection*, they were both seized'.[22] Here Diemerbroeck distinguishes this case of conceit from an external 'infection' yet his commentary reads:

How wonderful the Strength of Imagination is, we have experienced in many Persons, for that by the Motions of the Mind it frequently works Miracles ... [and] thus in these two Gentlewomen through the continual and constant Cogitation caused by the Preceding Fear, that Idea of the Small Pox, so strongly Imprinted in their Minds, and thence in their Spirits and Humours, begat therein a disposition and Aptitude to receive the Small Pox.[23]

This use of 'receive' leads us to think that Diemerbroeck concludes that mental perturbation had left the sisters vulnerable to contagious infection from outside, but his commentary immediately proceeds to another case of smallpox, which being supposedly prompted by a dream suggests a particularly subtle appeal to a configuration explanation.

Diemerbroeck had once visited a 'Noble *German*, who Dreamt that he was drawn against his Will to visit one that was Sick of the Small Pox, and was very much disfigur'd'; being 'unable to drive the dream out of his thoughts', after three weeks this nobleman fell 'into a fever and was pepper'd with the Small Pox'.[24] While this last case in particular might invite a modern, perhaps Freudian interpretation couched in terms of contagious emotions or hysteria, my concern in unearthing these puzzling histories from the archive is not with pursuing retrospective diagnoses, but rather with comprehending how these early commentators struggled to accommodate such reports within their own largely imaginative models already morally freighted with ultimately punitive notions of contagion and pollution. To this end, it is important to distinguish – initially at least – between claims that fear could exacerbate a poor prognosis in an individual case, and the larger – and to modern minds more 'unscientific' claim – that smallpox was actually being contracted and/or triggered by the imagination.

The first of these beliefs informed the popular practice of removing any mirrors from the vicinity of a smallpox patient. This was to prevent recovery being undermined by the trauma of self-alienation that overcame victims who caught sight of the horrendous distortion confluent pustules rendered to their features and which threatened permanent disfigurement through scarring. Professional medical commentators endorsed such precautions; writing on the management of smallpox in 1718, Dr John Woodward thought that the physician's 'Masterpiece, and chief Care is to raise the Fancy, steer and rightly rule the Passions, and continually keep up the Hopes of the Patient.'[25] Claiming that unless one learns this 'great Art' of calming the smallpox patient, 'the best Medicines, directed with the utmost Wisdom, in the Small-Pox ... will prove generally ineffectual'

Woodward goes so far as to claim that "tis certain there are greater Numbers hurryed out of Life by the Disorders brought on by fright, Surprize, Apprehension, the bustle and indiscreet Shew of Concern by Relations, Friends, and those about the patient, than by the Malignity of the Disease'.[26] His brief physiological explanation for this vulnerability is couched in terms of humoral imbalance, while specifically invoking the pyrolic spasms of nausea:

> The reason of which will be evident to those who are rightly inform'd of the Contrivance of the Body of Man: and know that the Stomach, which is the Fountain of those Principles that supply, form, and raise the Small-Pox, is likewise the Seat of the Passions. Now every unseasonable Rouseing of them must needs disturb the Oeconomy, and regular Egress of those Principles; upon which Oeconomy, and Regulation, the Event of the Disease depends.[27]

'The Passions' – a key term in eighteenth-century psychology – formed one of the rather misleadingly termed 'six non-naturals' of Galenism (the others were 'airs', 'diet', 'exercise', 'repletion and evacuation' and 'sleep and wakefulness') and as such were an essential consideration in disease aetiology and preventive 'Regimen'.[28] Even as smallpox was increasingly being recognized as communicable through infection by air-borne disease-specific particles, the role of the Passions remained crucial, accounting for why some individuals were more vulnerable than others. In a list of types and conditions rendering individuals vulnerable to infection by measles and smallpox, Fuller for example, includes those who have 'strong Fancies' and who are 'apt to fall into frights and Terrors' and anyone 'when the Spirits are confused, beat down, suppress'd, and put by their natural Functions and Operations; as by Frights, Terrors, Grief, strong Imaginations etc.'.[29] Fuller was claiming that 'the Hysteric and Hypochondriac' whose 'Spirits are not able to oppose the assaulting Enemy' are vulnerable to external attack from contagious 'variolous' particles carried in the air; but crucially he also asserts there 'have been very numerous Instances of people that have got the Small Pox (but not the Measles, or any other Sorts that I have heard of) by mere Fancy and Fear'.[30] Thus the risk of a perturbation of the Passions overwhelming the imagination not only underscored concerns over the effects of fear, but clearly helped to account for some otherwise baffling cases of smallpox acting at a distance.

Claims that smallpox in particular could be prompted by fear or 'conceit' no doubt owe much to the disease's gruesome visual symptoms, but also suggest a vestigial adherence to earlier, related concepts of contagion through *fascination* or occult *sympathy* (as discussed by Nancy Frelick and Clause Gagnon in the present volume). Certainly, the practice of encouraging smallpox victims to remain indoors or wear veils or masks in public was informed by the belief that the disease could be triggered by shock or fear. Of several such cases known to Fuller one of 'the most unaccountable' was that of a young man, 'who being scared with seeing one that lately had it, was taken ill upon the spot ... and had them come out upon the very next day'. Emphasizing that this was not a simple case of invasive particles, Fuller thought this case 'beyond all Rule and Precedent; for there was no Time for Assimilation or Concoction of the matter before Expulsion'.[31] The rapid onset in this case posed a challenge to Fuller's usual humoral understanding of the disease as an eruptive fever with a regular symptom pattern (like many, Fuller equated these stages with a process of crisis and elimination which he terms, assimilation, concoction, eruption, augmentation and maturation).[32] Although Fuller's account attempts to reconcile old and new ideas by describing the triggering of 'innate seeds' (what he terms 'ovula') by contagious disease-specific particles within the mechanistic vocabulary of corpuscularian ('atomistic') matter theory, he is unsure whether triggering by the imagination amounts to true contagion:

when a Person is taken with a thorough panic and Fright, and thinks of nothing but Infection, that extraordinary Perturbation and Terror may form the Spirits into such Species, and create such an Alteration of the Particles of the Body, as will directly and peculiarly act upon the latent *ovula* as effectively as an actual Contagion might do.[33]

He certainly suggests that the shock of seeing a confluent victim acts as the trigger for the 'ovula' lying dormant in the blood and as if seeking to cover all possibilities asserts that 'NATURE, in the first compounding and forming of us, hath laid into the Substance and Constitution of each something, equivalent to *Ovula*, of various Kinds, productive of all the Contagions, venomous fevers, we can possibly have as long as we live', and this in turn explains why, when 'all men have in them those specific sorts of Ovula which bring forth Small pox', all are 'liable to them'.[34] Fuller also betrays the underlying moral and religious assumptions behind the idea of smallpox corruption as a maternal inheritance, for having moulded the inherited 'innate seed' theory to produce his own conjectural explanatory edifice, he then buttresses this with biblical authority:[35]

Because these Ovula are of different Kinds; and every one of these Kinds is as essentially different from all the rest, as Eggs of different Fowl are from each another; therefore every sort of these Ovula can produce only its own proper Foetus, as it is said, 1 Corinth. xv. 38. *To Every Seed its Own Body*; and therefore the Pestilence [i.e. Bubonic Plague] can never breed the Small Pox, nor the Small-Pox the Measles ... any more than a Hen can a Duck, a Wolf a sheep, or a Thistle, Figs; and consequently, one Sort cannot be a preservative against any other.[36]

Fuller's adherence to a model of specific disease entities might invite interpretation as a small step in the direction of modern germ theory, but his
typically hybrid explanations owe little to empirical microscopic observation and everything to a rhetorical embroidering of older accounts. The
most striking feature is Fuller's imaginative elaboration upon the sexualized
meanings already implicit in terms such as 'breeding', 'seminaria' or 'seeds',
as he asserts that the ovula 'always lie quiet and unprolific, till impreg-
nated, and therefore these distempers seldom come without Infection;
which is as it where the Male, and the Active Cause'.[37] Surely it is here, in
this wholly conjectural, sexualized aetiological scenario, that we glimpse
the cultural pressures shaping Fuller's otherwise arbitrary adoption of the
term 'ovula' over 'seed' (or 'seminaria'): it simply reinforces the passive
female origins of the innate seeds while enabling him at least to equate the
external triggers with manly activity! Borrowing Nutton's remarks on the
essentially rhetorical value of Fracastoro's contagious 'seeds' hypothesis,
like many similar explanatory accounts, Fuller's simply represented 'a
philosophical luxury for the intellectual practitioner'[38] but had no therapeutic impact.

And where do the smallpox 'ovula' originate? In answering, Fuller simply
rehearses the traditional idea of impurities derived from the maternal blood
which, he argues, remain dormant in the body like other poisons such as
syphilis and rabies. More tellingly, Fuller endorses the underlying religious
implications in this concept of 'innate seeds' by paraphrasing an 'excellent
passage' from St. Severinus (d. 482):

DISEASES (saith he) have, as well as Plants and Animals, their proper
seeds in their Way and Manner. God hath (ever since the Fall of Adam)
created for the Punishment of man the seeds of evil, as well as of good
things. Our Bodies are the soil where they are sown: they grow up
and bring forth certain Distempers, every one according to its several
natures; each sort hath its peculiar and like symptoms; every Sort of
plant hath its Fibres and Figure.[39]

While Fuller attempted to marry modern medical theory with established
religious doctrine other physicians were more sceptical. Discussing the
causes of smallpox in a treatise of 1696 the physician Gideon Harvey notes
how it has been the 'universally received' opinion 'that there is a *Labes*, or
taint and impurity inherent in the Maternal Blood that gives nutrition ...
to the parts of the Foetus ... of which impurity, nature, at some uncertain
time after Birth, doth discharge and purify all the parts and juices of the
Body, by throwing it out into Measles or Small-Pox' and that it has been
assumed that 'this original *labes* is synonymous to original Sin' so, as a consequence, all mankind must at some stage in their lives undergo 'that purgatory' of either measles or smallpox (or both) as 'one of the most Ancient

Diseases of Lost Paradise'. But he concludes that the idea of smallpox being 'derived from the foulness of the menstruous Blood on the fluid, or solid parts of the Infant, is scarce possible' because, he argues, some people do not have smallpox until they are very old by which time any active 'tainted ... particles' would have long lost their malignancy.[40]

Although other Restoration and early Georgian physicians either rejected or ignored the maternal seed theory, none the less, as Miller observes, 'the notion ... was a very persistent one',[41] not least because it provided a reasonable explanation for why, when smallpox became epidemic in a particular district, not everyone succumbed. It also accorded with the common observation, reflected in Fielding's novel, that you had smallpox only once. Moreover, as Felicity Nussbaum has recently analysed, smallpox continued to be represented in punitive moral narratives as the disease which providentially threatened to disfigure and thus bring social death to vain young women; smallpox scarring was interpreted as a pointed reminder that all disease is the direct consequence of Eve's proud and undisciplined imagination.[42] Thus in *Inoculation; or Beauty's Triumph* (1768), the poet Henry Jones supplicates a popular inoculator:

> Oh! Hadst thou power to purge the darker Passions
> From the human Breast, with moral medicine,
> And inoculate the Soul; couldst thou, SUTTON,
> Quick kill the Seeds of each Distemper there,
> Of each irruptive fever, that deforms
> The maker's Image in th'immortal Mind,
> And blots bright beauty in the outward Frame,
> With marking, deep degrading Spots, those banners
> Of frail Defect, those legacies of EVE,
> That give th' angelic Face the Lye,
> And bring the fatal Apple to our View.[43]

Seeming to anticipate the putrefaction of the grave, smallpox disfigurement served as a lasting emblem of the underlying corruption of a fallen human nature, responsibility for which could be traced back to a transgressive femininity. Citing this and related literary representations, Nussbaum overlooks the extent to which such moralizing religious perspectives were supported by misogynist medical theories, which where themselves reinforced by Classical, Judeao-Christian and Islamic taboos concerning women and blood pollution.[44] While some physicians continued to claim that smallpox was a maternal inheritance, poets invariably personified 'Variola' as a monstrous female.[45] A Medusa-like 'Fury' with lethal looks, 'Variola' epitomized the female grotesque which, in Bakhtin's influential formulation, is 'unfinished, outgrows itself [and] transgresses its limits'.[46] This deep-rooted imaginative association between smallpox as maternal pollution and monstrous growth is made overt in *An Elenchus of Opinions Concerning the Cure of the Small Pox* (1661), when its author, the Royal physician Tobias Whitaker, exploits an established but false etymological linkage between *menstrua* and *monstrum* to create the telling neologism in the epithet 'Maternal Menstruosity' when discussing the neonatal origins of the 'seeds' of smallpox.[47] This purely figurative linkage leads us straight back to contemporary smallpox histories invoking fear; specifically a sub-set of cases in which mothers reportedly 'imprint' the disease onto the foetus after seeing a smallpox victim or being overtaken by fearful imaginings.

Again we have Fuller, this time reporting the case of a gentlewoman who was accidentally brushed by some soil when passing a Sexton digging a grave: on returning home she was 'most terribly frighten'd, fancying some Corps that had dy'd of the Small-Pox had broke out upon her; upon which she fell in Travail, brought forth a Child full of them; both Mother and Child dy'd. So the Distemper bred upon both from nothing at all but a mistaken Conceit.'[48] The 'Case of a Lady, Who Was delivered of a Child, which had the Small Pox appeared in a day or Two after its Birth', printed in the *Philosophical Transactions* of the Royal Society in 1749, describes an incident dating from 1700 when a pregnant woman encountered someone full of smallpox across a 40-foot courtyard; this leaves Dr Cromwell Mortimer, the report's author, baffled by 'how the Imagination only, affected by the Disagreableness of the sight, should convey the Infection to this Child in this case...especially as there was no Fright or Surpize'.[49] Fuller thought as 'strange a Thing as breeding of the Small-Pox by the force of fear and Fancy' was no more to be wondered at than the common knowledge that 'longing-women' can 'by the pure workings of the Imagination, form the Spirits into such Ideas, figures, and Species, as to Imprint marks upon their Foetus in the Womb'.[50] As Donald Beecher elaborates in his Afterword to this volume, this specific endorsement of traditional concerns with 'imprinting' during conception and gestation by seventeenth-century physicians amounts to a psychosomatic mechanism of 'contagion' and one that continued to be invoked to account for so-called 'monstrous births' well into the Georgian era.[51] The specific notion of smallpox being triggered by maternal 'conceits' was perhaps inevitable when the porosity of pustulate flesh pictured forth such a monstrous disruption of the integrity of the normally sealed boundaries of the body. Scholars were encouraged by existing medical prejudice and religious precept to blame the source of this contamination on that most primal exception to bodily integrity: the umbilical linkage between the embryo and its mother.

Traditionalist medical historians largely disregarded the early-modern claims for contagion by fearful conceits or maternal imprinting discussed above.[52] Rosenberg claimed that 'eighteenth-century physicians did not understand the nature of the "virus" that was passed from individual to individual during inoculation', yet 'it was clear that the epidemics of this great killer were caused by a specific, reproducible "matter"'.[53] But Miller had already shown that – however illogical in hindsight – not only did configuration explanations persist into the age of inoculation, but for many the increasing empirical evidence that inoculated smallpox left the patient free of the disease for life merely served to confirm that it grew from an inherited 'seed' susceptible to an artificially induced purging. Moreover, this theory also helped to allay religious scruples opposing inoculation, notably those prompted by the much discussed claim of the Revd. Edmund Massey that Satan was the first inoculator when he inflicted Job with 'sore boils'.[54] If smallpox was merely caused by a seed already present in the body then, it could be argued, inoculation did not amount to the evil introduction of a new disease. As we have seen, Fuller's 1730 treatise acknowledges the force of the imagination, but it also bears an appendix defending inoculation against Massey's charges of blasphemy (though, in the present context, it is worth noting that Fuller's list of exceptional circumstances in which the operation is inadvisable includes 'such as are extremely fearful, fanciful, Hysteric or Hypochondriac').[55] The inherent danger of being frightened by the sight of a confluent victim of natural smallpox also formed part of the pro-inoculation argument; as late as 1779 a pseudonymous pamphleteer was arguing for the advantages of inoculation on these very grounds.[56] Posing no immediate challenge to an innate seed model, inoculation did not necessarily nor immediately undermine a popular belief that smallpox could be either triggered or contracted by 'conceits'.

As this discussion makes evident, Sarah Fielding's 1753 portrayal of Camilla's concerns that her disturbing letter may have put Mrs Ratcliff's life in jeopardy was in keeping with a popular belief endorsed by medical opinion. But in the event Camilla had no need to feel responsible because 'she knew that Mrs Ratcliff had long ago had that Distemper, and had visible marks of it on her face; though, in order to have an Opportunity of making herself of Consequence by her affected Frights and Fears, she insisted ... that they were only the Marks of the Chicken-pox' (386). 'Mrs Ratcliff's case', it emerges, is simply a selfish ploy to ensure that her husband does not become attached to his godson Peter and make him his heir; a case of hypochondria in the pejorative sense of manipulative self-delusion. Thus Fielding's satirical portrait of a woman affecting vulnerability to smallpox through the mere words in a letter relied on a complex manipulation of persistent medical, religious and rhetorical associations between the imagination, femininity and the risk of contagion by conceit.

Acknowledgement

Research for this essay was supported by a Williams Clark Memorial Library Fellowship, UCLA (2001) and a British Academy AHRB Leave Grant (2004).

Notes

1 First published in 1744; this episode appears in Book VII, chapter X of 'Volume the Last' issued in 1753. Citations, hereafter in brackets, to Sarah Fielding *The Adventures of David Simple*, edited by Malcolm Kelsall (London: Oxford University Press, 1969).

2 J. R. Smith, *The Speckled Monster: Smallpox in England, 1670–1970, with particular reference to Essex* (Chelmsford: Essex Record Office, 1987), pp. 179–82; F. Fenner [et al.], *Smallpox and its Eradication* (Geneva: World Health Organization, 1988).

3 G. Miller, *The Adoption of Inoculation for Smallpox in England and France* (Philadelphia: University of Pennsylvania Press, 1957), p. 241.

4 *Ibid.*, p. 242.

5 *Ibid.*

6 *Explaining Epidemics and other Studies in the History of Medicine* (Cambridge: Cambridge University Press, 1992), p. 295.

7 *Ibid.*, p. 295; see also R. Porter, in *Western Medical Tradition, 880 BC to 1800*, ed. L. I. Conrad, M. Neve et al. (Cambridge: Cambridge University Press, 1995), pp. 407–8.

8 Miller, *The Adoption of Inoculation*, p. 256.

9 D. E. Shuttleton, 'A Culture of Disfigurement: Imagining Smallpox in the Long Eighteenth Century', in *Imagining and Framing Disease in Cultural History*, ed. G. S. Rousseau with M. Gill, D. Haycock and M. Herwig (London: Palgrave, 2003), pp. 68–91.

10 D. R. Hopkins, *Princes and Peasants, Smallpox in History* (Chicago and London: University of Chicago Press, 1983), pp. 22–3; S. Watts, *Epidemics and History: Disease, Power and Imperialism* (New Haven and London: Yale University Press, 1997), pp. 84–111.

11 Hopkins, *ibid.*, pp. 32–3. The debate over treatment intensified in the 1660s when Thomas Sydenham adopted corpuscularian theories to argue for an innovatory 'Cold method'.

12 Often attributed to Arabic sources by medieval and Renaissance European writers, Jarcho can only find the idea in the writings of 'Haly Abbas'; S. Jarcho, *The Concept of Contagion in Medicine, Literature, and Religion* (Malabar, FL: Krieger Publishing Co., 2000), p. 23.

13 Miller, *The Adoption of Inoculation*, pp. 242–56.

14 V. Nutton, 'The Seeds of Disease; an Explanation of Contagion and Infection from the Greeks to the Renaissance', *Medical History*, 27 (1983) 1–34.

15 As professor of physic and anatomy at Utrecht, Diemerbroeck (1609–1674) taught visiting British medical students and the English translation of his useful overview of late seventeenth-century opinion on the history, causes and treatment of smallpox was readily available as part of his popular *The Anatomy of Human Bodies ... To which is added a particular treatise of the small-pox and measles ... Translated [from original Latin] by William Salmon* (1689; rpt London: W. Whitwood, 1694).

16 *Ibid.*, p. 30.

17 *Ibid.*

18 *Ibid.*

19 London: C. Rivington and S. Austen, 1730, p. 189.

20 *Ibid.*

21 Diemerbroeck, *The Anatomy of Human Bodies*, p. 29.

22 *Ibid.*; my emphasis.

23 *Ibid.*

24 *Ibid.*

25 *The State of Physick: and of Diseases; with an inquiry into the causes of the late increase of them; but more particularly of the small-pox* (London: T. Horne and R. Wilkin, 1718), p. 69. For medical concerns with fear, see A. Luyendijk-Elshout, 'Of Masks and Mills: The Enlightenment Doctor and his Frightened Patient' in *The Languages of Psyche: Mind and Body in Enlightenment Thought*, ed. G. S. Rousseau (Berkeley and Oxford: University of California Press, 1990).

26 Woodward, *ibid.*, p. 69,

27 *Ibid.*, pp. 69–70.

28 See W. Riese, introduction to *Galen on the Passions and Errors of the Soul*, trans. P. W. Harkins (Columba: Ohio State University Press, 1963); and G. Sill, *The Cure of the Passions and the Origins of the Novel* (Cambridge: Cambridge University Press, 2001).

29 Thomas Fuller, *Exanthematologia, or an Attempt to Give a Rational Account of Eruptive Fevers, especially of the Measles and Smallpox* (1730), p. 104.

30 *Ibid.*, pp. 104-5.

31 *Ibid.*, p. 188.

32 *Ibid.*, p. 263.

33 *Ibid.*, p. 149.

34 *Ibid.*, p. 175.

35 Fuller talks of 'peculiar venemous Particles' that find a fit with the ovula and which themselves consist of 'rigid, infrangible, and unalterable Atoms, so subtle, pointed edged, and perhaps indented, crooked [and] barbed ... as to be ... wholly destructive to the Spirits, Blood and solids of Man'. He mentions that the microscopist Leewenhoek 'thought no Kingdom in Europe contained so many men, as he saw animalcules in the Seed of an Oyster', yet regarding his smallpox 'particles', Fuller concedes that 'their real particular, geometrical Figures, measures, and mechanic manner of exerting their Powers' must forever remain 'undiscoverable to us in this our...State of imperfection' (pp. 179, 181).

36 *Ibid.*, p. 175.

37 *Ibid.*

38 Nutton, 'The Seeds of Disease', p. 14.

39 Fuller, *Exanthematologia*, p.178. Source in Severinus untraced.

40 G. Harvey, *A Treatise of the Small-pox and the Measles; describing their Nature, Causes and Signs* (London: W. Freeman, 1696), pp. 7–10. Harvey's own views on the origins of smallpox add a typical racist slur to the usual misogyny when he suggests that 'the Ancient Arabian Physicians ... observing the promiscuous converse of their women with men, fell into a notion, that their Wombs must necessarily attract thence a great fowlness, which most certainly they did' and seeing 'their spurious Issue were surprised at some blotches, crusty Pimples, Ulcers, and pains of Limbs...which to me appears rather a Distemper of the Great Pox, than the Small' (p. 8).

41 Miller, *The Adoption of Inoculation*, p. 256.

42 F. Nussbaum, *The Limits of the Human: Fictions of Anomaly, Race and Gender in the Long Eighteenth Century* (Cambridge: Cambridge University Press, 2003), pp. 109–132.

43 H. Jones, *Inoculation; or Beauty's Triumph: a Poem in Two Cantos* (Bath: C. Pope, 1768). (1768), 7–8.

44 For a psychoanalytic approach, see Julia Kristeva, *Powers of Horror: An Essay on Abjection* (New York: Columbia University Press, 1982), esp. pp. 71ff.

45 For literary examples, see the final chapter of R. Anselment, *Realms of Apollo: Literature and Healing in Seventeenth-Century England* (Newark: Univ. of Delaware Press, 1995); and Shuttleton, 'A Culture of Disfigurement', pp. 76-80, and Guy Poirier in this volume.

46 M. Bakhtin, *Rabelais and his World*, trans. H. Iswolsky (Cambridge, MA: MIT Press, 1968), p. 27.

47 *An Elenchus of Opinions Concerning the Cure of the Small Pox Together with Problematicall Questions Concerning the Cure of the French Pest* (London: Nathaniel Brook, 1661), p. 8.

48 Fuller, *Exanthematologia*, pp. 189–90.

49 *The Philosophical Transactions and Collections, to the end of the year MDCC, abridged [etc.]* (London: W. Innys, R. Ware, J. and P. Knapton [et al], 1749), 46, pp. 233–4. Later works, such as George Pearson's *Observations on the Effects of Variolus Infection on Pregnant Women* [offprinted] from the *Medical Commentaries*, Vol. XIX, p. 213 (London, 1794), concerned with whether inoculation could cause *in utero* infection no longer countenance imprinting, but such beliefs persisted at a popular level.

50 Fuller, *Exanthematologia*, p. 189.

51 The Enlightenment medico-philosophical context is discussed in G. S. Rousseau, 'Pineapples, Pregnancy, Pica, and Peregrine Pickle', in *Tobias Smollett: Bicentennial Essays Presented to Lewis M. Knapp*, ed. G. S. Rousseau and P.-G. Boucé (Oxford: Oxford University Press, 1971), pp. 79–109; P.-G. Boucé, 'Imagination, Pregnant Women, and Monsters, in Eighteenth-century England and France', in *Sexual Underworlds of the Enlightenment*, ed. G. S. Rousseau and R. Porter (Manchester: Manchester University Press, 1987), pp. 86-100; M.-H Huet, *Monstrous Imagination* (Cambridge, MA: Harvard University Press,1993); and D. Todd, *Imagining Monsters: Miscreations of the Self in Eighteenth-Century England* (Chicago and London: University of Chicago Press, 1995), with further related references, p. 283, fn 26.

52 Miller briefly notes the role of the Passions in some accounts (248), but 'conceits' find no place in her attempt to give a pattern to what she admits is a confusing picture.

53 Rosenberg, *Explaining Epidemics*, p. 295, n. 4.

54 Miller, *The Adoption of Inoculation*, pp. 260–1.

55 Fuller, *Exanthematologia*, p. 414,

56 [Anon.] *A Letter to J. C. Lettsom M.D. F.R.S. S.A.S. &c. occasioned by Baron Dimsdale's Remarks on Dr. Lettsom's letter to Sir Robert Barker ... upon general inoculation. By an uninterested spectator* (London: J. Murray, 1779), pp. 22–3, discusses fear as trigger in the then fashionable language of nervous sensibility.

An Afterword on Contagion

Donald Beecher

The word 'contagion' contains a buried metaphor pertaining to 'touch'. But the notion has been generalized to express all manner of pathogenic transmission through proximity, and then generalized again to express moral contamination, imitative emotions or the psychology of crowds. Through such analogical applications, the history of contagion becomes even more extensive, one that relates not only to the best scientific and philosophical explanations from the ancients to the early moderns concerning the spread of diseases, but, by extension, to an analysis of the psychodynamics of groups. Given that microbiology belongs only to the last two centuries, earlier thinkers were challenged to account for contagion according to their 'received' philosophies of nature, or in terms of what they presumed to see and verify prior to an understanding of microorganisms. Consequently, they had no choice but to turn to the language of correspondences, occult and spiritual forces, environments and temperaments, poisons, vapours, stares and the polluting touch. But when these operations were applied to the transfer of passions and ideas it was no longer for a lack of understanding of the microbiological world, but of the emotional and cognitive mechanisms whereby minds copy passions and belief structures in seemingly spontaneous, subconscious and often destructive ways. These are socio-psychological phenomena merely resembling pathogenic operations; the relationship would appear to be one of pure metaphor. Yet the alignment between contagious diseases and transmissive emotions has been so close in Western culture as to blur the line between material cause and figurative application. In the pages to follow, there is room only for a few circumscribing examples – touchstone stories that illustrate some of the more liminal or culminative moments in the evolution of the 'idea' of pathological causation and its behavioral applications.

As an idea, a theory of agency, a narrative of events whereby patterns of disease have been replicated in healthy bodies, contagion has never been reserved exclusively to the field of medicine. Because the mechanisms of remote transmission were largely beyond material explanation, yet clearly

243

causal, such operations were barely less metaphorical than when applied to more figurative chains of 'epidemic' transmission, whether the hysteria of crowds, the blight of sin, religious *enthusiasmos* or the rapid spread of pernicious ideas such as heresy or rebellion. It was precisely the mystery in contagion, together with its generic plasticity as a working model of transmission, that made it serviceable in explaining any non-medical transfer of energies, such as emotions, or mental states. These extensions of contagion logic tended to draw even the moral and emotional phenomena to which they were applied back into the circle of medical analysis, in contrast to our own tendency to expose the earlier uses of metaphor in order to assess the spontaneous communication of ideas and emotions in socio-psychological terms. In short, reimagining their philosophical spaces regarding the transmission of actual diseases is perhaps easier than segregating modern from earlier thinking concerning infectious ideas and emotions. For the former, microbiology is the watershed; for the latter, 'contagion' remains an operative principle even today. The first two examples to follow will reveal the problem.

Regarding the case of the young women of Milesia who, in a collective state of emotional frenzy, decided to end their lives by hanging themselves, we may well ask whether contagion applies in more than a metaphorical sense. Plutarch recounts the story in the section on the bravery of women in the Moralia, an anecdote which, by-the-by, entered into medical history.[1] We would presume that the 'infection' passed from victim to victim as a form of imitative behaviour or as a transfer of emotions. Such conduct is fully recognized in the recent work of Hatfield, Cacioppo and Rapson,[2] although the phenomenon still retains some of its causal mystery. But for earlier thinkers, such 'contagious' reactions required more tactile explanations to account for the transfer of states and symptoms. Not surprisingly, then, the baffled city magistrates of Milesia, at the prospect of losing a generation of women, sought an explanation not only in the visitation of the angry gods, but in pestilential conditions present in the environment that were afflicting these women. Plutarch goes on to say that such spontaneous madness could originate only in a melancholy disorder of the brain, arising from a derangement of the humours in accordance with the constitutions peculiar to young virgins. That derangement was traced in turn to a contagious miasma – a poisoned or contaminated condition of the air.

This analysis is crucial in profiling the generic sense of disease transmission among the ancients. A common condition had afflicted these women because they all suffered the same delusions and sought the same cure in suicide. But the foetid or polluted air was not the disease itself, nor did it convey the disease from one person to another. Rather, it created the pathogenic conditions whereby the same disease arose spontaneously in each young woman because they all shared the same constitutional vulnerability. Once the malaise had corrupted the imagination, however, the condition was no longer environmental, but a melancholy disorder. The cure would be psychological. The city fathers simply told the women that henceforth the cadavers of the dead would be paraded through the streets of the town, naked. Where reasoning had failed, shame prevailed. Such 'methodical' cures would, in fact, become general among later medical practitioners where diseases of the imagination were concerned, as in the case of erotic melancholy, with its perverted phantasms. Avicenna was among the first to recommend that old ladies be hired to defame the beloved and dispraise her beauty in order to dislodge her image from the patient's mind.[3]

Vivian Nutton points out, in terms borne out by the example of the women of Milesia, that the ancient Greeks did not have a word corresponding to the modern sense of contagion as the direct transfer of pathogenic agents.[4] What they understood by 'the transmission of disease' was the catalysing of the diseased state when specific effusions, poisons, excrements, affluxions or miasmas came into contact with particular constitutions, temperaments and humours. Nevertheless, there is a nascent sense of 'touch' implicit here, in so far as the exposure to immediate or remote pollutants was a precondition to the epidemic. Intuitively, some form of propinquity or proximity had to be a constant factor in the replication of diseases, infectious behaviour or ideas. Hence, contagion pertains more specifically to those transmissions made close to hand through actual contact, as signified by its origin in the word *tangere*. Thomas Lodge would define the word many centuries later (in 1603) as 'an evil qualitie in a bodie communicated vnto an other by touch', including contact with lepers or those who died of the plague, the ingurgitation of poisons or other corrupted substances such as rotting flesh, the transfer of bodily fluids (including osmosis), contact with demons or evil spirits, or congenital 'contact' – all of these opposed to the remote agencies of air and water, miasmas, vapours or general atmospheric conditions more strictly attributed to infection.[5] Nevertheless, this critical distinction was largely lost because of the compounding of causes in the description of diseases. Leprosy was contagious both by touch as well as by breathing the fetid air caused by stinking breath and rotting flesh, much as the plague was associated both with miasmatic effects and fear of direct contact with the afflicted. Boccaccio's remarkable account of the plague in Florence – that which sent the young aristocrats to the hills to tell the stories that formed *The Decameron* – includes the anecdote of the pigs that fell down dead after snuffling through the cast-off clothing of the victims.[6] The story illustrates the principle of contagion by touch, extended to *fomites* or the clothes and bedding of the deceased, which, according to Fracastoro, remained infectious for a considerable period of time.[7]

Although Plutarch succeeded in drawing the behaviour of Milesia's hysterical virgins back into the circle of medical analysis – behaviour that we would otherwise attribute to purely emotional contagion – there were manifestations in kind in later centuries of such a magnitude that medical diagnostics fell short of the etiological task. For example, the group psychodynamics of St Vitus' dance – a condition related to rheumatic fever – are discussed even now as a form of contagion, as subsequent quotations will demonstrate. Certainly it seemed so to late medieval observers, who witnessed the spread of emotional disorientation and frantic physical manifestations on an unprecedented scale whereby 'touch' could only have the tangibility of thought. So many participants surpassed the logic of toxic air and vulnerable constitutions. Only behavioural paradigms were in the air, compellingly infectious paradigms that slipped past defense mechanisms and invaded spirits.

But for later observers, there is little difference between the suicidal hysteria of the Milesian maidens and the manic dancers or carolers of Colbeck described by Robert Mannyng in *Handlyng Sunne* (1303). The story originated in Germany and was retailed by the mendicant friars for 60 years before it was first set down in chronicle form. In Mannyng's version, a small company of young revelers induce the priest's daughter Ave to join their group. They begin their dancing and chanting in the churchyard well before dawn, and continue to dance through matins and the morning mass, whereupon Ave's father, interrupted in his office, stopped the service to chastise and finally to curse them for their disobedience. His 'magical' words compelled them all to dance a twelvemonth before their release, which in fact they did, without food or rest. Pilgrims came to see them, blaming the priest who had cast the spell. After the requisite time, the revelers entered the church and collapsed. Then, after three days, they departed, leaving Ave dead on the floor.[8]

'The Carolers of Colbeck' is a fictionalized account of the St Vitus' dance that swept through Germany and the surrounding areas following the years of the Black Death. Presumably, so much loss and despair had led to a kind of collective frenzy, and so 'the dancing mania began and spread like contagion'.[9] Participants would hold hands, just as the Colbeck dancers did, leaping into the air for hours, chanting, crying or seeing visions – impervious to all efforts to stop them until they collapsed from their own fatigue. The phenomenon persisted for many decades, 'propagated by the sight of the sufferers, like a demoniacal epidemic' in the words of J. F. Hecker.[10] Officials were concerned by the numbers – 500 in Cologne, 1,100 in Metz – so that priests were appointed to conduct the rites of exorcism, or men were sent in to swathe them around the waists until the writhing subsided. Paracelsus, in later years, recognizing the value of shock therapy for what was clearly a mental-emotional disorder, had them immersed in cold water, placed in solitary confinement, and subjected to enforced fasting.[11] Whether Sydenham's chorea – another name for St Vitus' dance – or the homiletic preoccupation with death, culminating in the *danse macabre* representations in the fourteenth and fifteenth centuries, had roles to play in perpetuating the dance mania remain moot points. What is clear is that transmissive emotion was the predominant factor, and that contagion, materially or metaphorically, was the best default explanation of its replicative powers.

The anomaly of our present work is that we are only allowed to tell these stories as they were understood of yore, with or without intimations of replicating 'seeds', with or without presentiments of the transmission of emotion. Thus it is to be recorded that the cure of choice for the dancers was to send them as pilgrims to the shrine of St Vitus, an early martyr whose apocryphal life was associated with epilepsy, and whose reputed remains had been transferred to the Abbey of Corvey in Westphalia in 836. Yet even for modern observers, contagion remains the categorical term for the dancing mania – a measure of the degree to which our own thinking is still attached to the metaphor where ostensibly occult causation is involved. Emotional susceptibility and behavioural conformity are familiar experiences, but their mechanisms are imperfectly understood. Diagnostics would lead the medieval mind to melancholy, demons or heresy, making them the centre of our study, but if we put them aside, we are still left with contagion theory pertaining to fear, superstition, conformity, empathy or gestural imitation whereby volition is activated in the absence of informed cognition. The authors of *Emotional Contagion* (1994) have sought explanations for these reactions in the latest of clinical and speculative terms, yet, tellingly, they too resort to metaphor in the title of their book.

Where the emotions are concerned, contagion may linger in modern discourse because the precise means whereby mind states are transferred to by-standers in hard causal terms remain at least partially undisclosed. To be sure, the power of empathy whereby we replicate in ourselves what others are feeling has received extensive investigation. But why we have the mental competence to empathize is less clear. Presumably, however, emotional transfer has to do with selected, adaptive traits that favour survival in groups. As Horace Kallin has pointed out, 'laughing and weeping are both contagious',[12] which is to say, both states demand emotional responses in others, or at least make others sense their alienation if they are unable to respond to such claims. If these gestures have value, then, it is in their demands on a community to confirm them. Such forms of expression are contagious because they are hard-wired reactions based on our vulnerability to empathy, demonstrated through those limbic responses that signal back to the group. Weeping and laughing are adaptive gestures, not because they prepare the body for action, but because they are nearly impossible to dissimulate. In Darwinian terms, those who were less efficient in emotional signalling failed to pass on their traits. Of course, these competencies arose when group fortunes were restricted to dealing with predators, food supply or imminent loss through separation and death. Crying with others creates a debt that would be repaid at the time of one's own peril. Thus, emotional synchronicity became a powerful drive in the confirmation of group solidarity, making the 'contagion' factor deep-seated, dynamic and real. Cultural and religious institutions and practices, and extended community sizes, needless to say, have subsequently provided new and sometimes bizarre contexts for the fashioning of these emotional demands. The women of Milesia provide a clear example, the manic dancers of the Black Death years another; they are, for us, cases of 'emotional contagion'. For earlier thinkers, however, the combined medical and demonological discourses ambiguously hovering between material and metaphorical analysis not only grey the margins, but potentially create an epistemological anxiety around our own sense of the word 'contagion'.

One of the stranger adaptations of 'tactile' contagion is epitomized by William Harvey's theory of conception. This English physician, renowned for his pivotal treatise on the circulation of the blood, also took on the vexing question of genetic imprinting from parent to offspring. To shape the question for him: how do the traits of the noble rooster make their way through a hen and an egg to once again produce a noble rooster? His solution to the problem was contagion theory, again based on remote touching; it was an omni-purpose paradigm. Intercourse was responsible for the transmission of 'this transitory thing, which is neither to be found remaining, nor touching, nor contained, as far as the senses inform us, and yet works with the highest intelligence and foresight, beyond all art'. If the sperm could not be imagined in terms of genetic coding, reverse engineering nevertheless necessitated that it possess a quality of intelligence, occult in its operations – one that might work in the female like a miasma, thereby constituting the unique configurations of the offspring. Fecundity, that most ancient of miracles, also fell under the expediencies of paradigm transfer that only contagion and infection could explain – Harvey employs both words.[13] Conception, for him, is an infection according to its earliest meaning, which is the absorption of liquids by osmosis according to mysterious mechanical processes; it is an act of immersion or infusion. With Harvey, then, we have a further occurrence of the word, far from the propagation of pandemics, but perhaps not so far from the metaphorical sense of information replicating itself in a distant host by occult means. Such an analysis may represent movement forward from the sixteenth-century preoccupation with the role of the imagination in determining the traits of the offspring, but it reveals at the same time the perdurability of the contagion model of proximate transmission to account for hereditary traits, still sub-material, yet a 'thing' more substantial than a constellation of conditions meeting humours.

Could it be feasible, then, that even the condition of the mother's imagination at the time of conception was a form of contagion? The case seems clear. The literature on the subject includes treatments by Jean Fernel, Luiz

Mercado, Levinus Lemnius, Johannes Wier and Jean Liébault,[14] but all of them tending in the common direction retailed in Ambroise Paré's *Treatise on Monsters*.[15] In the ninth chapter, he tells the story of the birth of a girl-child covered with shaggy hair. At the foot of the mother's bed was a picture of Saint John the Baptist dressed in the skins of wild animals. By traditional medical reasoning, the image had lodged itself in the mother's imagination at the time of conception – the critical moment at which form is imprinted on matter. Hence the imaginative faculty itself, notoriously subject to pollution and humoral corruption, could become the agent of contagion, stamping upon the unborn child the images in the mother's mind. Paré was a man of his age in espousing this doctrine, much as he was in believing that the plague was caused by a corruption of the air that might be dealt with allopathically through the carriage of nosegays of angelica, or in revisiting the Levitican notion that children conceived during the menses or through contact with menstrual blood would be subject in later life to leprosy, scurvy, gout and scrofula. All were based on contagion theory. Corrupt materials produce corrupt conditions as a self-fulfilling prophecy; causal symmetry so dictates. Hence, nothing could be less surprising than the employment of contagion theory to explain the transmission of unwanted traits in progeny. But this will not be the limit of the mind's role in generating and transmitting contagion in its ambiguous location between the materiality of the body and the immateriality of pure thought – between pathogenesis and spiritual causation – as subsequent examples will reveal. But before proceeding to those complex matters, I want to cite a further case of touch contagion, viewed from an entirely different perspective.

St Catherine's spiritual crusade to raise the stakes in the race for self-mortification might seem remote from the matter in hand. Yet contagion, or more precisely the fear of contagion by touch, is at the centre of her bid to gain complete mastery over worldly concerns. Her strategy involves things fetid, polluted, excremental, gangrenous, septic or venomous. St Catherine, in attendance on an aged woman suffering from a suppuration of the breast, vomited at the rupture of the abscess. At that moment a new path to sanctity rushed to her mind, for she bent down to breathe in the noxious vapours for a long period in an act of protracted humility. Little matter that the patient came to despise Catherine for making spiritual theatre out of her miserable condition. When, on a second visit, Catherine was again nauseous, she gathered up the pus from the wound and drank it as a means to overcome the devil. Her victory in the race was manifest, for according to the account, Christ himself appeared to her in a dream, acknowledging that her mastery of disgust had surpassed his own, thereby earning her the privilege of sucking directly on the wound in his side – Eucharist without transubstantiation.[16] Her spiritual exploitation of the dreaded fears of contagion is a reminder that contagion has always carried connotations of repugnance, and that disgust as a primitive limbic response has had its role to play in the prophylactic strategies interposed between phobia and sepsis. In this way, contagion becomes a powerful cultural matrix, compelling societies to define by belief structures and behavioural patterns the thresholds of disgust on an item-by-item basis in all matters pertaining to food and drink, ambient air, excrements, rotting substances and diseased bodies. Catherine of Siena turned to these instincts as markers of the world to be mastered by the spirit. But there is a second and equally important sense in which contagion is at play. Revulsion is a primal response with high adaptive value in emotionalizing conditions in the environment deemed noxious, pathogenic or deleterious to the health. Emotions concerning fear of pollution are themselves contagious in groups, revealing yet again a phenomenon, not unlike that which afflicted the Milesian girls, which is analytically complex, and most closely approximated in contagion theory metaphorically applied. The St Catherine anecdote, in its way, also illustrates the sliding relationship between literal and metaphorical contagion.

Returning to the conversion of the foetus into a monster through the powers of the image-polluted imagination, we find ourselves in the theatre of cognition as the mechanism of congenital contagion. The slippage towards metaphor is clearly in the making because it is nearly impossible for us to credit hereditary traits to the preoccupations of the imagination. Yet, according to the medical precepts of his age, Paré does not cross the line. So where, then, is the equator? Jean Starobinski reasoned that emotions attach themselves to words that name them, so that words may not only convey ideas and images but feelings by the semantic powers invested in them. By merely naming the emotion the 'efficacy' of the word 'helps to fix, to propagate, to generalize the emotion which it represents'.[17] Once again, a link is established between contagion theory and the psychodynamics involved in the transmission of all forms of emotions, moral qualities or ideas that seem to contain their own powers of replication. Denis de Rougemont takes us to the heart of metaphor in examining the spread of Nazi ideology among his students in the 1930s when he was a visiting professor in Germany. The phenomenon was simply beyond explanation in rational terms. He too could only fall back on a theory of contagion with all the benefits of microbiology, for his analogy was at the cellular level where health and happiness belong to the single unit with its two nuclei and their protein synthesizing capacities – until the cell is invaded by a virus. All seems well at first, because the foreign body allows itself to be digested and assimilated, but only to re-emerge as the 'soul' of the host by altering its genetic code. The host then destroys itself in the production of many new copies of the virus, which eventually explode through the walls and commence their reproductive forays in neighbouring cells.[18] Viral assault is the Ur-contagion, the phenomenological paradigm for all transmissions of subversive information, whether the host is a computer or a human memory. Wherever replication takes place, there parasites can exploit the mechanism to their own advantage. Totalitarianism can take over a healthy system, alter its belief structures by degrees, then collapse the system without killing the host, making his mind the carrier of the contagion. According to the metaphor, ideas are pathogens of the mind; as subsequent paragraphs will reveal, it is an idea with a future. But are we in the realm of pure metaphor?

Evil knowledge as an infection of the mind is by no means a recent figure. St Paul was tormented by the degree to which sinful action seemed to bypass all systems of control – insidious ideas that captured the will as a form of pollution. Pliny the Younger in his *Epistles* (X. xcvi. 9) evoked contagion to account for the spread of Christianity. Stephen Gosson held that theatre plays had such seductive powers over the imagination that to see them was to part corrupted. 'Vice is learned with beholding, sense is tickled, desire pricked, and those impressions of mind are secretly conveyed over to the gazers, which that plaiers do counterfiet on the stage. As long as we know our selves to be flesh, beholding those examples in theaters that are incident to flesh, wee are taught by other mens examples how to fall. And they that come honest to a play, may depart infected.'[19] Bodies themselves, conceived as sacred places, as 'temples of the living God', were always subject to profanation in a perfect meeting of medicine and metaphor.

These several examples reveal the ongoing interplay between transmissive phenomena in the social world and medical analysis. They tease our thoughts to the extent they blur the distinctions between mind and body, idea and pathogen, moral and physical corruption, natural and supernatural agents, or the occult transmission of diseases and the contagious energy of groups. As stated earlier, emotions are contagious, and adaptively so, because they are the physical markers of truth wherein we can read the 'embodied' thoughts of others in manifestations they are powerless to dissemble. Yet all such applications, along with the many others pertaining to sin, heresy or crowd frenzy, remain essentially descriptive rather than prescriptive. They are the informing anecdotes in the history of an idea that does not yet exceed the sum of its parts. But arguably there are two salient moments in Western intellectual history in which contagion-as-metaphor achieved the potency of an *ideé force* – an idea offering to explain more pervasive dimensions of the human condition – the first at the end of the fifteenth century, the second in the last two decades of the twentieth. Both systems were concerned with the psyche's capacity to be polarized by invasive information. The first concentrated on the role of devils or demons as the agents of transmission. The second styled the information chunks themselves as viruses, benign or malign according to their ideological coding. Both systems sought to shake off all metaphorical connotations by advancing themselves in empirical and phenomenological terms. These are the two instances in which the metaphorical dimensions of contagion theory achieve the status of complete paradigms.

It would be historically cavalier to say that the Dominican monks Heinrich Kramer and James Sprenger, precisely in the year of our Lord, 1486, released on the world a new paradigm of contagion in their treatise the *Malleus maleficarum*.[20] The work, in fact, followed hard on the papal bull *Summis desiderantes affectibus* of Innocent VII in which he anatomized the demonic world of incubi and succubi, amatory charms, spells and incantations, and related preternatural forces able to pervert the natural course of human sexuality. The dyad between the operations of the body and the spirit world has figured in Christian thought from its inception. But these two monks turned the implicit into an explicit account of the extensive agency of demons and devils in conveying or inciting diseases. Demons had malign intent, but they were not the diseases any more than they were sins to be 'caught'. Rather, they had access to the body through all the normal diagnostic channels, enabling them to incite the conditions of sickness, both physical and mental. It was through the systems of the body that they contaminated the faculties of imagination and judgement, leading to all manner of perverted phantasms presented to the appetites and will. To be sure, the will remained free – that was a fundamental rule set by God. But the devils had powers to stretch men in torments to rival Job's. Disease was to be Satan's means for perpetuating his war against humankind, so that all the contaminants, putrefactions and vaporous contagions became the elements of demonic mediation. In this regard, it hardly matters whether these agents were spiritual (the Fourth Lateran Council) or corporeal (Cajetano), because in either form they were proximate, unobstructed by the conventional barriers of matter, possessed alterative powers, and could insinuate their effects into the very matter of thought itself. According to Martin Del Rio, human consciousness could be reached entirely by natural causes unknown to man.[21]

The insidiousness of the *Malleus* analysis is that demonic volition made contagion 'organic' and alive. Demons were not viruses, but they had the intelligence of creatures able to will harm to their hosts. Throughout the following century, many of the finest minds succumbed to the logic of these writers. Devils became an efficient cause of any disease of the imagination, including philocaption, through the workings of the melancholy humours which were, according to St Jerome, 'the devil's bath'. What should be concluded concerning this construction of disease in which supernatural agents arrange and abet the causal circumstances in full anticipation of their ends? Is this metaphor or materiality in action? We are increasingly hard pressed to say, depending on where we place our own historical imaginations. The vision of reality created by Kramer and Sprenger no longer distinguished between material and spiritual causes, even though, in pro forma fashion, they allowed that such diseases as erotomania could be due to the wandering of the eyes, rather than to the work of devils exciting inner perceptions and humours, or of 'necromancers and witches with the help of devils'.[22] But ultimately, spiritual and pathological contagion become one, thereby collapsing metaphor to material cause in a Manichean world embedded in the body; the invisible forces of evil had become omni-pathogenic agents. This epitomization of a latent idea served to diminish the role of medicine in the treatment of the diseases of the soul, generally, because demonic agency was the several of the ecclesiastical establishments, which, to combat contagion, imposed penitence, exorcism, and the purgatives of the inquisition.

Arguably, there have been no other theories of 'metaphorical' contagion that have achieved such focus and influence. But a potential candidate arose from the work of Richard Dawkins in *The Selfish Gene* where he proposes that bits of cultural information, much like genes, carry identifying codes, and that these travel among minds in the form of 'memes'.[23] Douglas Hofstadter, Henry Plotkin and Daniel Dennett seized on the idea, seeing in the meme the equivalent of cultural DNA. Memes were discrete packets of information with the replicating potency of viruses – viruses that used the minds of their hosts for their capacity to evolve and spread ideas. Denis de Rougement might have claimed founder's status had he pushed his metaphor of the virus of totalitarianism into a system of contagion. Memetics simply carries the paradigm shift to its ultimate conclusions through Plotkin's study of the evolutionary basis of learning in *Darwin Machines and the Nature of Knowledge* (1993), and Dennett's exploration of replicating memes in *Consciousness Explained* (1991).[24] The paradigm shift is, quite simply, to think, not of individuals inventing and expounding ideas, but of information in its signature chunks behaving like autonomous organisms in an adaptive and selective environment, building up cultural markers around themselves as they achieve dominance in mind communities. A mind virus is an entity in the world – an idea, a jingle, a proverb, a gesture, a musical phrase, a doctrine, a group craving to commit suicide or dance till they drop, a seductive proposition appealing to any disposition of mind that will endorse and reproduce it. Now just as the devil, by contagion theory, became the agent in a somatopsychic economy of moral combat, the meme, by the logic of viral contagion, becomes the determinant of all cultural consensus and change. One can speak of 'good' and 'bad' memes, but this is merely in keeping with their own capacities to appeal and survive. Richard Brodie in *Virus of the Mind* (1996) tackles the question straight on, whether meme transmission is causally empirical in the terms in which it is expressed. 'I'm *not* saying this is the Truth. I'm *not* saying this is what Really Happens. I'm *not* saying this is the Only Way or The Right Way to look at the mind.'[25] But he is also not saying that it is merely metaphor. Rather, it is something in-between like a scientific model, a way of looking at things. He then goes on to list the humoral system and Newtonian physics as models which, in their respective times, were employed as truths. By their own theories, memeticists are compelled to release their model to the world on the same competitive basis that pertains to memes in general. Models are all that we have ever possessed to explain causation, and new models of contagion theory simply challenge us to pitch our imaginations into the future in the same way that we are challenged to cast them into the past. The simple gist of it is that our thoughts are not our own, they are 'caught'. Brodie has great hopes for memetics as 'the secret code of human behaviour, a Rosetta Stone finally giving us the key to understanding religion, politics, psychology, and cultural evolution'.[26] It may or may not get that far as a mass paradigm, but ideas with an evolutionary mission to survive in the minds of their hosts, replicate and spread, is a metaphorical twist to contagion theory that commands a certain attention.

The dissonances between the history of contagion, the ideas of contagion, and the metaphors of contagion will continue to create liminal zones where reason must struggle with interrelationships. Is it an act of contagion that one vibrating string placed near another of the same length and pitch should make the second also vibrate? Is it contagion that one person laughing should set a second one laughing, or that yawning should spread yawning, or that one person weeping at the happy ending should induce others to weep? Is contagion simply the body's incapacity to resist any form of altering influence from doctrines to emotional spasms? Are not all of these reactions part of the 'profoundly pluralistic outlook on contagion in the pre-modern world'?[27] Is it a contagion of the spirit when Pentecostals speak in tongues or break out in holy laughter? When Aristotle suggested that tragedies involved plots with recognition and reversal scenes built around persons of a particular stature and nature as the preconditions for achieving in the limbic systems of all spectators a sensation akin to pity for those whose fortunes are diminished unto death, and fear for ourselves, is the formula itself a mechanism of contagion whereby many are infected with the same set of emotions? The metaphor can, perhaps, be overplayed, losing its force in generic applications beyond custom. Touch, transmission, contagion, scale: where does the order of metaphor begin and end?

Perhaps just one or two more examples of boundary confusion, and then a halt.

Both will invoke a smaller scale of claims, and both will return us to the world before microbiological analysis. The gonococcus bacterium is the pathogen in the disease carrying its name, characterized by an inflammatory discharge from the urethra or vagina. It is highly infectious and is transmitted sexually, which act may be styled the most intimate form of touch. Gonorrhoea was rampant in all levels of society throughout Europe in the eighteenth century. Socially, it was a condition that called for a degree of discretion and secrecy where reputations were at stake, yet a condition openly acknowledged and treated through a variety of medical procedures. To be sure, the bacterium was unknown, but the emissions left no doubt of infection, arising after short intervals of time – six to seven days. Hence, partners might be identified and accusations leveled. By the same token, contagion was never more within the control of human volition, yet no less rampant in a social milieu of desire, loss of self-control, subterfuge and dissimulation, and the commercialization of sexuality. From a pathological perspective, this contagion was a contact sport involving the transfer of bodily fluids. But without microbiological explanations ready to hand, the contagion factor of this disease was quintessentially social and moral, and hence reapproaching the metaphorical. For in almost seriocomic fashion, the victims carried themselves to their own contamination, resulting in discovery, suffering, treatment, vows, guilt and recidivism; a lack of moral resolve was the efficient cause.

If the history of this disease were to receive its profile in a single exemplum, it would begin and end in the journals of James Boswell, who, in 1782, became the ninth Laird of Auchinleck. He is best known to the literary world for his *Life of Samuel Johnson*. Boswell's achievements must also include what may prove a lifetime record for cases of the clap, some nineteen documented instances during a 35-year period, at least twelve of them demonstrably new, rather than recrudescent, according to the careful analysis of William B. Ober.[28] In December 1762, he resorted to a 'safe girl' for company, an Edinburgh actress recently separated from her husband. At her first yielding, Boswell found himself impotent, and so they delayed for ten days. Following the assignation he created a journal entry that is too frank in its detail to repeat here, although he boasts that she was madly in love him and that 'five times' he was 'fairly lost in supreme rapture'. It came with a price, for six days later he 'began to feel an unaccountable alarm of unexpected evil; a little heat in the members of my body sacred to Cupid, very like a symptom of that distemper with which Venus, when cross, takes it into her head to plague her votaries. But then I had run no risks. I had been with no woman but Louisa; and sure she could not have such a thing.'[29] Still unsure of the symptoms, he went so far as to ask Louisa if she was infected, dallying with her again on that same occasion. But within a few more days the inception of his third bout with the disease was beyond doubt. So ended his hopes for 'a winter's safe copulation'. He was thereafter confined to his lodgings with a severe case of urethritis, concerning which he kept a close account! He was put on a low-calorie diet, given electuaries for his bowels and subjected to a seance of bloodletting, although in this case he was spared the insertion of 'medication into the urethra by syringe'.[30] He then went to accuse the poor girl, who believed in her innocence, although she confessed to an infection three years prior, from which she had been symptom-free for some 15 months. Ober explains, however, that women can carry the disease in latent form as a low-grade endocervicitis for quite some time.[31] Poor Boswell. Yet for the next 30 years, he would drink to excess and haunt the brothels in cities as far south as Venice, where the courtesans were again cruel to him in kind. The man bore his guilt, his vows and his folly to the end, rarely indulging in self-pity. He would marry a woman who was nine times pregnant and who would leave him five children to raise at the time of her death in 1789. Yet Boswell managed never to pass his condition to her; we may well ask how. The larger question is why a man of such standing and literary talent would carry this philandering demon with him for so many years, and let his condition degenerate into the predictable urethral stricture with all the attendant complications of uraemia that would carry him to his grave? Those are questions for his biographers. Yet there is little question that a deep-seated conflict with his father, doubts of virility and the powerful heritage of Church of Scotland guilt and expiation had much to do with Boswell's sexual demons. His case is raised here, not only as a chapter in the social history of contagion, but as a further illustration of the multiple dimensions of the 'idea' of contagion, for while clap was the epitome of tactile contamination, it was also an epidemic of social risk calculation against the drives of the flesh, and a moral contagion sent to philanderers as the scourge of God.

If, to this point, the nucleus of a definition holds, at least where pathogenic transmission is concerned, what are we to say of those phantom symptoms emanating from a morbid imagination assaulted merely by the fear of contagion? I confess to the playfulness of the question. But there is nothing like real symptoms without empirical cause to take us to the heart of the matter concerning contagion as fact and metaphor. Hypochondriasis has been a medical meme for all ages, but it appears to have enjoyed a copycat apogee in the seventeenth century – or such is the impression left by the satirists attacking the profiteering physicians who exploited the condition. What are we to say of patients in perfect health who impose states of disease on themselves through the powers of their own imaginations, and then resort to all manner of irrelevant cures, believing deeply in their efficacy, against all evidence, and in the profession that administers them? Hypochondriasis is so many conditions at once: a phobic reaction to the fear of unspecific contagion; a fashionable self-indulgence; an obsession with phantom symptoms and their cures; a metaphorical condition without aetiological substance; and a veritable disease originating in the corruption of the melancholic humours. In its multifaceted presentations, this condition represents nearly every aspect of contagion already discussed, including the demonic. Yet, because the patient complains of one set of symptoms, while the physician treats only the condition that makes the patient imagine such things, we discover through the satirist's eyes the cross-purposes that lock these 'players' in a deaf dialogue and a meaningless regimen. Yet the strange fact remains, for contemporary practitioners, that the phantom conditions resulting from hypochondriasis arose in a real obstruction of the spleen by distempered blood, backing up to the liver, according to the etiology of John Hill.[32] In this way, phantom symptoms derive from a real disease that we would consider a phantom in its own right. With hypochondriasis, contagion becomes a metaphor for a metaphor.

Admittedly Argan, in Molière's *Le malade imaginaire*, is a literary character and no case study *per se*, yet the author asseverates by his example that the age was much taken up with semi-public 'doctoring' in relation to the most socially fashionable *malaises*.[33] Moreover, scholars of medical history know that Argan suffers from a clinical condition spelled out without irony in the medical treatises of the age, as examined in masterful detail by Patrick Dandrey in *Le 'Cas' Argan*.[34] At the play's opening, he is alone on stage counting out the receipts for an astonishing variety of treatments, all of which he considered good value, even as the playwright calls attention to their hyperbolical number and their lavish costs: enemas to soften, moisten and refresh the entrails; more enemas with a double catholicon of rhubarb, honey and other ingredients to sweep, wash and clear the lower belly, not to mention soporific juleps, fresh cassia and oriental senna for further purgation, plus treatments for purging bile, then treatments for flatulence, then more preparations to clarify and refresh the blood, a cordial and preservative potion made of the most fashionable ingredients including bits of bezoar stone, with still more to follow. But what was Purgon, the physician in attendance, treating: the symptoms of the imaginary disease, or the initial cause of the melancholy delirium? The point is never clear, for Argan's disease is eternally diagnostic and imaginary at once.

On the one hand, rarely are we allowed to forget that the mind is its own source of contagion, touching the body by autosuggestion with its own false apprehension of sickness. It is the phobia of pathological imminence. In such a mind state, the immunological strategies of the race have never appeared more frangible, although none of the threat is real. On the other hand, by the medical logic of the age, hypochodriasis, even as a disease of the imagination, is entirely material and pathological. It differs little, in fact, from related diseases of melancholy, such as erotic love, which likewise, by medical consensus, begins with the adustion of black bile in the hypochondries, sending its smudgy vapours by the spinal column up to the brain where the imagination is corrupted in accordance with its compulsive preoccupations. The diagnostics of such conditions is necessarily circular – a hermeneutic loop of sorts – for thought is the provocation of perturbations in the body, while such perturbations are the cause of the incessant fixations. Such an analysis allowed for a variety of causes internal, external, efficient, procatarctic, remote, astrological and involving all the major seats of disease in the body including those of the emotions and of the reason, thereby permitting every conceivable homeo- or allopathic cure, from diet and distraction to the surgical and pharmaceutical, to be packaged into a long and costly regimen.

How then does the modern observer describe the contagion of hypochondriasis? It is variously the fear 'meme' communicated from person to person in an environment of pathological concerns, a form of self-contagion through the powers of the mind to simulate symptoms. For Argan's contemporaries, it was a pollution of the imagination through the adustion of biles, and, by remote causes, provoked by environmental contaminants. By seventeenth-century logic, the mind was incapable of such delirium without the corruption of the hypochondries. Yet we maintain a sense of interference, not from our inability to understand the bizarre logistics of adustion and polluting vapours, but from the scepticism regarding hypochondriasis on the part of Molière himself, who called it *'imaginaire'*. The 'something' which Argan manifested was an idea, a transmissive one, and a meme of astonishing potency – a reminder that all diseases of the imagination are metaphorical, yet entirely material once they are linked to physical causes. The contagion is hence social, psychological, and physiological together. We may surmise, at the same time, a subliminal intent on the part of Argan to expose himself to all the attention-gaining, socially prioritizing measures of a medical regimen. That he underestimated the empathy and credulity of his entourage is merely a complicating factor. Finally, his penchant for pharmaceuticals is a reminder that the terms and conditions of contagion had created a parallel world in all the counteractive measures and ingredients known to medical philosophers, surgeons and apothecaries. But now to the most dangerous part: a summary that exchanges stories for inferences.

Contagion is that part of diagnostics concerned with the transmission of diseases or disease-inducing conditions whereby identifiable pathological states are seen to replicate themselves in different hosts. Paradoxically, for the earliest theorists, this entails a constellation of pathogenic circumstances, rather than the actual transfer of infectious entities. Undoubtedly, humoral thinking compromised their understanding of exopathic agency because the humours themselves were so often invoked in endopathic terms. Nevertheless, early observers did recognize by degrees that identical conditions were passed by causal mechanisms from person to person through direct contact or through carrier agencies. After 1550, infection from a distance became the model for all metaphors pertaining to the non-empirical transfer of influences, sensations, or conduct networks in ways that obscured the line separating fact and metaphor. It may prove that ideas, emotions, and pathogens are ultimately on an equal footing – all are causally and phenomenologically 'real'. In a word, then, contagion is the measure of deleterious 'information' that slips past the immune system, the thresholds of the limbic system, or the mind's resistance to change.

Contagious diseases ravaged the ancient and early modern worlds, raising insistent questions about origins and transmission. Matters of such urgency attracted the best and most inventive minds, their theories forming traditions of authority that would settle into western consciousness as frames of analysis. But there were anomalies on all fronts. Leprosy, ostensibly the epitome of contagion, was explained by physicians in the thirteenth century as a disease of excessive melancholy bile acting like a poison upon the entire body.[35] André Du Laurens confessed to the role of the eyes in attracting lovers, yet the blood vapours carrying the species from image to beholder were also toxic and apt to contaminate the blood; love itself was a miasma.[36] And Hell itself, that ever-present environment of the damned, could spout its own infectious exhalations to the peril of both body and soul – or so the vivid imagination of Hamlet would have it: 'When churchyards yawn, and hell itself breathes out / Contagion to this world.'[37] But the final word goes to Thomas Lodge, from his 1603 *Treatise on the Plague*: 'For very properly is he reputed infectious, that hath in himselfe an evil, malignant, venemous or vitious disposition, which may be imparted and bestowed on another by touch, producing the same and as daungerous effect in him to whom it is communicated.'[38] In this quotation the principle of exopathy is clearly recognized in the direct passage of the plague from soma to soma. Yet the disease itself is grounded in spiritual metaphor, for the buboes were known as 'God's tokens', the wages of sin, while the plague, itself, is named for the *plaga* or stripes received by transgressors.[39] When diseases spread the gods are angry; they are the cause of all things. Touch is the foundation of contagion theory, but in a world order in which the body is the theater of a moral and spiritual economy, that which touches will generate its own allegory.

Notes

1 Plutarch, *The Philosophie commonlie called the Morals*, trans. P. Holland (London: Arnold Hatfield, 1603).
2 E. Hatfield, J. Cacioppo and R. Rapson, *Emotional Contagion* (Cambridge: Cambridge University Press, 1994).
3 J. Ferrand, *A Treatise on Lovesickness*, ed. D. Beecher and M. Ciavolella (Syracuse: Syracuse University Press, 1990), pp. 314, 326.
4 V. Nutton, 'Did the Greeks Have a Word for it?', in *Contagion: Perspectives from Pre-Modern Societies*, ed. L. Conrad and D. Wujastyk (Aldershot: Ashgate, 2000), pp. 137ff.
5 T. Lodge, *A Treatise of the Plague* (London: Edward White and N.L., 1603), n.p.
6 G. Boccaccio, *The Decameron*, trans. E. Hutton (New York: Heritage Press, 1940).
7 S. Jarcho, *The Concept of Contagion in Medicine, Literature, and Religion* (Malabar, FL: Krieger Publishing Company, 2000), p. 56.
8 *Middle English Literature*, ed. A. Brandl and O Zippel (New York: Chelsea Publishing, 1965), pp. 156–8.

9 H. Klawans, *Newton's Madness: Further Tales of Clinical Neurology* (London: Headline Book Publishing, 1990), p. 237.

10 J. F. Hecker, *The Dancing Mania of the Middle Ages*, trans. B.G. Babington (1873; New York: Burt Franklin, 1970), pp. 1-2.

11 Hatfield et al., *Emotional Contagion*, p. 124.

12 H. Kallin, *Liberty, Laughter and Tears* (De Kalb, IL: Northern Illinois University Press, 1968), p. 50.

13 *Anatomical Exercises on the Generation of Animals* in *The Works of William Harvey M.D.* trans. R. Willis (London: Sydenham Society, 1847), p. 372.

14 Note in ed. of Ferrand, *A Treatise on Lovesickness*, pp. 423-6.

15 A. Paré, *On Monsters and Marvels*, trans. Janis Pallister (Chicago: University of Chicago Press, 1982).

16 W. I. Miller, *The Anatomy of Disgust* (Cambridge, MA: Harvard University Press, 1997), p. 151.

17 J. Starobinski, 'The Idea of Nostalgia', *Diogenes*, 54 (1967) 81.

18 D. de Rougement, *The Myths of Love*, trans. R. Howard (London: Faber and Faber, 1964), p. 154.

19 S. Gosson, *Playes Confuted in fiue Actions* ([London:] n.d., 1582), sig. G4.

20 H. Kramer and J. Sprenger, *Malleus maleficarum*, trans. M. Summers (London: John Rodker, 1928).

21 M. Del Rio, *Les controverses et recherches magiques*, trans. A. du Chesne (Paris: Jean Petit-pas, 1611); *Investigations into Magic*, trans. and ed. P. G. Maxwell-Stuart (Manchester: Manchester University Press, 2000).

22 Kramer and Sprenger, *Malleus maleficarum*, p. 51.

23 R. Dawkins, *The Selfish Gene* (Oxford: Oxford University Press, 1989).

24 H. Plotkin, *Darwin, Machines and the Nature of Knowledge* (Cambridge, MA: Harvard University Press, 1993); D. Dennett, *Consciousness Explained* (Boston: Little, Brown, 1991).

25 R. Brodie, *Virus of the Mind: The New Science of the Meme* (Seattle: Integral Press, 1996), p. 33.

26 *Ibid.*, p. 26.

27 *Contagion: Perspectives from Pre-Modern Societies*, p. xvii.

28 W. B. Ober, *Boswell's Clap and Other Essays* (New York: Harper and Row, 1988).

29 J. Boswell, *Boswell's London Journal, 1762–1763*, ed. F. Pottle (New York: McGraw-Hill, 1950), p. 149.

30 Ober, *Boswell's Clap*, p. 7.

31 *Ibid.*

32 J. Hill, *Hypochondriasis: A Practical Treatise* (1766), ed. G. S. Rousseau (Los Angeles: Clark Memorial Library, 1969), p. 7.

33 Molière, *Le malade imaginaire*, ed. R. P. L. Ledésert (London: George Harrap, 1969).

34 P. Dandrey, *Le 'Cas' Argan: Molière et la maladie imaginaire* (Paris: Klincksieck, 1993).

35 F.-O. Touati, 'Contagion and Leprosy: Myth, Ideas and Evolution in Medieval Minds and Societies', in *Contagion: Perspectives from Pre-Modern Societies*, p. 187.

36 A. Du Laurens, *A Discourse of the Preservation of the Sight; of Melancholike Diseases; of Rheumes, and of Old Age*, trans. R. Surphlet (London: Felix Kingston for Ralph Lacson, 1599), 34v.

37 W. Shakespeare, *Hamlet* (New York: Signet, 1963), III.ii.396–8.

38 Lodge, *Treatise on the Plague*.

39 M. Healy, 'Anxious and Fatal Contacts: Taming the Contagious Touch', in *Sensible Flesh: On Touch in Early Modern Culture*, ed. E. D. Harvey (Philadelphia: University of Pennsylvania Press, 2003), p. 23.

General Bibliography

This list contains works of interest not cited in the notes; conversely, the notes include references not retained here.

Primary sources

Angelis, G. B. de. *Racconto d'alcuni de' molti miracoli operati da S. Francesco Saverio in Napoli nel tempo della pestilenza* (Roma, c.1660).

Astruc, J. *Dissertation sur la contagion de la peste. Où l'on prouve que cette maladie est véritablement contagieuse* (Toulouse: J. J. Desclasson, 1724).

——. *Dissertation sur l'origine des maladies épidémiques et principalement sur l'origine de la peste* (Montpellier: J. de Martel, 1721).

d'Aumont, A. 'Épidémie' in *Encyclopédie ou dictionnaire raisonné des sciences, des arts et des métiers*. Eds D. Diderot and J. d'Alembert (Paris: Briasson, David, Le Breton, Durand; then Neufchâtel: S. Faulche, 1751–65) V, pp. 788–9.

Binet, É. *Consolation et réjouissance pour les malades et personnes affligées* (Paris, 1617).

——. *Remèdes souverains contre la peste et la mort soudaine*. Ed. P.-L. Combet (1628; rpt Grenoble: Jérôme Million, 1998).

Bodin, J. *De la démonomanie des sorciers* (1587; n.p.: Gutenberg Reprint, 1979); *On the Demon-Mania of Witches*. Trans. R. A. Scott, ed. R. A. Scott and J. L. Pearl (Toronto: Centre for Reformation and Renaissance Studies, 1995).

Boguet, H. *Discours exécrable des sorciers* (Marseille: Laffitte Reprints, 1979); *An Examen of Witches*. Trans. E. A. Ashwin, ed. Rev. M. Summers (Bungay, Suffolk: John Rodker, 1929).

Briefve Institution pour preserver et guerir de la peste (Paris: Nicolas Buffet, 1545).

Burton, R. *The Anatomy of Melancholy*. Eds T. C. Faulkner, N. K. Kiessling and R. L. Blair (Oxford: Clarendon Press, 1989–97).

Chalmel de Viviers, R. (Raimundus a Vinario) *De peste libri tres*. Ed. J. Dalechamps (Lyon, Guil. Rouille, 1552).

Cureau de la Chambre, M. *Les Charactères des passions* (Paris: J. d'Allin, 1663).

Dante, A. *Divine Comedy*. Trans. John D. Sinclair, 3 vols (New York: Oxford University Press, 1976).

——. *Vita Nuova*. Trans. Mark Musa (Bloomington: Indiana University Press, 1973).

Defoe, D. *A Journal of the Plague Year*. Ed. P. R. Backscheider (1722; New York and London: W. W. Norton, 1992).

Del Rio, M. *Les Controverses et recherches magiques*. Trans. A. Du Chesne (Paris: Jean Petit-pas, 1611); *Investigations into Magic*. Trans. and ed. P. G. Maxwell-Stuart (Manchester: Manchester University Press, 2000).

Despars, J. *Avicennae Canon (liber III) cum Jacobus de Partibus* (Lyon: J. Trechsel, 1498).

Descartes, R. *Traité des passions* (1649; rpt Paris: Bourgois, 10/18, 1965); *The Passions of the Soul*. Trans. and ed. S. Voss (Indianapolis and Cambridge: Hackett Publishing Co., 1989).

Diemerbroeck, I. van. *The Anatomy of Human Bodies ... To which is added a particular treatise of the small-pox and measles ... Translated [from original Latin] by William Salmon* (London: W. Whitwood, 1694).

Du Laurens, A. *Discours de la conservation de la veue: des maladies melancoliques, des catarrhes et de la vieillesse* (Tours: Jamet Mettayer, 1594); *A Discourse on the Preservation of the Sight: of Melancholike Diseases; of Rheumes, and of Old Age*. Trans. R. Surphlet (1599; Oxford: Humphrey Milford, Oxford University Press, 1938).

Evelyn, J. *The Diary of John Evelyn*. Ed. E. S. De Beer, 11 vols (Oxford, 1955).

Ferrand, J. *A Treatise on Lovesickness*. Eds D. Beecher and M. Ciavolella (Syracuse: Syrcuse University Press, 1990).

Ferrier, A. *Remedes preservatifs et curatifs de la peste* (Paris, 1562).

Ficino, M. *Commentary on Plato's Symposium on Love*. Ed. S. Jayne (Dallas: Spring Publications, 1985).

———. *The Three Books on Life*. Trans. C. Kaske and J. Clark (Binghamton: Medieval & Renaissance Texts & Studies, 1989).

———. *Consilio contra la pestilenza* (Firenze, 1481).

Fracastoro, G. *L'Anima*. Ed. and trans. E. Peruzzi (Firenze: Le Lettere, 1999).

———. *De sympathia et antipathia rerum liber unus. De contagione et contagiosis morbis et curatione libri III* (Venice: Giunta, 1546; reeds Venice, Scoto, 1546; Lyon: Nicolas Bacquenoys, 1550; Lyon: Bacquenoys for Gazeau, 1550; Lyon: De Tournes and Gazeau, 1554).

———. *Fracastoro's Syphilis*. Ed. and trans. G. Eatough (Liverpool: Francis Carins, 1984).

Fregoso, B. *Contramours: Anteros ou Contramour de Messire Baptiste Fulgose, jadis Duc de Gennes* (Paris: Martin le Jeune, 1581).

Fuller, Thomas. *Exanthematologia, or an Attempt to Give a Rational Account of eruptive fevers, especially of the measles and smallpox* (London: C. Rivington and S. Austen, 1730).

Gosson, S. *Playes Confuted in fiue Actions* (London, 1582).

Grillando, P. *Tractatus de hereticis, et sortilegiis* (Lugduni: apud Jacobum Giuncti, 1536).

Guidon, J. *Traicte et remedes contre la peste* (Paris, 1545).

Harvey, G. *A Treatise of the Small-pox and the Measles; describing their Nature, Causes and Signs* (London: W. Freeman, 1696).

Harvey, William. *Anatomical Exercises on the Generation of Animals, in the Works of William Harvey M.D.* Trans. R. Willis (London: Sydenham Society, 1847).

Hecquet, Ph. *Le Naturalisme des convulsions dans les maladies de l'épidémie convulsionnaire* (Soleure, 1733).

———. *Traité de la peste* (n.p.: Cavelier Fils, 1722).

Hery, T. de, *La Methode curatoire de la maladie venerienne, vulgairement appellee grosse Verolle, et de la diversité de ses symptomes* (Paris, Jean Dehauy, 1660).

Hill, J. *Hypochondriasis: A Practical Treatise*. Ed. G. S. Rousseau (1766; Los Angeles: Clark Memorial Library, 1969).

Kornmann, H. *Linea amoris, sive commentarius in versiculum glossae visus* (Francofurti: typis M. Beckeri, 1610).

Kramer, H. and Sprenger, J. *The Malleus Maleficarum*. Trans. M. Summers (New York: Dover, 1971 [Rpt of the 1928 Rodker ed. with the addition of the 1948 introduction]).

Jaffray, A. *The Diary of Alexander Jaffray*. Ed. J. Barclay (Aberdeen, 1856).

Jaucourt, L. de. 'Peste', in *Encyclopédie ou dictionnaire raisonné des sciences, des arts et des métiers*. Eds D. Diderot and J. d'Alembert (Paris: Briasson, David, Le Breton, Durand; then Neufchâtel: S. Faulche, 1751–65), XII, pp. 452–7.

———. de. 'Vérole, petite', in *Encyclopédie ou dictionnaire raisonné des sciences, des arts et des métiers*, XVII, pp. 79–83.

Jones, H. *Inoculation; or Beauty's Triumph: a Poem in Two Cantos* (Bath: C. Pope, 1768).

Joubert, L. *Traité de la peste*. Trans. G. Des Innocents (Toulouse: Lertout, 1581).

——. *Traité du ris* (1579; rpt. Geneva: Slatkine, 1973); *Treatise on Laughter*. Trans. and ed. G. D. de Rocher (Alabama: University of Alabama Press, 1980).

La Rochefoucauld. 'De l'origine des maladies', in *Œuvres complètes*. Ed. L. Martin-Chauffier (Paris: Gallimard, 1950) *Réflexions diverses*, XII, pp. 375–6.

Le Loyer, P. *IIII Livres des spectres ou apparitions et visions d'esprits* (Angers: Georges Nepveu, 1586). Republished in 1605 under the title *Discours et histoires de spectres* (Paris: Nicolas Buon, 1605); *A Treatise of Spectors or straunge Sights, Visions and Apparitions appearing sensibly unto men* (London: Mathew Lownes, 1606).

Lemnius L. (Lemne Liévin). *Les Occultes merveilles et secrets de nature* (Paris: Pierre du Pré, 1567).

A letter to J. C. Lettsom, M.D. F.R.S. S.A.S. &c. occasioned by Baron Dimsdale's Remarks on Dr. Lettsom's letter to Sir Robert Barker ... upon general inoculation. By an uninterested spectator (London: J. Murray, 1779).

Malebranche, N. *De la recherche de la vérité*. Ed. A. Minazzoli (1674; Paris: Presses Pocket, 1990); *The Search After Truth*. Trans. T. M. Lennon and P. J. Olscamp (Columbus: Ohio State University Press, 1980).

Marescot Michel, *Discours veritable sur le faict de Marthe Brossier de Romorantin, pretendue demoniaque* (Paris: Mamert Patisson, 1599); *A True Discourse, Upon the Matter of Martha Brossier of Romarantin, pretended to be possessed by a Devill*. Trans. A. Hartwel (London: J. Wolfe, 1599).

Mead, R. *A Discourse on the Plague* (1744; New York: AMS Press, 1978).

——. *A Short Discourse Concerning Pestilential Contagion and the Methods to be Used to Prevent it* (London: R. Smith and S. Buckley, 1720).

Ménuret de Chambaud, J.-J. *Essai sur l'action de l'air dans les maladies contagieuses* (Paris: Société Royale de Médecine, 1781).

——. 'Miasme', in *Encyclopédie ou dictionnaire raisonné des sciences, des arts et des métiers*, X, pp. 484–5.

Moine, J. *Advis sur ce temps contagieux* (Paris: A. Taupinart, 1628).

Molinet, J. 'Ballade de la maladie de Naples', in *Les faictz et dictz de Jean Molinet* (Paris: Société des Anciens textes français, 1937).

Montaigne, M. de. *Les Essais de Montaigne*. Ed. P, Villey (Paris: Presses Universitaire de France, 1965); *Essays of Michel de Montaigne*. Trans. C. Cotton. Ed. W. C. Hazlitt (1685; Project Gutenberg Release 3600; http://onlinebooks.library.upenn.edu/webbin/gutbook/lookup?num=3600, 1/10/2004).

Paré, A. *Des monstres et des prodiges*. Ed. J. Céard (Geneva: Droz, 1971); *On Monsters and Marvels*. Trans. and ed. J. L. Pallister (Chicago: The University of Chicago Press, 1982).

——. *Traité de la peste, verolle et rougeolle, avec une breve description de la lepre* (Paris: André Wechel, 1568).

Paulet, J.-J. *Le Secret de la médecine révélé ou Préservatif contre la petite vérole et l'inoculation, mémoire pour servir de suite à l'Histoire de la petite vérole* (Paris, 1768).

Pearson, G. *The Medical commentaries* (London: 1794).

Pilet De La Mesnardiere, H. *Traitté de la melancholie, sçavoir si elle est la cause des effets que l'on remarque dans les possédées de Loudun* (La Flèche: Martin Guyot et Gervais Laboé, 1635).

Planis Campy, D. de. *La Verolle recogneue, combatue et abbatue sans suer, et sans tenir chambre, avec tous ses accidens* (Paris: Nicolas Bourdin, 1623).

Rabelais, F. *Oeuvres complètes*, 2 vol. (Paris: Garnier, 1962); *The Five Books and Minor Writings*. Trans. W. F. Smith (London: Alexander P. Watt, 1893).

'Les Sept Marchans de Naples' in *Recueil de poésies françoises des XVe et XVIe siècles*. Ed. A. de Montaiglon, II (Paris: P. Jannet, 1855).

Sigogne, *Les Œuvres satyriques du Sieur de Sigogne* (Paris: Bilbiothèque des Curieux, 1920).

Taxil, J. *Traicté de l'epilepsie* (Lyon: Robert Renaud, 1603).

'Le Triumphe de Dame Verolle', in *Recueil de poésies françoises des XVe et XVIe siècles*. Ed. A. de Montaiglon, IV (Paris: P. Jannet, 1856).

Viau, T. de. *Œuvres complètes*. Ed. G. Saba (Paris: Nizet, 1979).

Valleriola, F. *Observationum medicinalium libri sex* (Lugduni: apud Antonium Candidum, 1588).

Venel, G. F. 'Contagion', in *Encyclopédie ou dictionnaire raisonné des sciences, des arts et des métiers*, IV, p. 110.

Veyries, J. de. *La genealogie de l'amour divisée en deux livres* (Paris: Chez Abel l'Angelier, 1609).

De Vigo, J. *De Vigo en françoys. S'ensuit la pratique et cirurgie de tres excellent docteur en medecine Maistre Jehan de Vigo, nouvellement translatee de latin en françoys* (Lyon: Benoist Bounyn, 1525).

Voltaire. 'Sur l'insertion de la petite vérole', in *Mélanges*. Ed. J. Van den Heuvel (Paris: Gallimard, 1961), pp. 28–32.

Weyer, J. *Witches, Devils, and Doctors in the Renaissance*. Trans. J. Shea (1564; Binghamton, NY: Medieval & Renaissance Texts & Studies, 1991).

Whitaker, T. *An Elenchus of Opinions Concerning the Cure of the Small Pox Together with Problematicall Questions Concerning the Cure of the French Pest* (London: Nathaniel Brook, 1661).

Woodward, T. *The State of Physick: and of Diseases; with an inquiry into the causes of the late increase of them; but more particularly of the small-pox* (London: T. Horne and R. Wilkin, 1718).

Yvelin, P. *Response à l'examen de la possession des religieuses de Louviers* (Evreux: Jean de la Vigne, 1643).

Critical and historical studies

Collective works

Contagion and Infection. Ed. A. Weinstein. Special issue of *Literature and Medicine*, 22, 1 (2003).

Contagion: Historical and Cultural Studies. Eds. A. Bashford and C. Hooker (London and New York: Routledge, 2001).

Contagion: Perspectives from Pre-Modern Societies. Eds L. I. Conrad and D. Wujastyk (Aldershot: Ashgate, 2000).

Emotional Contagion. Eds. E. Hatfield, J. Cacioppo and R. Rapson (Cambridge: Cambridge University Press, 1994).

Epidemics and Ideas. Essays on the Historical Perception of Pestilence. Eds T. Ranger and P. Slack (Cambridge: Cambridge University Press, 1992).

Eros & Anteros: The Medical Traditions of Love in the Renaissance. Eds. D. Beecher and M. Ciavolella (Ottawa: Dovehouse, 1992, University of Toronto Italian Studies 9).

Medicine from the Black Death to the French Disease. Eds J. Arrizabalaga, A. Cunningham and R. French (Aldershot: Ashgate, 1998).

L'Origine de la syphilis en Europe, avant ou après 1493? Actes du Colloque international de Toulon, 25-28 novembre 1993. Eds. O. Dutou, G. Pálfi, J. Bérato and J.-P. Brun (Paris: Errance editions, 1994).

The Regulation of Evil: Social and Cultural Attitudes to Epidemics in the Late Middle Ages. Eds A. Paravicini Bagliani and F. Santi (Firenze: Società Internazionale per lo Studio del Medioevo Latino, 1998; Micrologus' Library, 2).

Sensible Flesh: On Touch in Early Modern Culture. Ed. E. D. Harvey (Philadelphia: University of Pennsylvania Press, 2003).

The Western Medical Tradition, 880 BC to 1800. Eds. L. I. Conrad, M. Neve [et al.] (Cambridge: Cambridge University Press, 1995).

Other secondary sources

Anselment, R. The Realms of Apollo: Literature and Healing in Seventeenth-Century England (Newark: University of Delaware Press, 1995).

Babb, L. The Elizabethan Malady: A Study of Melancholia in English Literature from 1580 to 1642 (East Lansing: Michigan State University Press, 1951).

Baldwin, P. Contagion and the State in Europe, 1830–1930 (Cambridge: Cambridge University Press, 1999).

Barstow, A. L. Witchcraze: A New History of the European Witch Hunts (San Francisco: Pandora, 1994).

Baumgartner, L. and Fulton, J.-F. A Bibliography of the poem Syphilis sive Morbus Gallicus by Girolamo Fracastoro of Verona, (New Haven, CT: Yale University Press, 1935).

Beecher, D. 'The Essentials of Erotic Melancholy: The Exemplary Discourse of André Dulaurens', in Love and Death in the Renaissance. Eds. K. R. Bartlett, K. Eisenbichler and J. Liedl (Ottawa: Dovehouse Editions, 1991).

——. 'The Lover's Body: The Somatogenesis of Love in Renaissance Medical Treatises', Renaissance and Reformation, 17, 1 (1988) 1–11.

——. 'Marguerite de Navarre's Heptameron and the Received Idea: The Problematics of Lovesickness', in International Colloquium Celebrating the 500th Anniversary of the Birth of Marguerite de Navarre April 13 & 14, 1992 Agnes Scott College. Ed. R. Reynolds-Cornell. (Birmingham, AL: Summa, 1995), pp. 71–8.

——. 'Quattrocento Views on the Eroticization of the Imagination', in Eros & Anteros: The Medical Traditions of Love in the Renaissance, pp. 49–65.

Beecher, D. and M. Ciavolella. 'Jacques Ferrand and the Tradition of Erotic Melancholy in Western Culture' in J. Ferrand, A Treatise on Lovesickness. Trans. and eds. D. Beecher and M. Ciavolella (Syracuse: Syracuse University Press, 1990), pp. 1–202.

Biraben, J.-N. Les Hommes et la peste en France et dans les pays européens et méditerranéens, 2 vols (Paris: Mouton, 1975–76).

Boucé, P-G. 'Imagination, Pregnant Women, and Monsters, in Eighteenth-century England and France', in Sexual Underworlds of the Enlightenment. Eds. G. S. Rousseau and R. Porter (Manchester: Manchester University Press, 1987), pp. 86–100.

Brodie, R. Virus of the Mind: The New Science of the Meme (Seattle: Integral Press, 1996).

Buisine, A. 'Casanova: bonheurs de la vérole', Magazine littéraire (July–August 2000) 38.

Cabanis, Cl. 'La peste de 1720 à Avignon d'après cinq livres de raison avignonnais' (Thèse de Université de Paris, 1971).

Cairns, F. 'Fracastoro's Syphilis, the Argonautic Tradition and the Ætiology of Syphilis', Humanistica Lovaniensia, 43 (1994) 246–61.

Campbell, A.M. The Black Death and Men of Learning (1931; New York: AMS Press, 1966).

Carmichael, A. 'Contagion theory and practice in XVth century Milan', Renaissance Quarterly, 44 (1991) 213–56.

Coopland, W. *Nicole Oresme and the Astrologers* (Liverpool: University Press, 1952).

Céard, J. 'The Devil and Lovesickness: Views of 16th Century Physicians and Demonologists', in *Eros & Anteros: The Medical Traditions of Love in the Renaissance*, pp. 33–47.

Céard, J. 'Folie et démonologie au XVIᵉ siècle', in *Folie et déraison à la Renaissance*, (Brussels: Université de Bruxelles, 1976).

Céard, J. *La Nature et les prodiges*, (1977; Geneva: Droz, 1996).

Certeau M. de. *La Possession de Loudun*, (1970; Paris: Gallimard-Julliard, 1990 Archives); *The Possession at Loudun*. Trans. M. B. Smith (Chicago: University of Chicago Press, 2000).

Ciavolella, M. *La Mallattia d'amore dall'Antiquità al Medievo* (Rome: Bulzoni, 1976).

Ciavolella, M. 'Eros and the Phantasms of *Hereos*', in *Eros and Anteros: The Medical Traditions of Love in the Renaissance*, pp. 75–85.

Clifton, J. D. *Images of the Plague and other Contemporary Events in 17th Century Naples* (UMI Dissertation, 1989).

Crisciani, C. 'Oro potabile fra alchimia e medicina. Due testi in tempo di peste', in *Storia e fondamenti della chimica. Atti del VII Convegno Nazionale*. Ed. F. Calascibetta (Roma: Accademia Nazionale delle Scienze, 1997), pp. 83–93.

Crisciani, C. and M. Pereira. 'Black Death and Golden Remedies: Some Remarks on Alchemy and the Plague', in *The Regulation of Evil: Social and Cultural Attitudes to Epidemics in the Late Middle Ages*, pp. 7–39.

Dandrey, P. *Médecine et maladie dans le théâtre de Molière* (Paris: Klincksieck, 1998).

——. '*Moralia & medicinalia*. Cadastre, semences et moissons (1977–1997)', *XVIIe siècle*, 202 (1999) 37–53.

——. 'De la pathologie mélancolique à la psychologie de l'autosuggestion: l'herméneutique de la sorcellerie au XVIIe siècle', *Littératures classiques*, 25 (1995) 135–59.

Darmon, P. *La variole, les nobles et les princes: la petite vérole mortelle de Louis XV* (Brussels: Complexe, 1989).

Derrida, J. 'La pharmacie de Platon', in *La dissémination* (Paris: Seuil, 1972) pp. 70–197.

Droz, E. and A. C. Klebs. *Remèdes contre la peste* (Paris: Droz, 1925).

Dubois, E. 'How to Face the Plague: Some Seventeenth-century Practices', *Newsletter of the Society for Seventeenth-Century Studies*, 2 (1980) 29–36.

Ehrard, J. 'Opinions médicales en France au XVIIIe siècle. La peste et l'idée de contagion', *Annales: Economie, sociétés, civilisations*, 12 (1957) 46–59.

Esquirol, J. E. D. *Mental Maladies: A Treatise on Insanity*. Intro. R. de Saussure (1838; New York: Hafner Publishing, 1965).

Favre, Robert. 'Vers une politique de la santé publique', in *La mort dans la littérature et la pensée françaises au siècle des Lumières* (Lyon: Presses Universitaires de Lyon, 1978), pp. 244–71.

Fenner, F. [et al.]. *Smallpox and its Eradication* (Geneva: World Health Organization, 1988).

Foa, A. 'The New and the Old: The Spread of Syphilis (1494–1530)', in *Sex and Gender in Historical Perspective*. Eds E. Muir and G. Ruggiero (Baltimore: Johns Hopkins University Press, 1990).

Folger, R. *Images in Mind: Lovesickness, Spanish Sentimental Fiction and Don Quijote* (Chapel Hill: North Carolina Studies in the Romance Languages and Literatures, 2002).

Forrester, J. 'Contracting the Disease of Love: Authority and Freedom in the Origins of Psychoanalysis', in *The Anatomy of Madness: Essays in the History of Psychiatry*. Vol. 1. Eds W. F. Bynum, R. Porter and M. Shepherd (London: Tavistock, 1985).

Foucault, M. *Les anormaux. Cours au Collège de France. 1974–1975* (Paris: Gallimard/ Seuil, 1999).

——. 'Médecins, juges et sorciers au XVIIe siècle', in *Dits et écrits I, 1954–1975* (Paris: Gallimard, 2001, Quarto), pp. 781–95.

French, R. *Canonical Medicine: Gentile da Foligno and Scholasticism* (Leiden: Brill, 2001).

——. *Medicine Before Science* (Cambridge: Cambridge University Press, 2003).

Gentilcore, D. *Healers and Healing in Early Modern Italy* (Manchester: University of Manchester Press, 1998).

Goldstein, J. '"Moral Contagion": A Professional Ideology of Medicine and Psychiatry in Eighteenth- and Nineteenth-Century France', in *Professions and the French State, 1700–1900*. Ed. G. L. Geison (Philadelphia: University of Pennsylvania Press, 1984).

Gordon, B. L. *Medieval and Renaissance Medicine* (London: Peter Owen, 1959).

Gordon, D. 'The City and the Plague in the Age of Enlightenment', *Yale French Studies*, 92 (1997) 67–87.

Gouldsblom, J. 'Les Grandes épidémies et la civilisation des mœurs', *Actes de la recherche en sciences sociales*, 68 (1987) 8–9.

Grant, E. 'Medieval and Renaissance scholastic conceptions of the influence of the celestial region on the terrestrial', *Journal of Medieval and Renaissance Studies*, 17 (1987) 1–23.

——. *A Source Book in Medieval Science* (Cambridge: Cambridge University Press, 1974).

Gruner, C. O. *A Treatise on the Canon of Medicine of Avicenna* (New York: Kelley, 1970).

Harris, J. G. *Foreign Bodies and the Body Politic: Discourses of Social Pathology in Early Modern England* (Cambridge; Cambridge University Press, 1998).

Healy, M. 'Anxious and Fatal Contacts: Taming the Contagious Touch', in *Sensible Flesh: On Touch in Early Modern Culture*, pp. 22–38.

——. 'Defoe's Journal and the English Plague Writing Tradition', *Literature and Medicine*, 22, 1 (2003) 25-44.

Heffernan, C. F. *The Melancholy Muse: Chaucer, Shakespeare and Early Medicine* (Pittsburgh: Duquesnes University Press, 1995).

Hildesheimer, F. *Le bureau de la santé de Marseille sous l'Ancien Régime: le renfermement de la contagion* (Marseille : Fédération historique de Provence, 1980).

——. *Fléaux et société: de la Grande Peste au choléra, XIVe-XIXe siècle* (Paris: Hachette, 1993).

Hopkins, D. R. *Princes and Peasants, Smallpox in History* (Chicago and London: University of Chicago Press, 1983).

Huet, M.-H. *Monstrous Imagination* (Cambridge, MA: Harvard University Press, 1993).

Jacquart, D. and M. Micheau. *La Médecine arabe et l'Occident médiéval* (Paris : Maisonneuve et Larose, 1990).

Jarcho, S. *The Concept of Contagion in Medicine, Literature, and Religion* (Malabar, FL: Krieger Publishing Company, 2000).

Le Brun, J. 'Représentations du cancer à l'époque moderne (XVIIe–XVIIIe siècles)', *Prévenir*, 16 (1988) 9–14.

——. 'Les représentations du cancer dans les biographies spirituelles du XVIIe siècle', *Sciences sociales et santé*, II, 2 (1984) 9–31.

Le Guérer, A. *Les pouvoirs de l'odeur*, Paris, 1988; *Scent, the Mysterious and Essential Powers of Smell*. Trans. R. Miller (New York: Turtle Bay Books, 1992).

Levenstein, J. 'Out of Bounds: Passion and the Plague in Boccaccio's *Decameron*', *Italica*, 73, 3 (1996) 313–35.

Lindberg, D. C. *Theories of Vision from Al-Kindi to Kepler* (Chicago: Chicago University Press, 1976).

Lowes, J. L. 'The Loveres Maladye of Hereos', *Modern Philology*, 11, 4 (1914) 491–546.

Lund, R. D. 'Infectious Wit: Metaphor, Atheism and the Plague in Eighteenth-Century London', *Literature and Medicine*, 22.1 (2003) 45–64.

Maclean, I. *Logic, Signs, and Nature in the Renaissance. The Case of Learned Medicine* (Cambridge: Cambridge University Press, 2002).

Maire, C.-L. *Les convulsionnaires de Saint-Médard: Miracles, convulsions et prophéties à Paris au XVIIIe siècle* (Paris: Gallimard/Julliard, 1985).

Mandrou, R., *Magistrats et sorciers en France au XVIIe siècle* (Paris: Seuil, 1980).

——. *Possession et sorcellerie au XVIIe siècle, textes inédits* (Paris: Fayard, 1979, Pluriel).

Marcovich, A. 'Contagion et santé publique. Une représentation du lien social en Angleterre au XVIIIe siècle', *Sciences sociales et santé*, II, 2 (1984) 45–70.

Miller, G. *The Adoption of Inoculation for Smallpox in England and France* (Philadelphia: University of Pennsylvania Press, 1957).

Miller, W. I. *The Anatomy of Disgust* (Cambridge, MA: Harvard University Press, 1997).

Nelson, J. Ch. *Renaissance Theory of Love: The Context of Giordano Bruno's 'Eroici furori'* (New York: Columbia University Press, 1963).

Normand, S. ' Poison, maladie et métaphore dans le Dictionnaire universel', *Littératures classiques*, 47 (2003) 173–84.

Nutton, V. 'Did the Greeks Have a Word for it?', in *Contagion: Perspectives from Pre-Modern Societies*, pp. 137–62.

——. 'The Reception of Fracastoro's Theory of Contagion: The Seed that fell among Thorns?', *Osiris*, 2nd ser. 6 (1990) 196–234.

——. 'Seeds of Disease: An Explanation of Contagion and Infection from the Greeks to the Renaissance', *Medical History*, 27 (1983) 1–34.

Ober, W. B. *Boswell's Clap and Other Essays* (New York: Harper and Row, 1988).

Orlandi, A. 'Malinconia e antropologia nel *De intellectione* e nel *De anima* di Girolamo Fracastoro', in *Psicopatologia e filosofia nella tradizione veronese. Atti del seminario di studi.* Ed. L. Bonuzzi (Verona: sn, 1994), pp. 9–17.

Pagel, W. *Paracelsus: An Introduction to Philosophical Medicine in the Era of the Renaissance* (Basel; 2nd rev. edn, New York: Karger, 1982).

Palmer, A. '*Lux Dei*, Ficino and Aquinas on the Beatific Vision', *Memini* 6 (2002) 129–52.

Palmer, R. 'The Church, Leprosy, and Plague in Medieval and Early Modern Europe', in *The Church and Healing*. Ed. W.J. Sheils (Oxford: Blackwell, 1982).

Pellegrini, F. 'L'epidemia di *Morbus peculiaris* del 1546–1547 e il medico del Concilio del Trento', *Castalia*, 2 (1946) 271–78.

——. 'Frammento inedito di G. Fracastoro riguardante la pestilenza del 1534–1535', *Rivista di Storie delle Scienze Mediche e Naturali*, 26, 4a (1935) 253–9.

——. *Origini e primi sviluppi della dottrina fracastoriana del 'contagium vivum'*, (Verona, 1950).

——. *Scritti inediti di G. Fracastoro* (Verona: Valdonega, 1955).

——. *Trattato inedito in prosa di G. Fracastoro sulla sifilide* (Verona: Tipografia Veronese, 1939).

Pelling, M. 'The Meaning of Contagion: Reproduction, Medicine and Metaphor', in *Contagion: Historical and Cultural Studies*, pp. 15–38.

Peruzzi, E. 'Antiocccultismo e Filosofia naturale nel *De Sympathia & Antipathia rerum* di Girolamo Fracastoro', *Atti e Memorie dell'Accademia toscana di Scienze e Lettere. La Colombaria*, 45 [nuova serie 31] (1980) 41–131.

——. *La nave di Ermete. La cosmologia di Girolamo Fracastoro* (Firenze: Olschki, 1995).

Peter, J.-P. 'Le désordre contenu: attitudes médicales face à l'épidémie au Siècle des Lumières', *Ethnologie française*, 17, 4 (1987) 355–66.

Pigeaud, J. 'Reflections on Love-Melancholy in Robert Burton', in *Eros & Anteros: The Medical Traditions of Love in the Renaissance*, pp. 211–31.

Pomata, G. *Contracting a Cure: Patients, Healers, and the Law in early Modern Bologna.* Trans. by author (Baltimore: Johns Hopkins University Press, 1998).

Porzio, A. 'Immagini della peste del 1656', in *Civiltà del Seicento a Napoli* (Naples: Electa, 1984), 2, pp. 51–7.

Qualtiere, L. F. and Slights, W. W. E. 'Contagion and Blame in Early Modern England: The Case of the French Pox', *Literature and Medicine*, 22, 1 (2003) 1–24.

Quétel, C. *Le Mal de Naples. Histoire de la syphilis* (Paris: Seghers, 1986).

Randall, L. B. 'Representations of Syphilis in Sixteenth-Century French Literature' (Thesis, University of Arizona, 1999).

Rey, R. 'Contagion ou "constitution épidémique"', in *Naissance et développement du vitalisme en France de la deuxième moitié du 18ᵉ siècle à la fin du Premier Empire* (Oxford, Voltaire Foundation, 2000), pp. 270–319.

Robbins, R. H. *Encyclopedia of Witchcraft and Demonology* (New York: Crown Publications, 1959).

Rosenberg C. E. *Explaining Epidemics and other Studies in the History of Medicine* (Cambridge: Cambridge University Press, 1992).

Rousseau, G. S. 'Pineapples, Pregnancy, Pica, and Peregrine Pickle', in *Tobias Smollett: Bicentennial Essays Presented to Lewis M. Knapp*. Eds. G. S. Rousseau and P.-G. Boucé (Oxford: Oxford University Press, 1971), pp. 79–109.

Savio, P. 'Recerche sulla peste di Roma degli anni 1656–1657', *Archivio della Societa romana di storia patria*, 26, 3 (1972) 113–42.

Sies, R. *Das Pariser Pestgutachten von 1348 in altfranzösischer Fassung* (Pattensen/Han.: Wellm, 1977, Untersuchungen zur mittelalterlichen Pestliteratur 4, Würzburger medizinhistorische Forschungen Bd. 7).

Singer, C. and D. 'The scientific position of G. Fracastor (1478?–1553), with special reference to the source, character and influence of his theory of infection', *Annals of Medical History*, 1 (1917) 1–34.

Siraisi, N. 'Anatomizing the Past: Physicians and History in Renaissance Culture', *Renaissance Quarterly*, 53, 1 (2000) 1–30.

Smith, J. R. *The Speckled Monster: Smallpox in England, 1670–1970, with particular reference to Essex* (Chelmsford: Essex Record Office, 1987).

Sonnino, E. 'Di qui cominciò qualche terrore considerabile nella città di Roma: Popolazione e sanità del XVII secolo', in *Scienza e Miracoli nell'Arte del '600. Alle origini della Medicina Moderna*. Ed. S. Rossi (Milan: Electa, 1998), pp. 60–9.

Sournia, J.-C. 'Les Notions d'épidémie et de contagion dans les comportements sociaux', in *L'Origine de la syphilis en Europe* (Paris: Errance editions, 1994).

Stephanson, R. 'The Plague Narratives of Defoe and Camus: Illness as Metaphor', *Modern Language Quarterly*, 48 (September 1987) 224–41.

Stitziel, J. 'God, the Devil, Medicine, and the Word: A Controversy over Ecstatic Women in Protestant Middle Germany 1691–1693', *Central European History* 29/3 (1996) 309–37.

Temkin, O. *Galenism: The Rise and Decline of a Medical Philosophy*, (Ithaca, NY: Cornell University Press, 1973).

Thorndike, L. *A History of Magic and Experimental Sciences*, 3 vols (New York: McMillan, 1935).

Tilles, G. and Wallach, D. 'Le Traitement de la syphilis par le mercure. Une histoire thérapeutique exemplaire', *Histoire des Sciences Médicales*, 30, 4 (1996) 501-***.

Todd, D. *Imagining Monsters: Miscreations of the Self in Eighteenth-Century England* (Chicago and London: University of Chicago Press, 1995).

Touati, F.-O. 'Contagion and Leprosy: Myth, Ideas and Evolution in Medieval Minds and Societies', in *Contagion: Perspectives from Pre-Modern Societies*, pp. 179–202.

Watts, S. *Epidemics and History: Disease, Power and Imperialism* (New Haven and London: Yale University Press, 1997).

Viana, O. 'L'atto di ammissione del Fracastoro al Collegio medico di Verona', *Rivista di Storia Critica delle Scienze Mediche e Naturali*, 5 (1914) 382–3.

Vigarello, G. *Le propre et le sale. L'hygiène du corps depuis le Moyen Âge* (Paris: Seuil, 1985).

Wack, M. F. 'From Mental Faculties to Magical Philters: The Entry of Magic into Academic Medical Writings on Lovesickness, 13-17th Centuries', in *Eros & Anteros: The Medical Traditions of Love in the Renaissance*, pp. 9–31.

——. *Lovesickness in the Middle Ages: The* Viaticum *and its Commentaries* (Philadelphia: University of Pennsylvania Press, 1990).

Index